ISBN 978-1-5279-1068-3
PIBN 10910721

This book is a reproduction of an important historical work. Forgotten Books uses
state-of-the-art technology to digitally reconstruct the work, preserving the original format
whilst repairing imperfections present in the aged copy. In rare cases, an imperfection in
the original, such as a blemish or missing page, may be replicated in our edition. We do,
however, repair the vast majority of imperfections successfully; any imperfections that
remain are intentionally left to preserve the state of such historical works.

English
Français
Deutsche
Italiano
Español
Português

www.forgottenbooks.com

Mythology Photography **Fiction**
Fishing Christianity **Art** Cooking
Essays Buddhism Freemasonry
Medicine **Biology** Music **Ancient
Egypt** Evolution Carpentry Physics
Dance Geology **Mathematics** Fitness
Shakespeare **Folklore** Yoga Marketing
Confidence Immortality Biographies
Poetry **Psychology** Witchcraft
Electronics Chemistry History **Law**
Accounting **Philosophy** Anthropology
Alchemy Drama Quantum Mechanics
Atheism Sexual Health **Ancient History**
Entrepreneurship Languages Sport
Paleontology Needlework Islam
Metaphysics Investment Archaeology
Parenting Statistics Criminology
Motivational

BULLETINS

OF

AMERICAN

PALEONTOLOGY

(Founded 1895)

Vol. 75

No. 305

THE CORALS OF THE BRASSFIELD FORMATION (MID-LLANDOVERY; LOWER SILURIAN) IN THE CINCINNATI ARCH REGION

By

Richard S. Laub

1979

Paleontological Research Institution
Ithaca, New York 14850 U. S. A.

PALEONTOLOGICAL RESEARCH INSTITUTION

BULLETINS OF AMERICAN PALEONTOLOGY
and
PALAEONTOGRAPHICA AMERICANA

PETER R. HOOVER ———————————————————————EDITOR

Reviewers for 1979

PAUL K. BIRKHEAD
T. CRONIN
J. WYATT DURHAM
D. ECHOLS
A. MYRA KEEN
DAVID L. PAWSON
COLIN W. STEARN
NORMAN E. WEISBORD
V. A. ZULLO

A list of titles in both series, and available numbers and volumes may be had on request. Volumes 1 - 23 of *Bulletins of American Paleontology* have been reprinted by Kraus Reprint Corporation, Route 100, Millwood, New York 10546 USA. Volume 1 of *Palaeontographica Americana* has been reprinted by Johnson Reprint Corporation, 111 Fifth Ave., New York, NY 10003 USA.

Subscriptions to *Bulletins of American Paleontology* may be started at any time, by volume or year. Current price is US $25.00 per volume. Numbers of *Palaeontographica Americana* are priced individually, and are invoiced separately on request. Purchases for professional use by U.S. citizens are tax-deductible.

for additional information, write or call:

Paleontological Research Institution

The Paleontological Research Institution
acknowledges with special thanks
the contributions of the following individuals and institutions

PATRONS
($1000 or more at the discretion of the contributor)

JAMES A. ALLEN (1967)

ARMAND L. ADAMS (1976)

ATLANTIC RICHFIELD COMPANY (1978)

MISS ETHEL Z. BAILEY (1970)

CHRISTINA L. BALK (1970)

MR. & MRS. KENNETH E. CASTER (1967)

CHEVRON OIL COMPANY (1978)

EXXON COMPANY (1977 to date)

LOIS S. FOGELSANGER (1966)

GULF OIL CORPORATION (1978)

MERRILL W. HAAS (1975)

MISS REBECCA S. HARRIS (1967)

AMERICAN OIL COMPANY (1976)

MRS. WILLIAM B. HEROY (1969)

MR. & MRS. ROBERT C. HOERLE (1974-77)

RICHARD I. JOHNSON (1967)

J. M. MCDONALD FOUNDATION (1972, 1978)

MOBIL OIL CORP. (1977-1979)

N. Y. STATE ARTS COUNCIL (1970, 1975)

KATHERINE V. W. PALMER (to date)

CASPER RAPPENECKER (1976)

MRS. CUYLER T. RAWLINS (1970)

TEXACO, INC. (1978)

UNITED STATES STEEL FOUNDATION (1976)

MR. & MRS. PHILIP C. WAKELEY (1976 to da

INDUSTRIAL SUBSCRIBERS
(1979)

($250 per annum)

AMOCO PRODUCTION COMPANY

ANDERSON, WARREN & ASSOC.

ATLANTIC RICHFIELD COMPANY

CITIES SERVICE COMPANY

EXXON PRODUCTION RESEARCH COMPANY

EXXON COMPANY, U.S.A.

MOBIL EXPLORATION AND PRODUCING SERVI

SHELL DEVELOPMENT COMPANY

SUSTAINING MEMBERS
(1979)

($75 per annum)

MRS. AND MRS. JOHN HORN

TOMPKINS COUNTY GEM AND MINERAL CLUB

(continued overleaf)

Membership dues, subscriptions, and contributions are all important sources of funding, and allow the Paleontological Research Institution to continue its existing programs and services. The P. R. I. publishes two series of respected paleontological monographs, *Bulletins of American Paleontology* and *Palaeontographica Americana*, that gives authors a relatively inexpensive outlet for the publication of significant longer manuscripts. In addition, it reprints rare but important older works from the paleontological literature. The P. R. I. headquarters in Ithaca, New York, houses a collection of invertebrate type and figured specimens, among the five largest in North America; an extensive collection of well-documented and curated fossil specimens that can form the basis for significant future paleontologic research; and a comprehensive paleontological research library. The P. R. I. wants to grow, so that it can make additional services available to professional paleontologists, and maintain its position as a leader in providing Resources for Paleontologic Research.

The Paleontological Research Institution is a non-profit, non-private foundation, and all contributions are U. S. income tax deductible. For more information on P. R. I. programs, memberships, or subscriptions to P. R. I. publications, call or write:

Peter R. Hoover
Director

Paleontological Research Institution
1259 Trumansburg Road
Ithaca, New York 14850 U. S. A.
607-273-6623

BULLETINS

OF

AMERICAN

PALEONTOLOGY

(Founded 1895)

Vol. 75

No. 305

THE CORALS OF THE BRASSFIELD FORMATION (MID-LLANDOVERY; LOWER SILURIAN) IN THE CINCINNATI ARCH REGION

By

Richard S. Laub

August 31, 1979

Paleontological Research Institution
Ithaca, New York 14850 U. S. A.

Library of Congress Card Number: 79-90221

Printed in the United States of America
Arnold Printing Corporation
Ithaca, New York 14850

CONTENTS

LIST OF ILLUSTRATIONS

LIST OF TABLES

DEDICATION

To my wife and my parents

THE CORALS OF THE BRASSFIELD FORMATION (MID-LLANDOVERY; LOWER SILURIAN) IN THE CINCINNATI ARCH REGION

By

RICHARD S. LAUB

Department of Geology

Buffalo Museum of Science

Buffalo, New York

and

Department of Geological Sciences

State University of New York

Buffalo, New York

ABSTRACT

Fifty-four species of corals (29 rugose, 19 tabulate, 6 heliolitid) have been identified from the Brassfield Formation (limestone, dolomite and claystone of mid-Llandovery [Lower Silurian] age) in the Cincinnati Arch region of the United States. Included are one new rugosan genus (*Schizophaulactis*), four new rugosan species (*Streptelasma scoleciforme, Dinophyllum semilunum, Paliphyllum regulare, Pycnactis tenuiseptatus*), three new tabulate species (*Favosites densitabulatus, Catenipora favositomima, Syringolites vesiculosus*), and one new rugosan subspecies (*Paliphyllum suecicum* Neuman, 1968 *brassfieldense*). Coral assemblages of different, apparently environmentally controlled character are concentrated in the north and the southeast parts of the region.

The Brassfield coral fauna appears to have been largely restricted to the present Cincinnati Arch region during mid-Llandovery time. Closest affinities on the species level during this episode are with the corals of Anticosti Island (Canada). Eight to twelve species pre-date the mid-Llandovery, some existing as early as the Late Ordovician. These oldest members of the Brassfield coral fauna were found in a number of areas throughout the world, rather than in a single ancestral homeland. At least half the coral species survived the end of the mid-Llandovery, and spread into a large portion of the submerged continental areas of the world.

The geometry, facies distribution, environments and possible current directions and age of the Brassfield lithosome are considered, and current views on these subjects are summarized. Certain features of the Florida Keys area, including Rodriguez Bank and the French Dry Rocks, offer insight into the mode of origin of the Brassfield Formation and one extraordinary coral bed that it contains.

The relationships of coral species distribution to regional geography and sedimentary facies are discussed. With some exceptions, the assemblages found in the limestone and in the dolomite are distinct. Almost all species that occur in the clays of the higher Brassfield are found in the limestone as well. Because the clay occurs in close association with the limestone, and because there is evidence that many of the corals enclosed in the clay have been transported, it seems likely that these corals lived in the limestone-depositing environment.

Morphologic features of corals that are discussed include axial torsion, a potentially important phenomenon that has received little previous attention.

CORALS OF THE BRASSFIELD ORMATION
(MID-LLANDOVERY; LOWER SlURIAN)
IN THE CINCINNATI ARCH RGION

By

RICHARD S. LAUB
Department of Geology
Buffalo Museum of Science
Buffalo, New York

and

Department of Geological Scieres
State University of New Yor
Buffalo, New York

ABSTRACT

ty-four species of corals (29 rugose, 19 tabulate, heliolitid) have been
ed from the Brassfield Formation (limestone, domite and claystone of
andovery [Lower Silurian] age) in the Cincinni Arch region of the
States. Included are one new rugosan genus (*hizophaulactis*), four
igosan species (*Streptelasma scoleciforme, Dinophlum semilunum, Pali-
m regulare, Pycnactis tenuiseptatus*), three new tabute species (*Favosites
abulatus, Catenipora favositomima, Syringolites esiculosus*), and one
ugosan subspecies (*Paliphyllum succicum* Neumar 1968 *brassfieldense*).
assemblages of different, apparently environmenta' controlled character
oncer-ated in the north and the southeast parts of the region.

'h   field coral '   pears to have been ligely restricted to the
         ati Arch   ing mid-Llandovery me. Closest affinities
         evel duri   ode are with the coils of Anticosti Island
         ht to t·   es pre-date the mid-landovery, some ex-
         as the 7   ian. These oldest meners of the Brassfield
         re four   er of arras throughout he world, rather than
         estral   lea'   he coral spies survived the end
         landov   ea'   'rge portn of the submerged
         as of
         etry.   ents and ossible current direc-
         of '   nsidered, nd current views on
         s ar(of the brida Keys area, in-
         igu(insight into the mode
         th(nary coral bed that

         atic   onal geography and
         fa   sec   ' assemblages found
         stor   lo'   l species that occur
         of   ass   ne as well. Because
         ccu   soc   l because there is
         h·   cc   been transported.
                    ·   ng environment.
                    n   e axial torsion, a
                                       vious attention.

ABBREVIATIONS OF REPOSITORY INSTITUTIONS

AMNH: American Museum of Natural History, New York, New York, U.S.A.

BM(NH):   British Museum (Natural History), London, England

Cambridge: University of Cambridge Geological Museum, Cambridge, England

GS: Geological Survey of Great Britain, London, England

GSC: Geological Survey of Canada, Ottawa, Ontario, Canada

MCZ: Museum of Comparative Zoology, Harvard University, Cambridge, Massachusetts, U.S.A.

OSU: Ohio State University, Columbus, Ohio, U.S.A.

PIN: Paleontological Institute of Novosibirsk, Novosibirsk, U.S.S.R.

ROM: Royal Ontario Museum, Toronto, Ontario, Canada

SNIIGGIMS: Sibirskii Nauchno-Issledovatelskii Institut Geologii, Geofiziki i Mineralnogo Syrya, Novosibirsk, U.S.S.R.

UCGM: University of Cincinnati Geological Museum, Cincinnati, Ohio, U.S.A.

UM: Paleontological Institution, University of Uppsala, Uppsala, Sweden

UMMP: University of Michigan Museum of Paleontology, Ann Arbor, Michigan, U.S.A.

USNM: National Museum of Natural History, Smithsonian Institution, Washington, D.C., U.S.A.

YPM: Yale Peabody Museum, Yale University, New Haven, Connecticut, U.S.A.

INTRODUCTION

The end of Ordovician time saw a considerable change in the character of the world coral fauna. It is possible that this global change was tied to eustatic sea-level variations during a Late Ordovician ice age. Sheehan (1973) argues that the Ordovician-Silurian changeover in the North American brachiopod fauna was due to this same factor. Among the Rugosa, the dissepimentarium came into prominence (Hill, 1959a, pp. 151, 152) and several evolutionary bursts occurred (Hill, 1956, p. 254) during the Silurian. Among the Tabulata, the auloporids increased in importance, as did the halysitids. *Favosites*, with mural pores on the wall faces, became dominant over *Paleofavosites*, with pores in the corners, and *Halysites*, with interstitial tubes, outpaced *Catenipora*, which lacks these structures. Among the Heliolitida, the coenosteal tubes developed solid walls. The first-known true coral bioherms are found in Silurian rocks.

To date, nearly all our knowledge of Silurian corals is limited to the rock units of Upper Llandovery, Wenlock, and Ludlow ages, which are well represented in the Baltic region (especially Gotland), the British Isles, and the Falls-of-the-Ohio (Kentucky and Indiana). The Brassfield coral fauna bridges the gap between the Late Ordovician and this later Silurian record with a large, diverse fauna in which several important Silurian species make their first appearance but from which the Ordovician flavor has not yet faded.

ACKNOWLEDGMENTS

Numerous persons and organizations made valuable contributions to this study and its publication. The research project began as a doctoral dissertation problem in the Department of Geology, University of Cincinnati, Cincinnati, Ohio. The principal advisor was Prof. Kenneth E. Caster of that department who, along with Professors Wayne A. Pryor and William Dreyer (respectively, Departments of Geology and Biology of the University of Cincinnati), reviewed the manuscript and offered valuable criticisms. Much technical help and advice was given by Dr. William B. Harrison (Department of Geology, University of Western Michigan, Kalamazoo) and by Dr. William A. Oliver, Jr. (United States Geological Survey, Washington, D.C.). Thin-sectioning of some important type specimens was done by Mr. William Pinckney (United States Geological Survey, Washington, D.C.). Advice on various special problems encountered came from Dr. Kees de Jong, Department of Geology, University of Cincinnati; Prof. Edward J. Buehler, Department of

Geological Sciences, State University of New York at Buffalo; Dr. Alfred M. Ziegler, Department of Geophysical Sciences, University of Chicago; Dr. Don L. Kissling, Department of Geology, State University of New York at Binghamton; Dr. Klemens Oekentorp, Geological-Paleontological Institution and Museum, Westfälische Wilhelms-Universität, Münster in Westfalen, Germany; Dr. Stefan Bengtson, Department of Paleobiology, University of Uppsala, Uppsala, Sweden, and Dr. Robert N. Ginsburg, Rosenstiel School of Marine and Atmospheric Science, University of Miami, Miami Beach, Florida.

The quality of this paper has been enhanced by the editorial advice of Drs. Peter R. Hoover and Katherine V. W. Palmer of the Paleontological Research Institution, Ithaca, New York.

The extensive surveys of the scientific literature required by this study were made possible by the diligent cooperation of the library staffs of the University of Cincinnati Geology Department, the United States Geological Survey (Washington, D.C.), the Buffalo Museum of Science, and the State University of New York at Buffalo Geology Library. Grants from the Society of the Sigma Xi and the University of Cincinnati Geology Department helped finance the necessary field work.

Publication of this work was made possible through financial contributions by the University of Cincinnati Research Council, the University of Cincinnati Geology Department, the Buffalo Museum of Science, and the Department of Geological Sciences, State University of New York at Buffalo.

Numerous other persons, unmentioned above, have in various ways made contributions to this work. My failure to include them here in no way diminishes my gratitude for their kindness.

LOCALITIES OF BRASSFIELD EXPOSURES

Names used for localities and roads in this list are current as of 1973.

1. Type locality of Brassfield Formation, along abandoned cut of Louisville and Atlantic Railroad, between Panola and Brassfield, Madison County, Kentucky, starting from about 0.25 mile (0.4 km) west of right-angle bend in State Route 499 at Panola; also, exposure on north side of Route 499 about 0.5 mile (0.8 km) east of this right-angle bend of the road at Panola, and a third exposure, about 0.4 mile (0.64 km) further east, in Drowning Creek and along the

road (Rt. 499) going uphill east of the Creek. Virtually all specimens came from the second site. Panola Quadrangle; 39°30′25″N, 84°06′30″W.

2. About 6 miles (9.6 km) north of Manchester, Adams County, Ohio, on east side of Route 136. Manchester Quadrangle; 38°44′00″N, 83°36′45″W.

5. Exposure on north side of Route 52 in Adams County, Ohio, 10 miles (16 km) east of Manchester, Ohio. Concord Quadrangle; 38°41′10″N, 83°27′45″W.

7. Exposures in quarries belonging to Southwest Portland Cement Company and to Atlas Steel Company along Route 235, about 1 mile (1.6 km) east of the offices of the former company, near the juncture with Trebein Road, southeast of Fairborn, Greene County, Ohio. Yellow Springs Quadrangle; 39°48′45″N, 83°59′30″W.

7a. In vicinity of locality 7; largest working quarry of Southwest Portland Cement Company (as of 1972), about 3.5 miles (18,000 feet; 5.6 km) south of the main quarry office, along the quarry road. Yellow Springs Quadrangle; 39°46′30″N, 83°57′45″W.

9. Mine and quarry of Marble Cliff Company, near Lewisburg, Preble County, Ohio, 0.85 mile (1.36 km) west of juncture between Ohio Route 503 and Euphemia-Castine Road. Lewisburg Quadrangle; 39°51′45″N, 84°33′15″W.

11. Series of three outcrops on NW side of Route 41, from 2.5 to 3.0 miles (4.0 to 4.8 km) NE of West Union, Adams County, Ohio. West Union Quadrangle; 38°49′00″N, 83°30′30″W.

12. Route 41, where it crosses Ohio Brush Creek in Adams County, Ohio, 1.4 miles (2.24 km) SW of Jacksonville, Ohio. Peebles Quadrangle; 38°53′40″N, 82°27′10″W.

13. Todd Fork Creek, NE of Wilmington, Clinton County, Ohio. Top of section is about 350 yards (320 meters) downstream from Route 68; base and best exposure of section is in stream-cut and abandoned quarry further downstream, about 400 yards (366 meters) upstream from Center Road. Wilmington Quadrangle; upstream exposure is 39°28′45″N, 83°50′45″W; downstream exposure is 39°28′50″N, 83°51′20″W.

14. Aramco Steel Company quarry, Piqua, Miami County, Ohio. Piqua East and Troy Quadrangles; 40°07′30″N, 84°14′00″W.

15. Route 571, west of Interstate Route 75, and 0.3 mile (0.48 km) east of West Milton, Miami County, Ohio. West Milton Quadrangle; 39°57′55″N, 84°19′15″W.

16. Base of Wright Brothers Memorial, Dayton, Greene County, Ohio, at crest of bluff overlooking Route 444 and the Huffman Dam, immediately below U.S.G.S. benchmark. Fairborn Quadrangle; 39°47′45″N, 84°05′25″W.

19. Two miles (3.2 km) NE of Bryantsburg, Indiana, on Route 1150N, about 1.5 miles (2.4 km) from Route 421, near border between Ripley and Jefferson Counties, Indiana. This is the Brassfield "pinch-out". Rexville Quadrangle.

20. South of Belleview, Jefferson County, Indiana, on Route 250, 2.5 miles (4.0 km) from junction with Route 421. Canaan Quadrangle; 38°50′30″N, 85°21′15″W.

21. Tri-County Stone Company quarry, 4 miles (6.4 km) south of Cross Plains, Switzerland County, Indiana. Cross Plains Quadrangle; 38°53′45″N, 85°11′40″W.

22. Hanging Rock, 1 mile (1.6 km) north of Madison, Jefferson County, Indiana, along Route 7. Clifty Falls Quadrangle; 38°45′05″N, 85°23′45″W.

23. Stream-cut and outcrop at border between Highland and Adams Counties, Ohio, on Ohio Route 247. Belfast Quadrangle; 39°01′30″N, 83°35′15″W.

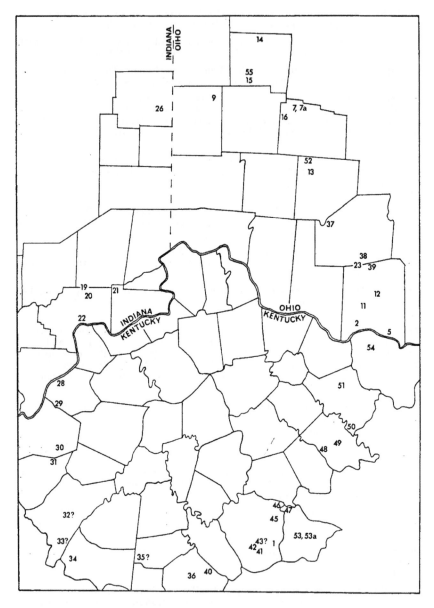

Text-figure 1. — Map showing locations of Brassfield Formation outcrops studied: Counties are outlined. North-south extent of map is 200 miles (320 km); east-west extent is 140 miles (224 km). ? = questionably Brassfield.

26. Top of Elkhorn Falls, just downstream from Indiana Route 227, about 2 miles (3.2 km) south of Richmond, Indiana town line, Wayne County, and just before Route 227 reaches Wolfe Road. New Paris Quadrangle; 39°46'40"N, 84°51'50"W.

28. Oldham County, Kentucky, near top of Ohio River bluff at "Y" junction of private roads to house and to river, north of Dunbar Hollow, on property of George Eggers. LaGrange Quadrangle; 38°27'40"N, 85°29'00"W.

29. Exposure on west side of Brownsboro Road, 0.35 mile (0.56 km) south of bridge over South Fork Harrod's Creek, Oldham County, Kentucky. Anchorage Quadrangle; 38°20'50"N, 85°33'40"W.

30. Road-cut on north side of Seatonville Road (Kentucky Route 1819), just above abandoned quarry on bluff of Floyd's Fork Creek, about 1 mile (1.6 km) west of Seatonville, Jefferson County, Kentucky. Jeffersontown Quadrangle; 38°08'10"N, 85°32'00"W.

31. Road-cut on east side of U.S. Route 31E, about 1.6 miles (2.6 km) SE of junction with Kentucky Route 44 in Mt. Washington, Bullitt County, Kentucky. Mt. Washington Quadrangle; 38°02'00"N, 85°32'00"W.

32. Abandoned quarry on south edge of Bardstown, Nelson County, Kentucky 0.6 mile (1.0 km) south of U.S. Route 31E on extension of 4th Street. Bardstown Quadrangle; 37°48'00"N, 85°28'10"W. [Possibly not Brassfield]

33. Small abandoned quarry on north side of Route 52, about 0.3 mile (0.5 km) north of north edge of New Hope, Nelson County, Kentucky. New Haven Quadrangle; 37°38'50"N, 85°32'10"W. [Possibly not Brassfield]

34. Road-cut on east side of Kentucky Route 527, on north edge of Raywick, Marion County, Kentucky. Raywick Quadrangle; 37°33'45"N, 85°25'50"W.

35. West bluff and east bank of stream in Webb Hollow tributary to Scrubgrass Creek, about 3.4 miles (5.4 km) southwest of Mitchellsburg, Boyle County, Kentucky. Parksville Quadrangle; 37°34'15"N, 84°59'30"W. [Possibly not Brassfield]

36. South bank of Neals Creek, just east of bridge and about 275 yards (252 meters) downstream from church, in Lincoln County, Kentucky. (locality 1 CONE of Foerste, 1906, pp. 147-148.) Halls Gap Quadrangle; 37°29'30"N, 84°39'15"W.

37. Abandoned quarry southwest of road junction at Sharpsville, Highland County, Ohio. Martinsville Quadrangle; 39°15'20"N, 83°45'05"W.

38. Natural exposure on northwest bank of a creek 0.7 mile (1.1 km) southwest of Belfast, Highland County, Ohio, where Ohio Route 485 crosses Rock Lick Creek. Belfast Quadrangle; 39°03'10"N, 83°32'30"W.

39. Natural exposure (near top) on south side of Ohio Route 73, 1.6 miles (2.6 km) west of Louden, Adams County, Ohio, where the highway crosses Flat Run. Sinking Spring Quadrangle; 39°01'50"N, 83°28'45"W.

40. Route 150, 3 miles (4.8 km) NW of Crab Orchard, Lincoln County, Kentucky, where road crosses Cedar Creek. Crab Orchard Quadrangle; 37°28'45"N, 84°33'10"W.

41. East side of Route 25, 1.6 miles (2.6 km) north of Berea, Madison County, Kentucky, where road crosses Silver Creek. Berea Quadrangle; 37°35'45"N, 84°16'55"W.

42. Interstate Route 75, 500 feet (152 meters) north of intersection (on overpass) with Route 595, about 3 miles (4.8 km) NW of Berea, Madison County, Kentucky. Berea Quadrangle; 37°36'00"N, 84°18'50"W.

43. Farristown, Madison County, Kentucky, where Route 1983 crosses Louisville & Nashville Railroad tracks. Berea Quadrangle; 37°37'25"N, 84°18'05"W. [*Possibly not Brassfield*]

45. Route 52 at Muddy Creek, 0.5 mile (0.8 km) west of Waco, Madison County, Kentucky. Moberly Quadrangle; 37°44'25"N, 84°09'15"W.

46. Type locality of Noland Member, Brassfield Formation, about 2 miles (3.2 km) east of College Hill intersection, on private road of the Armstrong Farm (as of 1972), where road ends in a gully after making a left and a right-hand turn, in Madison County, Kentucky. Palmer Quadrangle; 37°47'45"N, 84°05' 30"W.

47. South bluff of Red River on east side of Route 89, Estill County, Kentucky. Palmer Quadrangle; 37°49'00"N, 84°04'00"W.

48. Bluff on Steppingstone Road at Sugar Grove Church, just south of Interstate Route 64, southwest of Owingsville, Bath County, Kentucky. Preston Quadrangle; 38°06'10"N, 83°49'40"W.

49. Route 60, 1.6 miles (2.6 km) east of Owingsville, Bath County, Kentucky. Colfax Quadrangle; 38°08'45"N, 83°44'25"W.

50. South bluff of river valley and adjacent road-cut, along road about 0.6 mile (1.0 km) north of Colfax, Fleming County, Kentucky. Colfax Quadrangle; 38°13'20"N, 83°38'00"W.

51. Route 32, about 2 miles (3.2 km) NW of Goddard, Fleming County, Kentucky. Flemingsburg Quadrangle; 38°22'45"N, 83°38'40"W.

52. Lumberton, Clinton County, Ohio, at abandoned quarry on Wilkins property (as of 1973), about 1000 feet (305 meters) east of McKay Road, and 0.2 mile (0.3 km) south of where McKay Road crosses Anderson Fork Creek. Port William Quadrangle; 39°32'40"N, 83°50'20"W.

53. Kentucky Route 89, east side of road, about 0.3 mile (0.5 km) north of Estill Springs, Estill County, Kentucky, on the north outskirts of Irvine, Kentucky. (Estill Springs is no longer a town, but its site is marked by an historical marker on the west side of the road.) Irving Quadrangle; 37°42'30"N, 83° 58'40"W.

53a. Kentucky Route 89, west side of road, about 1 mile (1.6 km) north of Estill Springs (see comment under locality 53), on north outskirts of Irvine, Estill County, Kentucky. Irvine Quadrangle; 37°43'10"N, 83°58'55"W.

54. Outcrop on Kentucky Route 10, about 2.5 miles (4.0 km) east of Kentucky Route 57 in Tollesboro, Lewis County, Kentucky. Tollesboro Quadrangle; 38° 34'00"N, 83°32'30"W.

55. Ludlow Crush-Stone Quarry, about 1 mile (1.6 km) west of Ludlow Falls, Miami County, Ohio. West Milton Quadrangle; 39°59'30"N, 84°21'15"W.

THE BRASSFIELD LITHOSOME

The rocks of the Brassfield Formation have been recognized as distinct for nearly a century and a half. In his geological report on southwestern Ohio, Locke (1838, p. 239) referred to the "Flinty limestone" (a descriptive, not nomenclatural term) in Adams County, a name apparently pointing to the layers of chert common in the Brassfield there. The presence of ferrous oölite in the Brassfield caused earlier workers to refer it to the Clinton of New York (Owen, 1857, *fide* O'Donnell, 1967, p. 16; Orton, 1871, p. 145).

Foerste (1896) wrote that the differences between the fossil assemblages of the "Clinton" in Ohio and New York inclined him to regard the two as separate, and he suggested that the Ohio unit be called Montgomery Formation for "its typical development in Montgomery county, in Ohio" (p. 189). Later he found that this name was preoccupied and discontinued its use (*fide* O'Donnell, 1967, p. 17). In 1904, Foerste reported that several species found in the New York Clinton do not appear in the vicinity of the Cincinnati Arch until the post-Brassfield Osgood bed, further confirming his contention of the nonequivalency of the two "Clintons".

The term "Brassfield" for the unit as used herein was introduced in 1906 by Foerste (pp. 10, 27). The typical exposure is along the now-abandoned cut of the Louisville and Atlantic Railroad running between Brassfield and Panola in Madison County, Kentucky. The lower part, consisting of a few thick beds of "limestone" (dolomite), is best seen near Panola, while the middle and upper portions, consisting of numerous thinner beds of dolomite, often interlayered with thin beds of clay, is best seen near Brassfield.

The "type" Brassfield consists of these dolomites studied by Foerste along the railroad cut. Above them in this area lie interbedded dolomite and clay, which Foerste (1906, pp. 10, 27) labelled (in ascending order) Plum Creek Clay, Oldham Limestone, Lulbegrud Clay, Waco Limestone, and Estill Clay. He referred these five units to the "Crab Orchard division of Silurian" (p. 27), and wrote later (1935, p. 127) that it was W. M. Linney who in 1882 first used this term with Group status in a report on the geology of Lincoln County, published by the Kentucky Geological Survey. Rexroad, *et al.* (1965) assigned the lower four units, the Plum Creek, Oldham, Lulbegrud and Waco, to their Noland Formation, based upon the exposure at College Hill, Kentucky (locality 46 of this study).

O'Donnell (1967, pp. 38, 46) chose to regard the Noland as a member of the Brassfield Formation. Without making a judgement on the validity of this view, it has proven convenient to follow his usage in this study of the Brassfield corals. Hence, the "Brassfield Formation" as used here includes as a member the Noland Formation of Rexroad, *et al.* (1965).

GEOMETRY OF THE BRASSFIELD LITHOSOME IN THE
CINCINNATI ARCH REGION

The Brassfield Formation outcrops around the Cincinnati Arch in a narrow horseshoe pattern, elongate north-to-south, and opening southward. The latitudinal range is approximately from 20 miles (32 km) north of Dayton, Ohio, to 20 miles (32 km) south of Richmond, Kentucky. The longitudinal range is approximately from eastern Adams County, Ohio, to western Jefferson County, Indiana. The study area extends through southwest Ohio, southeast Indiana, and north-central Kentucky. Beyond this region, the Brassfield has been reported over a wide area, including Tennessee, Alabama, and Illinois (Berry & Boucot, 1970; Swartz, et al., 1942). Most reports of the Brassfield outside of the Cincinnati Arch area seem tentative, and these must be closely scrutinized before their true relationship to the type Brassfield can be understood.

It is likely that the Brassfield deposits originally extended over the Cincinnati Arch. This is suggested by isolated occurrences of Brassfield beds as inliers on the portion of the Arch circumscribed by the main outcrop. Foerste (1935, pp. 125, 169) reported such occurrences at Jeptha Knob near Shelbyville, and at Scrubgrass Creek near Mitchellsburg, both in Kentucky. Distribution patterns of the corals in this unit reinforce the contention that Brassfield sediments originally covered the Cincinnati Arch (see Text-fig. 3).

In this region, the Brassfield attains its greatest thickness (up to 15 meters) on the east side of the Arch between the latitudes of Cincinnati in the north and Lexington, Kentucky, in the south. It is considerably thinner on the west side of the Arch, especially in Indiana and northern Kentucky (with the exception of a thick outcrop near Cross Plains, Indiana, locality 21). The outcrops in this area are commonly less than a meter thick and in certain localities the Brassfield pinches out completely (in western Ripley and adjacent parts of Jennings and Decatur Counties, Indiana, according to Foerste, 1904, p. 326). Such a pinch-out may be seen at locality 19 of this study. O'Donnell (1967, fig. 6) showed the Brassfield thickening slightly eastward from this "Ripley Island pinch-out", and also to the west into the subsurface. While it is uncertain to what degree the westward thinning of the Brassfield reflects primary topography, as opposed to subsequent uplift of the Arch area, the reported thickening of this unit both east and west of the "Ripley Island" area suggests that the sea floor shoaled to the north and west, a conclusion also indicated by the coral and lithofacies distributions.

FACIES DISTRIBUTION

No attempt has been made herein to systematically sample the Brassfield Formation in order to analyze its lithology, or to examine it petrographically. O'Donnell (1967) has done such work, and should be referred to for details. Based upon field observations and the study of some thin-sections, however, I can make some broad statements concerning features of Brassfield sediments and their distribution. The terminology used here is that of Dunham (1962).

Four general lithofacies constitute the Brassfield lithosome. These include:

1. a dolomitic mudstone and wackestone facies
2. a pelmatozoan packstone and grainstone facies
3. a ferruginous pelletal or oölitic packstone and grainstone facies
4. a terrigenous(?) clastic facies of silty clay

Glauconite is common at certain horizons in the Brassfield, and might reasonably be included as a lithofacies, but insufficient information was gathered concerning its occurrence to warrant its inclusion.

Dolomite is nearly ubiquitous throughout the Brassfield in the area studied. It is uncertain to what extent it is primary, rather than a replacement product of micrite. Evidence for both possibilities was observed: In at least one instance (locality 47) shell fragments appear to be altered to dolomite around their peripheries. On the other hand, solitary corals are frequently found lying in a matrix of bioclastic limestone, their septocoels filled with dolomite rhombs. Even here, the possibility remains that the infilling dolomite was originally micrite. In only one thin-section was a possible occurrence of micrite observed, so that if the dolomite is entirely secondary, replacement must have been complete.

The bioclastic limestone seems to consist mostly of echinodermal fragments. Many specimens that appear to be entirely limestone prove, on being thin-sectioned, to consist of dolomite with greater or lesser proportions of shell fragments.

The bioclastic limestone contains numerous evidences of erosion, often well hidden, that indicate a longer period of deposition than the thickness of the deposit would imply. For example, at locality 7 (Fairborn, Ohio) a specimen of the colonial rugosan *Petrozium pelagicum* in a solid block of limestone proved, on close inspection, to have an erosional surface truncating several adjacent corallites and covered by the same sort of sediment as that below the erosional surface.

The ferruginous layers of the Brassfield are similar to those of the New York Clinton. They consist of bioclastic limestone (packstone and grainstone) containing smooth, rounded pellets or oöids. These layers are generally reddish, though not always markedly so. In some instances, the lithofacies was recognized by the presence of the characteristic rounded grains in thin-sections of ordinary looking limestone.

The terrigenous(?) clastic layers are not clearly understood. Their components were not analyzed, but they have a claylike consistency, are gritty to the bite, and effervesce in hydrochloric acid (perhaps a result of contamination by the surrounding lime). Their color is grey-green. Superficially they resemble the sediments of the Richmondian Elkhorn Formation, which in many localities directly underlies the Brassfield.

Possible explanations for the presence of this silty clay are that it represents a recurrence of the conditions that produced the Richmondian shales, or that it represents a reworking of sediments from eroded Richmondian units. A third possibility is that it is due in part to both these causes, and some evidence seems to support this idea. On the one hand, the clastics are frequently associated with scouring channels and with evidence of considerable turbulence (see below). On the other hand, no definite Richmondian fossils were identified from these clays, while they definitely do contain species found in other lithofacies of the Brassfield.

The following are some observations on the lithofacies distribution:

There is a general increase in the ratio of organic fragments (and thus, $CaCO_3$) to dolomite in the Brassfield in a northerly direction. While the southern localities in central Kentucky (*e.g.*, in the type area) are nearly all fine-grained dolomite with only a thin fossiliferous layer at the top (in the Noland Member), the northernmost localities, in the vicinity of Dayton, Ohio, are nearly all limestone, with dolomite occurring mostly toward the base and as scattered rhombs among the organic fragments of the limestone.

Because dolomite and limestone are frequently interlayered and even blend into single beds at some localities, it is not possible to mark a precise boundary between localities on the basis of the dominance of one or the other lithofacies. In general a line extending between the areas of localities 31 and 32 in the southwest, and localities 37 and 38 in the northeast would reasonably separate the more limy regions to the north from the more dolomitic regions to the south. A second, more southerly line extending from the area of localities 50 and 51 to that of localities 36 and 40 roughly separates the almost entirely dolomitic regions to the south from the somewhat more calcareous dolomites northward. (Text-fig. 2.)

With few exceptions, the terrigenous (?) clastic facies occurs either at the top or the bottom of the Brassfield Formation. Where the occurrence is basal, it appears to represent interfingering between the upper Richmondian shales and the Brassfield carbonates. This may be clearly seen at locality 5 in Adams County, Ohio.

The terrigenous (?) clastics of the upper Brassfield are well exposed at localities 46 (College Hill, Kentucky, type locality of the Noland Member) and 1 (east of Panola, Kentucky).

O'Donnell (1967, fig. 9) presented a fence-diagram based largely on subsurface data, showing the shales as thickest toward the south and east areas of the Arch and pinching out to the north and west. The sporadic development of this clastic facies in the Brassfield of the Cincinnati Arch apparently reflects the situation of this area on the border between the terrigenous clastic and the carbonate domains.

In northern localities (7a, 15, 55) the terrigenous (?) facies of the upper Brassfield is of considerable interest for several reasons. First, it occurs often in what seem to be scouring channels or large trough cross-beds near the top of the formation. One of these structures may be seen at locality 15 (West Milton, Ohio) on the south side of Route 571, where it appears in longitudinal profile running east-west.

The northern occurrences of this facies are noteworthy also for the great richness of the fossil coral assemblages they contain (in

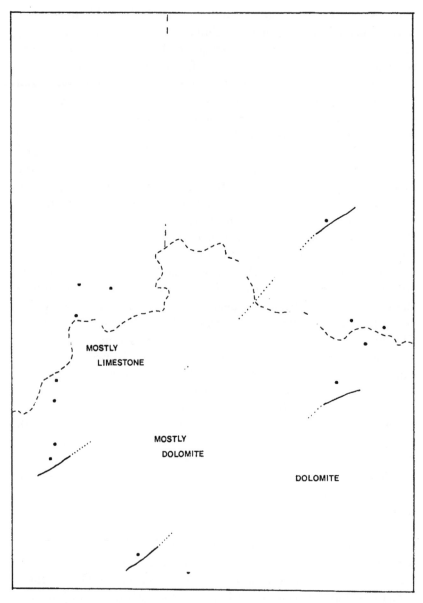

Text-figure 2. — Lithofacies map of the Brassfield Formation in the Cincinnati Arch Region. Boundaries are approximate; dots mark localities visited
(see Text-fig. 1).

contrast with the relatively barren southern clastics). Most of these corals occur also in the carbonate facies of the northern Brassfield, but a small number were observed only in the clastic facies (Table 5).

A clue to the nature of this facies is given by a striking layer found at locality 7a (Fairborn, Ohio) by Dr. William B. Harrison III during this study. This is one of the few localities in the north where the clastics are concentrated toward the middle, rather than the top of the Brassfield Formation. The bed in question is two feet (0.61 m) thick at its center, extending laterally (in outcrop) for some two hundred feet (61 m) before pinching out, and consisting of hundreds of colonial corals packed together like loaves of bread on a baker's shelf, with clay in the interstices. In addition to these, there are two species of solitary corals (one, *Dinophyllum semilunum*, sp. nov., has been found only in this bed). The colonies are worn, some apparently having been broken and abraded so that the cyclical spacing of the tabulae is manifested in parallel ridges on the truncated surfaces. A large number of the colonies lie upside-down in the bed, and the evidence of considerable turbulence, enough to transport and wear colonies as much as 0.4 meter in diameter as sedimentary cobbles, is unmistakable. Most of the colonial corals found in this bed are also found in the carbonates of the Brassfield outcropping in this region, and even occur immediately above the clastic bed in the same outcrop. A possible present-day homologue of this feature will be discussed in the following section.

The ferruginous facies of the Brassfield appears to be associated with a packstone and grainstone lithology. It is characterized, aside from its often striking red color, by the presence of smooth, rounded pellets or oöids. These particles sometimes appear in thin-sections even in the absence of the red color and probably represent the same environment as do their redder counterparts. The red color is most clearly seen in the northeastern localities (7a, Fairborn; 13, Todd Fork; 5, east of Manchester), but the presence of oöids or pellets in thin-sections from localities 37 (Sharpsville, Ohio), 21 (Cross Plains, Indiana), and 31 (Mt. Washington, Kentucky) suggest that the depositional environment of the ferruginous facies occurred there also, though the sediments are not red. Besides the presence of oöids or pellets, this facies is characterized by the omnipresence of calcareous fragments of organic origin and macrofossils, especially echinodermal fragments, bryozoans, solitary corals and orthocone cephalopods.

GEOGRAPHY OF THE BRASSFIELD SEA

Studies of dolomite and dolomitization in recent years indicate that this process is associated with shallow, intertidal, and even supratidal regions of the littoral (Murray and Pray, 1965, p. 2; Shinn, Ginsburg, and Lloyd, 1965). Penecontemporaneous dolomite may be seen today on Sugar Loaf Key in the Florida Keys.

O'Donnell (1967, fig. 11) traced the lithology of the Brassfield Formation for some distance away from the Cincinnati Arch and showed that the dolomitic rocks characteristic of the southern half of this area also occupy the entire region to the west of the Arch. If dolomite is characteristically associated with an environment near the shoreline, this should indicate that the shore ran west of the present Arch down to about the Ohio River. There it turned eastward across the Arch to reappear in the eastern outcrops. The dolomitic deposits of the southern area would represent an environment at or near sea-level with an embayment of shallow sea producing lime sediments.

Unfortunately, this montage is confused by the findings of Deffeyes, Lucia, and Weyl (1965), who reported dolomitization of Plio-Pleistocene sediments today by "refluxing" of magnesium-rich waters down through the permeable sediments of later years. Further, Murray and Lucia (1967) suggested similar secondary dolomitization, controlled by sediment texture in Mississippian beds of Alberta. This means, in terms of the above hypothetical geography, that we cannot be certain if the Brassfield dolomite represents near-shore conditions, or if it represents an area of finer carbonate sediment which was subsequently uplifted to near sea-level and dolomitized later.

Therefore, it seems wiser to use other criteria for clues to the topography of the Brassfield seafloor:

1. In the north and northwest, there is a sharp break between the Brassfield Formation and the underlying Richmondian beds. At several localities, pebbles from the underlying beds occur in the base of the Brassfield (localities 14 and 21), and in at least one instance

(at locality 21) there appears to be an angular unconformity below the Brassfield. Further south, on the eastern side of the Arch (locality 5) the transition from Richmondian to Brassfield beds appears more gradual, represented by a zone of interfingering Brassfield-like dolomite and Richmond-like shale. Gray and Boucot (1972) studied the fossil spores found in this transition zone and in the base of the Brassfield in this region (at locality 12) and concluded there was a paraconformity, recording a transition from near sea-level environment (with intermingling terrigenous and marine conditions), to a more constantly marine environment higher in the section.

2. The three coral biofacies detected in this study appear to occur in zones elongate in a northeast-southwest direction (Text-fig. 3), which conforms to the pattern shown in (1) above, of a gradient in profoundness-of-unconformity along a northwesterly line. The coral biofacies may roughly parallel the contours of the contemporary sea-floor.

3. The Brassfield is thickest to the east and south, and generally thinnest to the northwest, especially in southeast Indiana and northern Kentucky on the west side of the Arch. An actual pinch-out of the Brassfield may be seen at locality 19, near Bryantsburg, Indiana. O'Donnell (1967, fig. 6) showed these relationships in the form of an isopach map. He used subsurface data to extend the area over which the Brassfield's thickness is indicated, and he showed that this unit again thickens to the west of its area of minimum thickness in southeast Indiana. The presence of Brassfield deposits between "Ripley Island", the region of zero thickness, and the crest of the Arch, suggests that the "island" may indeed have been a topographically positive area, affecting Brassfield sedimentation, as stated by Rexroad (1967, p. 1). As the entire Arch area was uplifted subsequent to Brassfield deposition (O'Donnell, 1967, p. 121, said, "There is no relationship between the lithosomes [of the Brassfield] and the present Cincinnati Arch."), it is difficult to be certain to what extent the isopachs reflect original seafloor topography.

4. The proportion of bioclastic material increases toward the north and west, resulting in coarse-grained limestone dominating the north, and grading southward into fine-grained dolomite without

abundant shelly material. The contours of increasing liminess generally parallel the boundaries of the coral biofacies.

5. Large, transported coral colonies (in the clay layer at locality 7a), and apparent scouring channels or large cross-bed troughs (localities 7 and 15) suggest a high-energy environment in the northernmost area of the outcrop. This may indicate shallow water.

A possible modern homologue of the Brassfield sea may be found in the Florida Keys. Rodriguez Key consists of an *in situ* accumulation of calcareous biological debris, the result of a locally high level of biotic activity. It overlies lime-mud, similar to that currently forming in the restricted carbonate environment of Florida Bay (Turmel and Swanson, 1976). It is suggested that this sequence formed as a result of the post-Pleistocene marine transgression northward in this area. At an earlier stage of the transgression, the site of present Rodriguez Key experienced shallow, restricted marine conditions similar to Florida Bay today. As the water deepened, the increased circulation of an open-marine environment provided the food and oxygen necessary to support a large lime-producing community, whose remains have accumulated in Rodriguez Key. A similar stratigraphic sequence occurs in the northern localities of the Brassfield and is most clearly seen at localities 7 and 7a (Fairborn). A deposit of fine material (here dolomite, rather than micrite) is overlain by a large deposit of coarse bioclastic limestone, containing many features indicative of a high energy environment. It seems likely that this sequence reflects the same events as occurred at the site of Rodriguez Key. The Brassfield is generally regarded as a transgressive sequence, and this view is supported by the findings of Gray and Boucot (1972) and by my field observations. The dolomite of the lower Brassfield would represent (either as dolomite, or as dolomitized lime mud) shallow, restricted marine conditions which are succeeded (at least in the north) by more open-marine conditions suited to the support of a sufficiently large benthic community to produce the large shell piles seen in the north.

Another feature of the northern Brassfield may have its modern homologue in the Florida Keys (and not far from Rodriguez Bank). The French Dry Rocks, south of the Keys, are a shoal consisting of an accumulation of large coral fragments, a storm-deposit. The pieces of coral broken from the reef have become tangled forming a resistant structure. Perhaps this explains the origin of the layer of

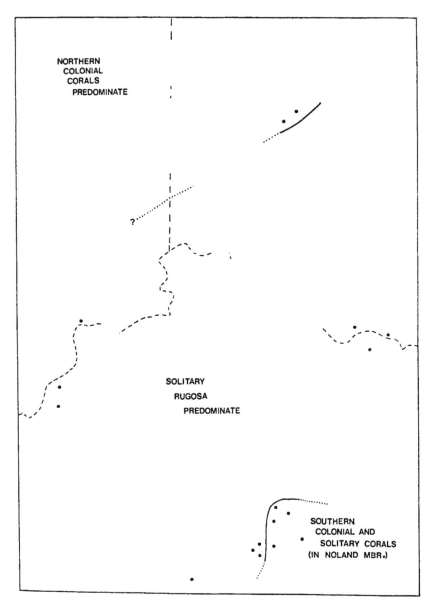

Text-figure 3. — Coral biofacies map of the Brassfield Formation in the Cincinnati Arch Region. Boundary lines of the biofacies are approximate; dots represent localities. See Table 3 for detailed analysis of species occurrence at each locality.

large, transported coral cobbles with interstitial clay (the clay-coral layer) at locality 7a (Fairborn). These coral colonies appear to have lived in the limestone environment and were probably swept from other areas into their final resting-place. The interstitial clay extends no higher in the section than the exact top of the coral layer, suggesting that its presence there was controlled by the coral cobbles. In other words, the "storm-beach" accumulation of coral cobbles served as a sediment trap for the clay particles.

Until the matter of the origin of the Brassfield dolomite is better understood, there seems little profit in speculating as to the location of the shore-line. The water energy regime is more easily understood, all signs pointing to the northern localities as having experienced the most rough conditions.

CURRENT DIRECTIONS

At several localities, cross-beds were found, as well as fossils whose forms and orientations appeared to hold information on bottom-currents in the Brassfield Sea. The fossils included solitary corals and orthocone cephalopods, which might indicate the *direction* in which the current moved (experiments showed that the aperture would tend to point downstream), and bryozoan twigs, which might indicate the *line* along which the current moved, without showing the direction. The most meaningful readings were taken on densely grouped fossils lying on about the same bedding-plane. Some individual fossil orientations were measured to see how they compared with the orientations of the assemblages.

All readings were made in the northeast quadrant of the outcrop area, from the Ohio River northward in Ohio to the Dayton region. Most of the readings indicated a northerly flow direction for the bottom currents. The orientation measurements were a by-product of this study and do not reflect a statistically rigorous sampling design. Consequently, the readings listed in Table 1, and presented graphically in Text-figure 4, should only be regarded as data to be considered in future, more carefully designed studies of directional indicators in the Brassfield Formation.

AGE OF THE BRASSFIELD FORMATION

O'Donnell (1967) and Berry and Boucot (1970) summarized the evidence used to determine the age of the Brassfield. In general outline, it is as follows:

Brachiopods and conodonts have proven the most useful tools in age-correlation work. Berry and Boucot (1970) used the designa-tions of Jones (1925) and Williams (1951 [brachiopods]) for sub-dividing the Llandovery. They divide the early Llandovery into five divisions (A_1 - A_5), the middle Llandovery into three divisions (B_1 - B_3), and the late Llandovery into six divisions (C_1 - C_6). Two brachiopod genera, *Platymerella* and *Microcardinalia*, are regarded as good index fossils by Berry and Boucot (1970, pp. 127-128). The former has a range of "pre-C_1" according to these authors, while the latter is C_1 - C_2. They regard the type Brassfield in Madison County, Kentucky, as being in the *Platymerella* zone, even though this genus is not found there, because northward, along strike, *Platymerella* oc-curs in Ohio beds assigned to the Brassfield that are of pre-*Micro-cardinalia* age.

Microcardinalia occurs in the Oldham limestone bed of the Noland Member (O'Donnell, 1967, p. 75). Rexroad, *et al.* (1965, p. 12) noted that, while one species of this genus occurs in the Old-ham in east-central Kentucky, another species of the same genus occurs at the top of the Brassfield beds immediately below the Noland at Dayton, Ohio, (one of the most northerly localities). They interpret this (p. 13) as an indication that the Brassfield trans-gressed northward.

Berry (1967) reported a new subspecies of the graptolite *Climacograptus scalaris* Hisinger from the Noland Member near Colfax, Kentucky (locality 50). Berry and Boucot (1970, pp. 190-191) noted that this graptolite belongs to a group which ranges from the early Late Ordovician into the early Late Llandovery, con-sistent with their estimate of a C_1 - C_2 age for the Noland in this area.

Rexroad (1967) studied the conodonts of the Brassfield, and concluded (pp. 12-14) that the Brassfield, with the Noland Member, is entirely within the period of Walliser's (1964, *fide* Rexroad, 1967) Bereich I (pre-A_1 to B_3), and specifically that it is entirely within the upper half of that period. Walliser's work was based on Silurian conodonts in central Europe (*fide* Berry & Boucot, 1970, p. 35 and Rexroad, 1967, p. 11). Rexroad further said (p. 14) that graptolite work indicates that the Brassfield Formation (not including the

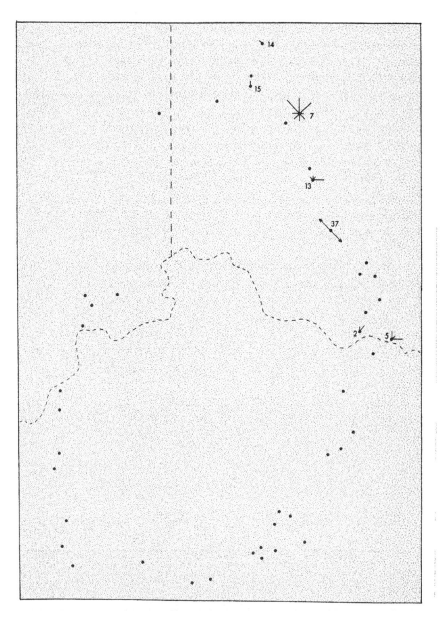

Text-figure 4. — Rose diagrams showing orientations of fossils at outcrops of the Brassfield Formation: Ray lengths are proportional to the number of readings falling in that directional segment. Patterns are based upon the data recorded in Table 1 but do not include the cross-bed readings from locality 5. The plot at locality 37 is based upon the alignment of numerous small, slender twiglike fragments from bryozoan zoaria on a single bedding-plane and indi-

Noland Member) falls within the European zones of *Monograptus cyphus, M. gregarius,* and *M. convolutus (i.e.,* European graptolite

Table 1. — Orientation Readings in the Brassfield.

Locality	Source of data	Orientations
2	3 horn corals lying together in lower Brassfield; calices face:	N 30° E; N; N 45° E
5	6 horn corals 3″ (7.6 cm) above base of Brassfield; calices face:	N 72° E; S 80° E; N 15° E; N 40° E; N 13° W; S 80° E
	4 Cross-bed dips in upper Brassfield, above the upper chert layer (52″-64″ or 1.3-1.6 m above chert)	N 80° E; N 10° W; N; S 25° E
7	22 orthocone nautiloids together on bedding plane within an area of about 1 square meter at top of Brassfield; apertures face:	N 10° W; N 55° W; N 25° E; N 50° E; N 55° W; N; N 15° E; N 55° W; S 25° W; S 30° E; S 35° W; S 30° E; N 40° W; S; N 40° W; S; N 20° E; N 30° E; N 25° E; N 85° W; N 85° W; E
13	3 horn corals in the ferruginous zone near top of Brassfield; calices face:	N 40° E; N 10° E; N 30° W
	3 orthocone nautiloids at same level; apertures face:	E; E; S 80° E
14	*H*orn coral, 106″ (2.7 m) above base of Brassfield; calice faces:	N 60° W
15	*H*orn coral, about 19 feet (5.8 m) above Brassfield base, in channel; calice faces:	N 20° W
37	Numerous twiglike bryozoan fragments on single bedding plane, in middle Brassfield (c. 11 feet (3.4 m) above quarry floor); elongation:	N 35° W — S 35° E

cates only the line along which the current appears to have moved on the sea floor at that point. The other readings are of conical shells (orthocone cephalopods and solitary corals) whose broad apertures empirically orient facing down-current. For these the bottom current direction is indicated.

zones 18 - 20, which include upper A_3 through B_3). This conclusion is based upon correlation of Rexroad's conodont zonation with the standard graptolite zonation.

Rexroad's dating of the Brassfield and Noland based upon conodonts thus produces a somewhat greater age than does the brachiopod work reported by Berry and Boucot. These latter authors, however (1970, p. 191), stated that Rexroad, in a personal communication in 1967, informed them that he found *celloni* Zone conodonts (C_1 through mid-C_3) in the uppermost Waco bed (of the Noland Member). This reconciles better the conodont and brachiopod data.

Hurley, *et al.* (1960, p. 1804), using K-Ar and Rb-Sr methods of age-measurement on glauconite from the Brassfield Formation in Adams County, Ohio obtained an absolute age of 410 million years for the unit.

MORPHOLOGICAL TERMS USED

The terms used to describe coral morphology in this work are, with a few exceptions, based upon Hill (1935; 1956, pp. F245-F251). The latter is a glossary of terms used in reference to rugosans, tabulates, heliolitids, and scleractinians (compiled by R. C. Moore, D. Hill and J. W. Wells), while the earlier work deals specifically with rugosan morphology. Exceptions to Hill's usages are italicized:

RUGOSA

The word *septocoel* is here defined as the space between two adjacent major septa, extending axially to the imaginary plane connecting the axial edges of these septa. Thus, the cardinal fossula (as seen in transverse view in the top portions of Pl. 15, fig. 18, and Pl. 16, fig. 1) consists of two wide septocoels on either side of the cardinal septum.

Neuman (1969, p. 5) used the term *septal lobe* to denote an extension of a septum into the axial region of the corallite which is bent in a sharply different direction from that of the main part of the septum. This is most clearly shown in Plate 21, figure 1. Hill (1956, p. F241) used a similar term, "axial lobe", for a different concept, and these should not be confused.

In corals where the tabularium and dissepimentarium slope steeply toward each other, a moat-like depression develops between them. Often this depression is shallowed by sub-horizontal plates, a large proportion of which are complete as seen in longitudinal sections of the corallite (Pl. 21, fig. 3; Pl. 22, fig. 14; Pl. 23, fig. 1). These are referred to here as *supplementary plates,* corresponding to the "outer series of tabulae" of Hill (1935, text-fig. 19 and p. 511).

TABULATA

Hill (1956, pp. F251, F311) used the term "trabecula" to denote the rodlike element of the septum in rugosans, and the spinose projections found in tabulates and heliolitids. As such blanket use of the term might be suggestive of homology of these features (in what are now regarded as three distinct groups), the more general term *spine* will be used in this work in the case of the tabulates and heliolitids, while Hill's usage will be retained for the rugosans.

In describing the halysitids, Hill and Stumm (1956, p. F469) used the term "microcorallite" for the subquadrate tubes located between the larger corallites in the chains of *Halysites* (Pl. 37, figs. 1, 2; Pl. 32, fig.2; Pl. 41, figs. 1, 2). As Etheridge (1904, p. 17) pointed out, other terms have been applied to these interstitial features. As it is not clear just what they represent, or even if they served a single purpose in various species, it seems best not to designate them as corallites, but to follow the usage of Etheridge (which he attributed to Nicholson) in calling them merely *interstitial tubes.*

The term *balken* refers to rodlike elements that occur in the space between corallites of the Halysitidae, parallel to the corallite axes. The rods occur as two pairs, each pair at the border between a corallite tube and the interstitial area. Hamada (1959, pp. 282-284) discussed and illustrated these features, and noted their potential as systematic tools.

HELIOLITIDA

In describing the heliolitids, Hill and Stumm (1956, p. F458) used the term "tabularium" for the large, tabulated tubes of the corallum, and "coenenchyme" for the tubular or vesicular matrix, separating the tabularia. In this work, the word *corallite* is substituted for "tabularium'", a practice found to be common in the literature. It is felt that the term "tabularium" as used by Hill and Stumm suggests an integral, rugosan-like relationship between these tubes and the intervening matrix of tubular or vesicular material which has not been demonstrated. (The reference here is to the common occurrence of an axial tabularium and a peripheral marginarium [which when vesicular, is referred to as a "dissepimentarium"] in single rugosan corallites.)

Hill stated (1956, p. F246) that "coenenchyme" refers collectively to the coenosteum (meaning "common hard-tissue") and the coenosarc (meaning "common soft-tissue"). As the fossils retain only the hard-parts of the original organisms, the term *coenosteum* is used in this work when referring to the intervening skeletal structure between the corallites of the heliolitids.

MORPHOLOGICAL OBSERVATIONS

The corals of the Brassfield Formation display several morphological peculiarities worthy of closer attention:

Cnidarians are often referred to as possessing radial symmetry. This radial symmetry is superficial. The common elongation of the mouth, the presence of a siphonoglyph at one end of the mouth as in the Octocorallia (Bayer, 1956, p. F168), and the pattern of development of the mesenteries in Recent representatives of this group all clearly point to a fundamental bilateral symmetry in the Cnidaria, upon which radiality has been superimposed (as in the Echinodermata).

In the Rugosa, the symmetry is clearly bilateral, if a careful examination is made of the septal pattern. The presence of a cardinal fossula reinforces this bilaterality. This feature also illustrates a second aspect of rugosan morphology which might be called "directedness".

The solitary Brassfield rugosans which bear distinct cardinal fossulae generally have smooth exteriors, and are curved, the fossula on the convex side of the corallite. This is probably related to the life position and feeding attitude of the animal. If the fossula may be presumed to mark an extension of the mouth or feeding apparatus, then probably the coral lay in the sediment on its convex side, with its oral disc oblique to, and touching the sea floor. This would bring the mouth region closer to the bottom and, presumably, to the principal source of the organism's food. Examples of this type of corallite may be seen in Plate 1, figures 1-3; Plate 2, figures 15-17; and Plate 3, figures 9, 10.

In the other major types of Brassfield solitary corals, the scolecoid, patellate, and talon-bearing forms, the life-position appears to have been different. The scolecoid forms (Pl. 6, figs. 16, 17) probably lay on the sediment under low-energy conditions on their sides, changing position either as occasionally rolled by gentle currents, or more likely, as a result of changes in the location of their center of gravity due to axial growth. In any case, this radial change in life-position would seem to preclude the formation of a specialized organ in association with a specific radial position.

The patellate forms (Pl. 4, figs. 5-7; Pl. 5, fig. 1) were generally broader than high, with explanate peripheries which encrusted the sediment surface to produce a stable structure. The oral disc was directed upward and was surrounded by the everted calice walls for an equal distance on all sides so that, again, the oral region was kept away from contact with the sediment surface, and no appreciable fossula formed.

The talon-bearing forms (Pl. 4, figs. 8, 9; Pl. 5, figs. 2, 3, and 6, 7) grew upward from the sediment, and were propped by lateral extensions of their basal portions (on the convex side when the corallite was curved). Probably, the oral disc was more-or-less horizontal, parallel to the sea floor and was certainly not in contact with it. These forms may have depended for their food on different sources than did those forms with mouths closer to the bottom.

There are three possible explanations of the relationship between curvature and radial position of the fossula in the first of these four types of corallites. (1) The position of the fossula may have been predetermined genetically, with the result that the curvature of the corallite followed suit. (2) The curvature of the corallite may have been predetermined genetically, so that the fossula was positioned to conform with the direction of curvature. (3) Both the orientation of the curvature and the position of the fossula may have been linked by some third factor. As this is a cardinal fossula, that is, one associated with the cardinal septum, it seems safe to eliminate the second possibility, because the cardinal septum is defined before the corallite is long enough for curvature to occur.

Minato (1961, text-figs. 20, 22) showed the ontogenetic stages of *Dalmanophyllum minimum* (Ryder), a solitary rugosan from the Silurian of Gotland. It is clear that curvature is convex on the cardinal side, but it is also seen that the strongest development of the cardinal fossula occurs at the point of greatest curvature of the corallite. This suggests that, while the cardinal septum surely pre-dates and probably determines the curvature, there may be a close tie between the position of curvature and the radial location of the fossula, both with regard to position and time. It is not yet possible to say which determines which.

The radial anisotropy of rugose corals is clearly defined. There is, however, another sort of anisotropy that has received less atten-tion. This is torsion of the axis ("axial vortex" of Hill, 1935, p. 512), in which the septa do not take the shortest possible route (as seen in transverse section) to the axis of the corallite. Rather, the septa bend laterally in the axial region of the corallite, producing a whirl-pool-like appearance (*e.g.*, Pl. 1, fig. 2; Pl. 16, fig. 1).

This torsional phenomenon is common among the rugosan corals, and while it has been cited in some generic and specific descriptions, it has received little attention for its own merit. Only Voynovskiy-Kriger (1971) has dealt with this subject in detail. His paper is preliminary in nature, but shows a background of con-siderable investigation by its author. Dealing specifically with rugose corals, Voynovskiy-Kriger gives a classification of torsional patterns and examples of species in which they occur. While it is possible that some of his torsional types will prove to be variants of one another rather than mutually exclusive, Voynovskiy-Kriger's classification seems to reflect some true, distinct patterns. He reinforces this by tracing the torsional patterns of his classification system from the Ordovician through the Carboniferous (not the Permian), show-ing that the various types seem to have replaced one another in a regular manner through time (pp. 19-21).

He also notes that torsion can appear very early in ontogeny, and that torsional type can change with the different growth stages of a corallite (p. 21).

Torsion appears to have some correlation with taxonomy, in the sense that it seems to be characteristic of certain taxa and not of others. The taxonomic level on which torsion is significant has not yet been determined. In the Brassfield corals, torsion was found in the following species:

Rhegmaphyllum daytonensis (Foerste) — Plate 15, figure 18; Plate 16, figure 12
Dinophyllum stokesi (Milne-Edwards and Haime) — Plate 1, figure 2
Dinophyllum hoskinsoni (Foerste) — Plate 12, figure 10; Plate 13, figure 5
Dinophyllum semilunum, sp. nov. — Plate 14
Schlotheimophyllum benedicti (Greene) — Plate 17, figures 2, 3, 5
Cyathactis sedentarius (Foerste) — Plate 38, figures 1, 2

The first five species belong to the Streptelasmatidae, suggesting that torsion was fairly common in this family during the middle Llandovery. All three of the species of *Dinophyllum* found during this study show torsion, while only one of the three species of *Schlotheimophyllum* shows it. Voynovskiy-Kriger noted that torsion in affected genera can be absent in some species, and in affected species it may be absent in some individuals. He also reported its occurrence in colonial forms, where it may be found in some corallites and not in others. Among the 29 corallites in the type suite of *Zaphrenthis charaxata* Foerste (1906) (USNM 87178) thirteen specimens had clearly visible calices, and eleven of these had axial whorls. Of nine additional specimens collected from the Ohio Brassfield Formation, all with clear calices, five show axial whorls. One specimen (UCGM 41938; Pl. 15, fig. 17-19) shows that torsion can occur at certain levels and not at others.

Voynovskiy-Kriger (p. 23) was impressed by the great predominance of the counterclockwise torsional direction in rugose corals (*i.e.,* looking down upon the calice floor from the distal end of the corallite, the axis has the appearance of a whirlpool rotating in a counterclockwise direction). All torsion in Brassfield corals is counterclockwise. Voynovskiy-Kriger cited only a few examples in which accounts in the literature seemed to indicate clockwise torsion (p. 22-23). Laub (1978), in a summary of Voynovskiy-Kriger's paper and his own findings, showed that these instances of supposed clockwise torsion were either questionable, or, in the case of Rominger's (1876) plates, definitely the result of printing error reversals. Ivanovskiy (1960) described a new species, *Dinophyllum breviseptatum,* and his figure shows the torsional direction of the corallite axis as clockwise (see discussion of *Dinophyllum stokesi*). I feel it likely that this figure will prove to have been reversed, or else that the original thin-section was made by mounting the sliced specimen upside-down, both of which are common errors and result in reversal of true torsional direction. If this is so, then it is possible that torsion in the Rugosa is universally counterclockwise.

It would be of interest to detect axial torsion among living corals, as then its significance might be determined through experiments with the living animal. An examination of the many figures provided by Wells (1956) turned up no clear cases of axial torsion in scleractinian corals.

This torsional phenomenon, geometrically, allows lengthening of the septa without an accompanying increase in corallite diameter. One might guess, then, that this is one mechanism by which the corallite could increase its surface area (or the absorptive area of its coelenteron) without an accompanying volume increase.

DISTRIBUTION OF CORALS IN THE BRASSFIELD FORMATION

The 54 species described here (29 rugosans, 19 tabulates, 6 heliolitids — see Table 2) represent the most prominent elements of the Brassfield coral fauna. More have been found but not yet analyzed, and many others doubtless await discovery.

The Brassfield Fm. dates from the mid-Llandovery, just before the abundant and diversified worldwide coral faunas characteristic of the later Silurian had developed, according to current views. Taken in the light of this historical context, the respectability of this fauna is substantiated in comparing it with described coral faunas from the classic rock units of the later Silurian. Stumm (1964, p. 10) reported 67 species of corals from the Louisville Limestone (Late Wenlock and Early Ludlow) at the Falls of the Ohio near Louisville, Kentucky. Amsden (1949) described a coral fauna of 41 species from the Brownsport Formation (Ludlovian) of western Tennessee.

GEOGRAPHIC DISTRIBUTION

The Brassfield Formation on the west side of the Cincinnati Arch is generally thin, with low coral diversity and abundance. Some exceptions are encountered, as at localities 30 (Seatonville, Kentucky) and 17 (near Bunker Hill, Indiana, which may not be Brassfield). Apparently, the combination of true primary shoalling in the vicinity of Ripley Island (see discussion under "Brass-field Lithosome"), and erosion during uplift of the Cincinnati Arch cloud the stratigraphic and geographic patterns of coral distribution in this area. Nonetheless, the species here are those found elsewhere in the unit, as shown in Table 3.

Table 2. — List of rugosan, tabulate, and heliolitid coral species in the Brassfield Formation of the Cincinnati Arch region.

RUGOSANS

Family Streptelasmatidae
  Streptelasma scoleciforme, sp. nov.
  Dinophyllum stokesi (Milne-Edwards & Haime), 1851
  Dinophyllum hoskinsoni (Foerste), 1890
  Dinophyllum semilunum, sp. nov.
  Dalmanophyllum linguliferum (Foerste), 1906
  Dalmanophyllum (?) *obliquior* (Foerste), 1890
  Rhegmaphyllum daytonensis (Foerste), 1890
  Schlotheimophyllum patellatum (Schlotheim), 1820
  Schlotheimophyllum benedicti (Greene), 1900
  Schlotheimophyllum ipomaea (Davis), 1887

Family Paliphyllidae
  Paliphyllum primarium Soshkina, 1955
  Paliphyllum suecicum Neuman, 1968 *brassfieldense,* subsp. nov.
  Paliphyllum regulare, sp. nov.
  Cyathactis typus Soshkina, 1955
  Cyathactis sedentarius (Foerste), 1906
  Protocyathactis cf. *P. cybaeus* Ivanovskiy, 1961

Family Tryplasmatidae
  Tryplasma radicula (Rominger), 1876
  Tryplasma cylindrica (Wedekind), 1927
  Tryplasma sp.

Family Calostylidae
  Calostylis spongiosa Foerste, 1906
  Calostylis lindstroemi Nicholson & Etheridge, 1878

Family Chonophyllidae
  Strombodes socialis (Soshkina), 1955

Family Arachnophyllidae
  Arachnophyllum mamillare (Owen), 1844
  Arachnophyllum granulosum Foerste, 1906
  Petrozium pelagicum (Billings), 1862
  Craterophyllum (?) *solitarium* (Foerste), 1906

Family Halliidae
  Pycnactis tenuiseptatus, sp. nov.
  Schizophaulactis densiseptatus (Foerste), 1906

Family Cystiphyllidae
  Cystiphyllum spinulosum Foerste, 1906

(Table 2 continued)

TABULATES

Family Favositidae
    Favosites favosus (Goldfuss), 1826
    Favosites discoideus (Roemer), 1860
    Favosites hisingeri Milne-Edwards & Haime, 1851
    Favosites sp. A
    Favosites densitabulatus, sp. nov.
    Paleofavosites prolificus (Billings), 1865b

Family Halysitidae
    Halysites catenularius (Linnaeus), 1767
    Halysites nitidus Lambe, 1899
    Halysites (?) *meandrinus* (Troost), 1840?
    Catenipora gotlandica (Yabe), 1915
    Catenipora favositomima, sp. nov.

Family Moniloporidae
    Cladochonus sp. A
    Cladochonus sp. B
    Cladochonus (?) sp. C

Family Syringoporidae
    Syringopora (?) *reteformis* Billings, 1858

Family Syringolitidae
    Syringolites vesiculosus, sp. nov.

Family Alveolitidae
    Alveolites labrosus (Milne-Edwards & Haime), 1851
    Alveolites labechii Milne-Edwards & Haime, 1851

Family Pachyporidae
    Striatopora flexuosa Hall, 1851

HELIOLITIDS

Family Proporidae
    Propora conferta Milne-Edwards & Haime, 1851
    Propora exigua (Billings), 1865b
    Propora eminula (Foerste), 1906

Family Heliolitidae
    Heliolites spongodes Lindström, 1899
    Heliolites spongiosus Foerste, 1906

Family Palaeoporitidae
    Palaeoporites sp.

On the eastern side of the Arch, the Ohio River provides a convenient line of reference. Twenty species were found only north of the River, while 24 were observed only south of this latitude. Only 10 species were identified, either definitely or questionably, from both north and south of the River (Table 4).

Remarkable changes in species diversity and abundance occur along this north-south transect. North of the Ohio River, especially in the vicinity of Dayton, Ohio, there are many localities with rich coral assemblages (especially 14, 55, 15, 7, 7a, 13, 37, 11 — see Text-

Table 3. — Occurrence of coral species in the Brassfield Formation at the localities visited. A dot indicates that the species was collected and identified from that locality by the author during this study. An *F* indicates that Foerste reported the species from that locality, but that it was not collected by the author. A *?* indicates that the identification by the author of that species from that locality is questionable.

1 2 5 7 8 9 11 12 13 14 15 16 19 20 21 22 23 26 28 29 30 31 32 33 34 35

Streptelasma scoleciforme
Dinophyllum stokesi
D. hoskinsoni
D. semilunum
Dalmanophyllum linguliferum
D. (?) obliquior
Rhegmaphyllum daytonensis
Schlotheimophyllum patellatum
S. benedicti
S. ipomaea
Paliphyllum primarium
P. suecicum brassfieldense
P. regulare
Cyathactis typus
C. sedentarius
Protocyathactis cf. *P. cybaeus*
Tryplasma radicula
T. cylindrica
Tryplasma sp.
Calostylis spongiosa
C. lindstroemi
Strombodes socialis
Arachnophyllum mamillare
A. granulosum
Petrozium pelagicum
Craterophyllum (?) solitarium
Pycnactis tenuiseptatus
Schizophaulactis densiseptatus
Cystiphyllum spinulosum
Favosites favosus
F. discoideus
F. hisingeri
F. densitabulatus
Favosites sp. A
Paleofavosites prolificus
Halysites catenularius
H. nitidus
H. (?) meandrinus?
Catenipora gotlandica
C. favositomima
Cladochonus sp. A
Cladochonus sp. B
Cladochonus (?) sp. C
Syringopora (?) reteformis
Syringolites vesiculosus
Alveolites labrosus
A. labechii
Striatopora flexuosa
Propora conferta
P. exigua
P. eminula
Heliolites spongodes
H. spongiosus
Palaeoporites sp.

fig. 5). Scattered among these localities are others with lower diversities.

From just north of the Ohio River and southward to roughly the latitude of Richmond, Kentucky, is a chain of localities which contains a monotonous fauna of low diversity. The only species likely to be found is the solitary rugosan, *Rhegmaphyllum daytonensis* (Foerste) though this species may be extremely abundant at some of these localities. *R. daytonensis* is the most cosmopolitan Brassfield coral species (see Text-fig. 5E).

South of this chain of localities is a small area containing sites of almost totally barren Brassfield. In these the top of the sections bears a coral fauna of even greater diversity than that encountered in the north. As in the north, these sites are surrounded by localities with relatively low coral abundance and diversity.

Much additional work is necessary for an understanding of the distribution of Brassfield corals and the environments in which they lived. A plausible model for the Brassfield Sea and its history can be obtained by analyzing the association of corals with sediment types and by using a possible modern environmental homologue such as Rodriguez Bank in the Florida Keys (see "the Brassfield Lithosome" and below).

CORAL-SEDIMENT ASSOCIATIONS

Table 5 shows the distribution of corals with respect to general type of sediment. All sediment types contain rugosans, tabulates, and heliolitids in roughly comparable proportions.

The fine-grained dolomite contains the most diverse assemblage, with twenty-nine species found only in this rocktype. Seventeen species occur only in the coarse-grained limestone. Four species occur in both the dolomite and limestone. Four species are restricted to the clay.

The distributions of corals with respect to sediment type (Table 5) and geography (Table 4) are closely correlated. The species that are characteristic of the southern area are there because of the dominance of fine-grained dolomite, while the northern species are apparently controlled by the restriction of coarse-grained limestone largely to that area. The corals possibly are not responding to the sediments themselves, but rather to the environments they represent.

Table 4. — Occurrence of Brassfield coral species. Headings refer to distribution with respect to the Ohio River, in the Cincinnati Arch Region.

NORTH ONLY

Dinophyllum semilunum
Dalmanophyllum (?) *obliquior*
Schlotheimophyllum patellatum
S. benedicti
S. ipomaea
Paliphyllum primarium
P. suecicum
P. regulare
Strombodes socialis
Petrozium pelagicum
Halysites catenularius
Catenipora gotlandica
C. favositomima
Cladochonus sp. B
Cladochonus (?) sp. C
Alveolites labechii
Striatopora flexuosa
Propora conferta
P. exigua
Palaeoporites sp.

SOUTH ONLY

Streptelasma scoleciforme
Dinophyllum stokesi
Dalmanophyllum linguliferum
Cyathactis typus
C. sedentarius
Protocyathactis cf. *P. cybaeus*
Tryplasma radicula
T. cylindrica
Tryplasma sp.
Calostylis spongiosa
Arachnophyllum mamillare
A. granulosum
Craterophyllum (?) *solitarium*
Pycnactis tenuiseptatus
Schizophaulactis densiseptatus
Cystiphyllum spinulosum
Favosites sp. A
Halysites (?) *meandrinus?*
Cladochonus sp. A
Syringolites vesiculosus
Alveolites labrosus
Propora eminula
Heliolites spongodes
H. spongiosus

NORTH & SOUTH

Dinophyllum hoskinsoni
Rhegmaphyllum daytonensis
Calostylis lindstroemi (occurrence questionable)
Favosites favosus
F. discoideus
F. hisingeri
F. densitabulatus
Paleofavosites prolificus
Halysites nitidus
Syringopora (?) *reteformis*

Text-figure 5. — Number of coral species at each locality in the Brassfield Formation:

A. Rugosan, tabulate and heliolitid corals; dot size is proportional to number of species found. In compiling this map, reliable data from Foerste's work were used. Several species questionably identified in the field were included where it was probable that they represented a species other than those already identified from the locality.

B. Rugose corals

C. Tabulate corals

D. Heliolitid corals

E. Distribution of the solitary rugose coral *Rhegmaphyllum daytonensis* (Foerste, 1890) [larger dots]. *R. daytonensis* is the most widely-occurring coral species in the Brassfield in the Cincinnati Arch Region.

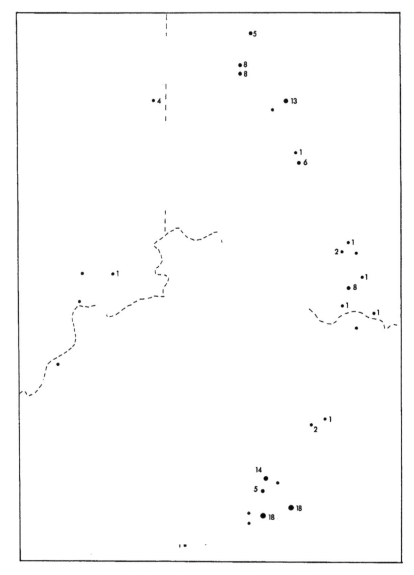

Text-figure 5A. — Number of rugosan, tabulate and heliolitid coral species at each locality in the Brassfield Formation. Dot size is proportional to number of species found.

ies at each locality in the

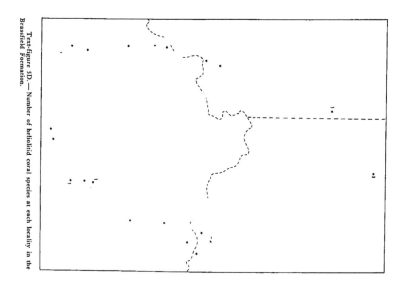

Text-figure 5C. — Number of tabulate coral species at each locality in the Brassfield Formation.

Text-figure 5D. — Number of heliolitid coral species at each locality in the Brassfield Formation.

Table 5. — Categorization of the Brassfield coral species with regard to the type of sediment with which they are associated. * = species in limestone or dolomite that also occur in clay.

DOLOMITE ONLY

Streptelasma scoleciforme
Dinophyllum stokesi
Dalmanophyllum linyuliferum
Paliphyllum primarium
Cyathactis typus
C. sedentarius
Protocyathactis cf. *P. cybacus*
Tryplasma radicula
T. cylindrica
Tryplasma sp.
Calostylis spongiosa
C. lindstroemi
Arachnophyllum mamillare
A. granulosum
Craterophyllum (?) *solitarium*
Pycnactis tenuiseptatus
Schizophaulactis densiseptatus
Cystiphyllum spinulosum
Favosites favosus
Favosites densitabulatus
Favosites sp. A
Halysites (?) *meandrinus?*
Cladochonus sp. A
Cladochonus sp. B
Syringolites vesiculosus
Alveolites labrosus
Propora eminula
Heliolites spongodes
H. spongiosus

LIMESTONE ONLY

Dalmanophyllum (?) *obliquior*
Schlotheimophyllum patellatum
S. ipomaea
Paliphyllum suecicum brassfieldense
**P. regulare*
**Strombodes socialis*
**Petrozium pelagicum*
**Favosites discoideus*
**Paleofavosites prolificus*
Halysites catenularius
H. nitidus
**Catenipora gotlandica*
C. favositomima
**Striatopora flexuosa*
**Propora conferta*
P. exigua
Palaeoporites sp.

CLAY ONLY

Dinophyllum semilunum
Schlotheimophyllum benedicti
Cladochonus (?) sp. C
Alveolites labechii

DOLOMITE & LIMESTONE

Dinophyllum hoskinsoni
Rhegmaphyllum daytonensis
**Favosites hisingeri*
Syringopora (?) *reteformis*

As with Rodriguez Bank, the fine-grained dolomite of most of the Brassfield is thought to have been the product of a restricted, near-shore environment, similar to Florida Bay today. Possibly these dolomites originally were lime-muds (micrites) and were subsequently dolomitized, but no definite pronouncements are warranted by available information. As the Brassfield Sea transgressed in a northerly direction, this facies supported a small number of coral species, virtually all of them solitary rugosans, few of which are identified and described in this study. They generally appear to be streptelasmatids, corals whose lack of a dissepimentarium makes them more akin to the Ordovician forms than to those of the later Silurian.

In places, as the water deepened during transgression and more open-marine conditions prevailed, the resulting improved circulation supported an increased benthic population, including bryozoans, corals, and pelmatozoan echinoderms. The skeletons of these organisms accumulated into "shell piles" similar to those at Rodriguez Key.

Clay found its way into the environment, and was sometimes stabilized, perhaps by cystoid roots and algae, into mounds such as those seen at locality 7 near Fairborn, Ohio (Kissling, 1973, p. 49; personal communication, 1974). The sources of this clay are uncertain but may have been the Richmondian shales to the north and west.

In these "banks" or "shell piles", coral life was abundant. Limestone bedding planes are covered with *Favosites discoideus* (Roemer), *Paleofavosites prolificus* (Billings), and other, less-abundant species. Here colonial forms dominated the fauna. Few if any seem to have cemented themselves to firm substrates (except presumably to shell fragments in the earliest stages of skeletal development), so that they were subject to transport by strong currents. Most of the massive forms maintained a low physical profile, offering a small cross-sectional area to the currents. Kissling (1973) described this lack of firm attachment, and even current-inspired vagrancy, in the case of *F. discoideus* (which he refers to as *F. favosus*) at Fairborn, Ohio and has compared its probable mode of life with that of the modern scleractinian coral, *Siderastrea radians* (Pallas) in the Florida Keys.

The remarkable "clay-coral" bed lies in the midst of these supposed "bank" sediments at locality 7a (Fairborn). It consists mostly of large colonial corals, a great proportion of them overturned and showing signs of considerable wear. These corals (*Paleofavosites prolificus* (Billings), *Catenipora gotlandica* (Yabe), *Propora conferta* Milne-Edwards and Haime, *Strombodes socialis* (Soshkina), *Favosites discoideus* (Roemer), and the solitary rugosans *Paliphyllum suecicum* Neuman *brassfieldense*, subsp. nov., and *Dinophyllum semilunum*, sp. nov.) lie in a matrix of clay and are stacked one above the other. The entire bed extends laterally for some 60 meters, and is about 0.6 meter in maximum thickness. In outcrop it is lens-shaped. The possibility that this bed represents a storm beach of corals swept from the bottom and piled into a wave-resistant structure comparable to the French Dry Rocks off of Tavernier Key, Florida Keys, has been discussed under "Geography of the Brassfield Sea".

The clay beds of the northern localities do not always contain such large and abundant coral specimens as are found in the clay-coral bed of locality 7a, but they do contain a much more concentrated assemblage of corals (especially at localities 15 and 55) than do the adjacent and surrounding limestones.

As shown in Table 5, ten of the fourteen species found in the clay are also found in the limestone. Only four of the species from the clay have yet to be found in another facies. Considering the evidence of transport and wear associated with the clay deposits at locality 7a (and at locality 15, in the form of scour channels) it seems likely that all or most of the corals found in the clay beds of the upper Brassfield were derived from the limestone facies.

In the southern region, especially at localities 1 (Panola), 46 (College Hill) and 53-53a (Irvine region), the top of the Brassfield (Noland Member) contains abundant shell material in dolomite (Pl. 6, fig. 1). Here is found the greatest diversity of coral species of the entire Arch, all in the uppermost few inches of the outcrops. A low-energy environment is suggested by several factors: Delicate fenestrate bryozoans are abundant; many of the solitary rugosans (*e.g., Tryplasma cylindrica* (Wedekind) and *Streptelasma scoleciforme*, sp. nov.) are scolecoid and must have lain free on their sides; several species (*e.g., Calostylis lindstroemi* Nicholson and Etheridge and *Protocyathactis* cf. *P. cybaeus* Ivanovskiy had small talons at their proximal extremities for attachment, which probably would not have supported them in rough water. (One specimen of *C. lindstroemi*, Pl. 6, figs. 16, 17, has a tiny talon, but the corallite is so long and scolecoid that it must have been attached only in its early stages.)

The Noland Member of the southern Brassfield is regarded as about coeval with the northern Brassfield in the vicinity of Dayton (see "Age of the Brassfield"). So, the diverse southern fauna would about correspond in age with that of the north. In support of this contention, some elements of the southern fauna were found near the top of the Brassfield in the north. Specifically, a single slab (UCGM 41859) from the upper Brassfield at locality 13 (Todd Fork near Wilmington, Ohio) contains *Favosites favosus* and *Calostylis lindstroemi*. This suggests some interfingering of the restricted, quiet environment inhabited by these species with the more open-marine environment of the northern fauna. It is uncertain why this southern fauna of the dolomites did not appear in the earlier transgressive sequence in the south though it may be due in part to its not yet having evolved.

THE ANCESTRY, GEOGRAPHIC EXTENT, AND FATE OF THE BRASSFIELD CORAL FAUNA

Knowledge of the major species components of the Brassfield coral fauna leads to consideration of the following questions:

1) Where did the corals come from? Did this fauna have a single ancestral homeland, or did its components immigrate from a number of places? How many Brassfield coral species existed prior to Brassfield deposition?

2) Over what area did the Brassfield coral fauna extend as a recognizable entity?

3) What became of these corals after the termination of Brassfield deposition in the Cincinnati Arch region?

In quest of answers to these questions, an extensive literature search was made of described Late Ordovician through Wenlockian (and some Ludlovian) corals of the world. About two hundred references were consulted, covering as many regions of the world as possible (Text-fig. 6; Table 6). All references used consisted of descriptions and figures (or, in the absence of figures, of descriptions sufficiently detailed to permit reasonable confidence of identification). Where possible, the specimens themselves were examined, though these represent a minority. In no instance was a faunal list, or the mere mention of an occurrence used as data in this compilation.

Works by Berry and Boucot (1970, 1972), Hamada (1958), Gignoux (1950) and Manten (1971) made it possible to define the ages of almost all occurrences sufficiently to give meaningful information on the history of these species before, during, and after Brassfield deposition (Tables 7, 8). The results were presented in an earlier paper (Laub, 1975) with geographic distributions of the species plotted on a map published by Briden *et al.*, (1973). In the present paper, maps supported by more recent studies of Ziegler, Hansen *et al.*, (1977) and Ziegler, Scotese *et al.*, (1977) are used. The former reference provides an Early Silurian (Llandovery) geography in the form of a "squashed" Mercator projection, used in Text-figs. 6-8. The latter reference provides a Middle Silurian (Wenlock) geography in the form of a true Mercator projection used in Text-fig. 9. Ziegler (personal communication, 1978) cautions that the position of China on both maps, in the absence of paleomagnetic data, is conjectural.

Calculations using paleomagnetic data from Hicken *et al.* (1972), done with the kind help of Dr. Kees deJong, gave a latitude of 27°S for the Cincinnati area during the Llandovery. This is in good agreement with the maps of Ziegler and his colleagues.

Pre-Brassfield distribution (Text-fig. 7). — Eight of the species appear to have existed prior to Brassfield time. Three more questionably predate the Brassfield. A twelfth species, belonging to the genus *Protocyathactis* Ivanovskiy, has not itself been reported in pre-Brassfield beds, but a similar, and possibly ancestral species, *P. cybaeus* Ivanovskiy, has. Of this total of 12 species, 8 arose in the Late Ordovician, while the remainder first appeared in the early Llandovery.

No single region appears to have been the homeland of these earliest Brassfield species, as they are found in North America, Europe, Asia, and possibly South America, with the Canadian maritime provinces and the Baltic area containing the greatest numbers.

None of the rugose species have been found in the pre-Brassfield rocks of North America. They occur in the Baltic area, the Siberian Platform, and possibly in Venezuela. On the other hand, except for occurrences in the Baltic region, the tabulates and heliolitids that later helped populate the Brassfield Sea are found only on the North American continent.

Distribution during Brassfield time (Text-fig. 8). — The current data indicate that about 42 of the 54 Brassfield coral species appeared first during Brassfield time. The fauna seems to have been restricted geographically, with only 8 species (2 of them questionably) occurring elsewhere during this time. These are found in the northern Great Lakes area, Manitoba, the Canadian maritime

Text-figure 6. — Map showing continental positions during the Early Silurian (Llandovery) (after Ziegler, Hansen, *et al.*, 1977, fig. 2). Localities for which faunal data were obtained for purposes of analyzing the history of the Brassfield coral fauna before, during, and after Brassfield time are marked by numbers which indicate the approximate number of references covering each region. Other references that do not apply to a specific area are not included (see Table 6). The position of China on this map is tentative.

Japan is located on the eastern end of the South China continent, marked "3". Girvan and most of Ireland lie just east of Greenland and south of the Equator, marked "4". England lies at the southwest point of the northwest Europe continent, marked "9". Kazakhstania includes south central Siberia. Siberia's position is inverted relative to the present, with northern Siberia here on the south side, parallelling the Equator. Afghanistan is a triangular block east of Iran-Arabia and just west of India. Tibetia and the Himalayas lie on a block marked "2", between North China and Australia. New Guinea is shown between Malaysia and Australia, on the Equator, and Tasmania is just east of Australia and Antarctica, marked "3".

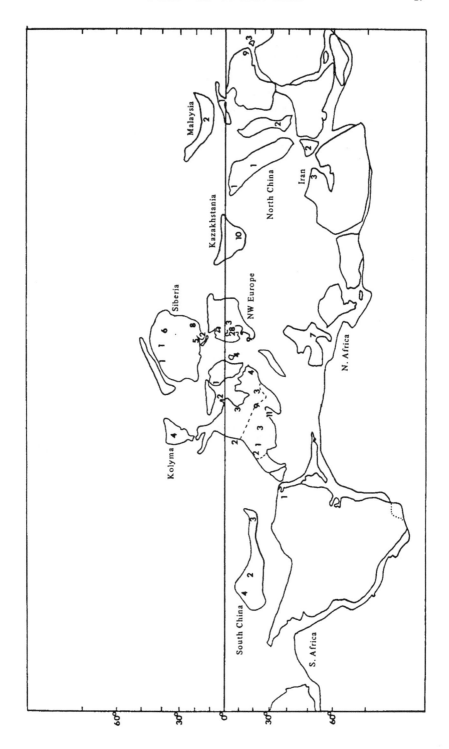

Table 6.—References used in analysis of the world-wide distribution of the Brassfield coral species before, during, and after the time represented by the Brassfield Formation in the Cincinnati Arch region.

NORTH AMERICA

Western Tennessee: Amsden (1949), Roemer (1860) — Sil.
Northern Great Lakes: Bolton and Copeland (1972) — Sil.; Rominger (1876) — Sil.; Hume (1925) — Sil.; Bolton (1957, 1966, 1968) — Sil.; Stumm (1963) — Ord.; Williams (1919) — Sil.; Sinclair (1955) — Ord.
Eastern Great Basin: Budge (1972) — Ord., Sil.
Southwest U.S.: Flower (1961), Hill (1959b) — Ord.
Northwest Ohio: Floyd, *et al.* (1972) — Sil.
Ohio, Indiana, Kentucky: [this study] — Sil.; Stumm (1964) — Sil.; Davis (1887) — Sil.; Kissling (1970, 1973) — Sil.
Canadian Maritimes: Northrop (1939) — Sil.; Twenhofel (1928) — Sil.; Shrock and Twenhofel (1939) — Sil.; Bolton (1972) — Ord. & Sil.
Canadian Cordillera: Norford (1962a, b) — Ord., Sil.; Wilson (1926) — Ord.
Maine, Quebec: Oliver (1962a, b) — Sil.; Stumm (1962)
Iowa: Philcox (1970) — Sil.
Illinois & Missouri: Savage (1913) — Sil.
Oklahoma: Sutherland (1965) — Sil.
Manitoba: Okulitch (1943), Nelson (1963), Sinclair (1955) — Ord.
Canadian Arctic: Teichert (1937) — Ord., Sil.; Cox (1937) — Ord.
Greenland: Troedsson (1928) — Ord., Sil.
New York State: Hall (1847) — Ord.; *H*all (1852) — Sil.; Oliver (1963) — Sil.

SOUTH AMERICA

Mérida Andes, Venezuela: Scrutton (1971)) — Sil.

EUROPE

Gotland: Tripp (1933) — Ord., Sil.; Lindström (1868, 1882a, 1896) — Sil. (possibly some Ord.) ; Jones (in part; 1936) — Sil.
Ireland: Kal'o and Klaamann (1965) — Ord.
Norway: Spjeldnaes (1961, 1964) — Ord.; Stasínska (1967) — Ord., Sil.; Kiär (1899) — Sil.; Scheffen (1933) — Sil.; Klaamann (1971) — Sil.?.
Sweden: Thorslund (1948) — Sil.; Stasínska (1967) — Ord., Sil.; Hisinger (1837-1841) ; Neuman (1968) — Ord.
Czechoslovakia: Prantl (1940) — Sil.; Pocta (1902) — Sil.; Prantl (1957) — Sil.; Galle (1968, 1973) — Sil.
Podolia: Bulvanker (1952) — Sil.; Rózkowska (1946) — Sil.; Sytova (1968) — Sil.
Girvan, Scotland: Nicholson and Etheridge (1878-1880) — Ord., Sil.; Wang (1948) — Ord., Sil.
Leningrad region: Cherkesov (1936) — Sil.
Germany: Weissermel (1894) — Sil.
Eastern Alps: Flügel (1956)
Estonia: Klaamann (1959, 1961a) — Sil.; Klaamann (1961b) — Ord.; Klaamann (1964) — Ord., Sil.; Kal'o (1956a, 1958a, 1961) — Ord.; Kal'o (1957, 1958b) — Ord., Sil.
England: Smith (1930a) — Sil.; Tomes (1887) — Sil.?; Alexander (1947) — Sil.; Colter (1956) — Sil.; Butler (1937) — Sil.; Jones (1936) — Sil.; Lonsdale (1839) — Sil.; Milne-Edwards & Haime (1850-1854) — Sil.; Sutton (1966) — Sil.
Miscellaneous Europe: Lindström (1882b) — Russia; Neuman (1969) — Ord. of Scandinavia; Stasínska (1967) northern Europe; Dybowski (1873, 1874) — Baltic region; Milne-Edwards & Haime (1851)

AFRICA

Kenya coast: Gregory (1938) — age uncertain

ASIA

Kazakhstan: Poltavtseva (1965) — Sil.; Sharkova (1964) — Sil.; Sultanbekova (1971) — Ord.
Lushan, Kueichow: Chen (1959) — Sil.; Tsin (1956) — Sil.
Northwest Korea: Shimizu, *et al.* (1934) — Sil.
Yunnan: Wang (1944, 1947) — Sil.; Grabau (1926) — Sil.; Fontaine (1964)
Inner Mongolia: Wu (1958) — Sil.
Kansu: Yu (1956) — Sil.; Chi (1935) — *possibly Devonian*
Afghanistan: Desparmet (1969) — Sil.; Lafuste and Desparmet (1970) — Sil.
East Iran: Flügel and Saleh (1970) — Sil.; Flügel (1962) — Sil.; Saleh (1969) — Sil.
Himalayas: Sahni and Gupta (1968) — Sil.; Cowper-Reed (1912) — Ord. & Sil.
Malaysia: Thomas (1963) — Sil.; Thomas and Scrutton (1969) — Sil.
Choltagh: Regnell (1941, 1961) — Sil.
Tuvan and Mongolian SSR: Chernyshev (1937b) — Sil.
Taimyr region, Norilsk: Chernyshev (1941a, 1941b) — Sil.; Ivanovskiy (1963) — Ord., Sil.; Zhizhina (1968) — Sil.; Zhizhina and Smirnova (1957) — Sil.
Uzbekstan: Dzyubo (1971) — Sil.; Lavrusevich (1971a) — Ord.; Lavrusevich (1971b) — Sil.
Siberian Platform (primarily the basins of the Podkamennaya Tunguska, and the Sukhaya Tunguska Rivers): Iskyul (1957) — Sil.; Ivanovskiy (1959a, 1960, 1962) — Sil.; Ivanovskiy (1961) — Ord.; Ivanovskiy (1959b, 1963) — Ord., Sil.; Soshkina (1955) — Ord., Sil.
Chukchee Penin., Kolyma River Basin: Kiryushina, *et al.* (1939) — Sil.; Preobrazhenskiy (1964, 1968) — Ord., Sil.; Rukhin (1938) — Sil.
Sayan region, Siberia: Naumenko (1970) — Sil.
Novaya Zemlya and region to south: Zhizhina (1969) — Ord.; Chernyshev (1938a) — Sil.
Salair Mts. and Gornogo Altay: Dzyubo (1960, 1962) — Ord.; Zheltonogova (1965) — Sil.; Cherepnina (1962) — Ord.; Cherepnina (1965) — Sil.
Tadzhikstan: Leleshus (1972) — Sil.; Lavrusevich (1965) — Sil.
Miscellaneous Asia: Orlov (1930) — Sil. of Ferghana; Yu (1960) — Ord. of China; Chernyshev (1938b) — U. Sil. of Letney River Basin

OCEANIA

New South Wales, Australia: Hill (1957) — Ord.; Hill (1940, 1954) — Sil.; Ross (1961) — Sil.; Webby (1971) — Ord.; Strusz (1961) — U. Ord., or L. Sil.; Webby and Semeniuk (1969) — Ord.; Webby (1972) — Ord.; Jones (1944)
Tasmania: Hill (1955) — Ord.; Hill (1942) — U. Ord.?, Sil.; Hill and Edwards (1941) — Ord.
Japan: Sugiyama (1940) — Sil.; Hamada (1956) — Sil.; Hamada (1958) — Ord., Sil.
New Guinea: Musper (1938) — Sil.?
Miscellaneous Oceania: Etheridge (1904, 1907) — Australia; Jones (1937) — Australian *Favosites*; Jones and Hill (1940) — Australian heliolitids

OTHER PERTINENT REFERENCES

Lindström (1899) — heliolitids of world
Tesakov (1965) Silurian corals of USSR
Ivanovskiy (1969) — U. Ord. to Dev., world-wide (*Tryplasmatidae* and *Cyathophylloididae*)
Yabe (1915) — *Halysitidae*, world-wide
Jones (1936) — *Favosites* of various regions
Hill and Jell (1970) — tabulates of various regions

Text-figure 7.— World distribution of Brassfield coral species prior to Brassfield deposition (pre-mid-Llandovery), using same base map as in Text-figure 6. Numbers of Brassfield coral species in each region are indicated.

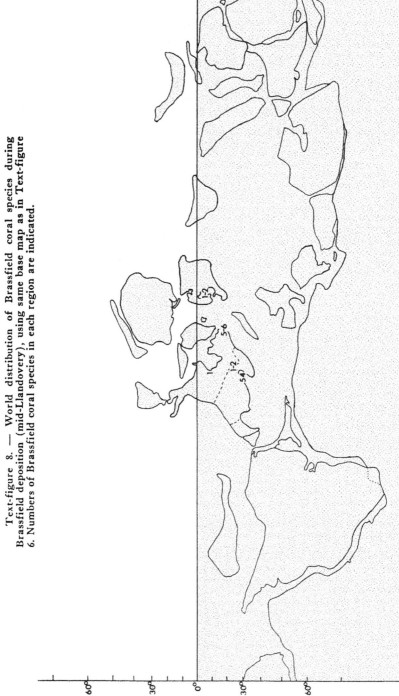

Text-figure 8. — World distribution of Brassfield coral species during Brassfield deposition (mid-Llandovery), using same base map as in Text-figure 6. Numbers of Brassfield coral species in each region are indicated.

provinces, and the Baltic area. Of these, the fauna of the Maritimes shows the greatest similarity to the Brassfield fauna.

Post-Brassfield distribution (Text-fig. 9). — At least 26, and possibly as many as 31 coral species survived the end of Brassfield deposition in the Cincinnati Arch region. These went on to populate a large portion of the epicontinental seas in later Silurian time, uniting with other species to form the well-known post-Brassfield coral faunas of Gotland, Britain, Bohemia, the Siberian Platform, the Far East, and eastern North America.

Several species reappeared in the general vicinity of what had been the Brassfield Sea. When the Louisville Limestone (Late Wenlockian and Early Ludlovian) was being deposited at Louisville, Kentucky, 7 or 8 Brassfield species were present: *Schlotheimophyllum benedicti* (Greene), *S. ipomaea* (Davis), *Halysites nitidus* Lambe, *Arachnophyllum mamillare* (Owen), *Heliolites spongodes* Lindström, *Favosites discoideus* (Roemer), *F. hisingeri* Milne-Edwards and Haime, and possibly *Dalmanophyllum linguliferum* (Foerste). Later, in the Early Ludlovian, the Brownsport Formation of western Tennessee was deposited in a sea inhabited by *Favosites discoideus*, *Favosites* sp. A, *Halysites meandrinus* (Troost), and possibly *Tryplasma radicula* (Rominger).

Current knowledge of the Ordovician and Silurian corals of western North America and South America is limited (though recent advances have been made in the former area). The presence of Brassfield coral species in the former, and possibly in the latter area marks them as promising regions for future investigations into the history of the fauna.

SYSTEMATIC PALEONTOLOGY

In this section the synonymies, besides serving their traditional purpose of summarizing the history of taxonomic names, also represent programs to the discussion section for each taxon. Thus, the prefix "Cf." in a synonymy indicates that the taxon in question will be compared with the taxon bearing this prefix.

Just before each species diagnosis is an indication of the "number of specimens examined". This is to give an idea of the quantity of material that I personally studied prior to composing a descrip-

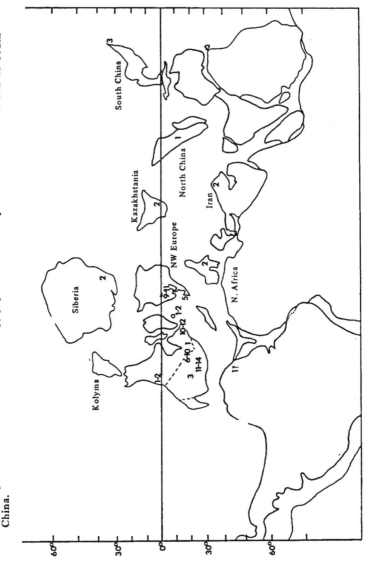

position of China on this map is conjectural.

The explanation for Text-fig. 6 serves as a guide to most of the unlabeled land masses on this map as well. Note that in this map, Japan is marked by a "3" on the north end of South China.

Table 7.— Chronologic ranges of the Brassfield coral species. X = presence; O = Late Ordovician; S = early Llandovery.

	Before Brassfield Time	Occurrence Synchronous with Brassfield outside the Cincinnati Arch Region	After Brassfield Time
RUGOSANS			
Family Feelasmatidae			
Streptelasma scoleciforme, sp. nov.			
Dinophyllum stokesi (Milne-Edwards & Haime), 1851			X
Dinophyllum hoskinsoni (?), 1890			
Dinophyllum semilunium, sp. nov.			
Dalmanophyllum ? ...m (?), 1906	X? (O) ?		X?
Dalmanophyllum (?) obliquior (Foerste), 1890			
Rhegmaphyllum ...is (Foerste), 1890	X? (S)	X?	X?
Schlotheimophyllum patellatum (Schlotheim), 1820			X
Schlothei ...m benedicti (Greene), 1900			X
Schlotheimophyllum ipomaea (Davis), 1887			X
Fmaily Paliphyllidae			
Paliphyllum primarium Soshkina, 1955	X (O)		
Paliphyllum suecicum ...in, brassfieldense subsp. nov.	X (O)		
Paliphyllum regulare, sp. nov.			X
Cyathactis ...tpus Soshkina, 1955			
Cyathactis sedentarius (Foerste), 1906			
Protocyathactis cf. P. cybaeus Ivanovskiy, 1961	X? (O)		
Family Tryplasmatidae			
Tryplasma radicula (Rominger), 1876			X
Tryplasma cylindrica (Wedekind), 1927			X
Tryplasma sp.			
Family Calostylidae			
...lis spongiosa Foerste, 1906			X
Calostylis lindstroemi Nicholson and Etheridge, 1878			X

	Before Brassfield Time	Synchronous with Brassfield Inside the Cincinnati Arch Region	After Brassfield Time
Family ...phyllidae			
Strombodes socialis (Soshkina), 1955		X	X
Family Anop...lidae			
Arachnophyllum mamillare (...), 1844			X
Arachnophyllum granulosum Foerste, 1906			
Petrozium pelagi...um (Billings), 1862		X	
Craterophyllum (?) solitarium (Foerste), 1906			
Family Hallii...de			
Pycnactis tenuiseptatus, sp. nov.			
Schizophaulactis densiseptatus (Foerste), 1906			
Family Cystiphyllidae			
Cystiphyllum spinulosum Foerste, 1906			
TABULATES			
Family Favositidae			
Favosites f...aus (...ls), 1826	X (S)	X?	X
Favosites i...dus (Roemer), 1860		X	X
Favosites hisingeri Milne-Edwards and Haime, 1851		X	X
Favosites sp. A			X
Favosites densitabu...lus, sp. nov.			
Paleofavosites prolificus (Billings), 1865b	X (O)	X	X
Family Halysi...de			
Halysites catenularius (Linnaeus), 1767			X
Halysites nitidus Lambe, 1899			X
Halysites (?) meandrinus (Troost), 1840?			X?
Catenipora gotlandica (Yabe), 1915	X (S)		X
Catenipora favositomima, sp. nov.			

	Before Brassfield Time	Synchronous with Brassfield in the Cincinnati Arch Region	After Brassfield Time
Family Moniloporidae			
...us sp. A			
...us sp. B			
...us (?) sp. C	X (O)		
Family Syringoporidae			
Syringopora (?) reteformis Billings, 1858			♦
Family Syringoli i de			
...us vesiculosus, sp. nov.			
Family ...			
Alveolites ...us (Milne-Edwards & Haime), 1851	X? (O)		X
Alveolites ...hii Milne-Edwards & Haime, 1851			X
Family Pachypori de			
Striatopora flexuosa Hall, 1851			X
HELIOLITIDS			
Family Proporidae			
Propora ...ta Milne-Edwards & Haime, 1851	X (O)	X	X
Propora exigua (Billings), 1865b	X (S)		X
Propora ...la (F...ste), 1906			X?
Family Helioli die			
Heliolites spongodes Lindström, 1899			X
Heliolites spongiosus ...rste, 1906			X?
Family Palaeoporitidae			
Palaeoporites sp.			

Table 8. — Brassfield coral species arranged geographically and chronologically: The middle column consists of occurrences in strata coeval with the Brassfield Formation (mid-Llandovery) but outside the Cincinnati Arch Region.
*=questionable occurrence.
†This form may be ancestral to *P. cf. P. cybaeus* from the Brassfield.

| | COEVAL WITH | |
PRE-BRASSFIELD	BRASSFIELD	POST-BRASSFIELD
(EASTERN UNITED STATES)		
P. conferta		*A. mamillare*
Cladochonus (?) sp. C		*C. spongiosa*
**A. labechii*		*S. benedicti*
		S. ipomaea
		**D. linguliferum*
		**T. radicula*
		H. nitidus
		**H.* (?) *meandrinus* ?
		F. discoideus
		F. hisingeri
		Favosites sp. A
		A. labrosus
		S. flexuosa
		H. spongodes
(NORTHERN GREAT LAKES)		
P. conferta	*S. socialis*	*T. radicula*
P. prolificus	**F. favosus*	*A. mamillare*
		D. stokesi
		**F. favosus*
		F. hisingeri
		P. prolificus
		**A. labrosus*
		**A. labechii*
		S. (?) *reteformis*
		**P. eminula*
(CANADIAN MARITIMES)		
P. conferta	*P. pelagicum*	*A. mamillare*
P. exigua	**F. favosus*	*D. stokesi*
P. prolificus	*F. discoideus*	**F. favosus*
F. discoideus	*F. hisingeri*	*F. discoideus*
	P. prolificus	*F. hisingeri*
	P. conferta	*P. prolificus*
		H. nitidus
		A. labechii
		S. flexuosa
		P. conferta
		P. exigua
		**P. eminula*

(Table 8 continued)

(BALTIC)

P. conferta	P. pelagicum	T. cylindrica
C. gotlandica	*R. daytonensis	S. patellatum
P. suecicum		*R. daytonensis
*D. linguliferum		F. discoideus
		F. hisingeri
		P. prolificus
		H. catenularius
		C. gotlandica
		P. exigua
		H. spongodes
		*H. spongiosus

(BRITAIN)

S. benedicti
C. lindstroemi
*D. linguliferum
F. discoideus
F. hisingeri
A. labrosus
A. labechii

(SIBERIAN PLATFORM)

P. primarium	S. socialis
P. cybaeus†	C. typus

(SW SIBERIA & IRAN)

S. patellatum
S. socialis
H. catenularius
C. gotlandica

(FAR EAST)

P. prolificus
(Japan)
H. catenularius
(Japan)
P. conferta
(Japan)
H. spongodes
(Kansu Prov., China)

(MISCELLANEOUS LOCALITIES)

P. conferta	P. conferta	T. radicula
(NE Canadian Arctic)	(Manitoba)	(Oklahoma)
P. prolificus		A. mamillare
(Manitoba)		(Iowa)
*R. daytonensis		S. socialis
(Venezuela)		(Lake Timiskaming in
		Canada)
		F. favosus
		(Iowa & Brit. Columbia)
		F. hisingeri
		(Bohemia)
		*H. nitidus
		(Brit. Columbia &
		Venezuela)
		H. spongodes
		(Bohemia)

tion of the species, and may be used as an index of how reliably the description reflects the variability of that species in the Cincinnati Arch region. The numbers indicated are commonly approximations, and represent both material I personally collected, and museum specimens considered to provide reliable data.

RUGOSA

Family **STREPTELASMATIDAE** Nicholson
in Nicholson and Lydekker, 1889

Nom. correct. Wedekind, 1927 (*pro* Streptelasmidae Nicholson *in* Nicholson and Lydekker, 1889)

Rugosa, consisting of solitary and fasciculate corals lacking a dissepimentarium; theca a septal stereozone; minor septa short; major septa dilated to varying degrees in early stages, meeting axially, but thinning and withdrawing from axis through ontogeny; in ephebic stage, major septa may intertwine in axis to form various types of axial structures, usually with stereomal deposits; tabulae usually domed axially, both complete and incomplete; cardinal side convex in almost all solitary species.

Genus **STREPTELASMA** Hall, 1847

1847. *Streptoplasma* Hall, p. 17.
1847. *Streptelasma* Hall, page facing p. 339.
1876. *Zaphrentis* Rafinesque, Rominger, pp. 140-142 (*partim*).
1927. *Dybowskia* Wedekind, p. 18 (*non* Dall, 1876, p. 46).
1940. *Brachyelasma* Lang, Smith, & Thomas, pp. 28, 55 (*pro Dybowskia* Wedekind, 1927, *non* Dall, 1876).

Type species. — *Streptelasma corniculum* Hall, 1847, p. 69 (by subsequent designation of Roemer, 1861, p. 19).

Type locality and horizon. — Lower Trenton Limestone (Upper Middle Ordovician)at Middleville, New York State.

(Table 8. — Miscellaneous localities continued)

H. catenularius of undetermined age in Australia
P. conferta in Silurian of Korea and Siberian Platform
F. discoideus in Silurian of Korea
T. radicula in Silurian of Great Basin (western U.S.)
Favosites sp. A in Silurian of Baffin Island (Canadian Arctic)
F. hisingeri in Llandovery of Southampton Island (Canadian Arctic)
A. labrosus in Silurian of Canadian Maritimes

Diagnosis. — Solitary corals lacking a dissepimentarium; theca a septal stereozone; cardinal side convex; septocoels open throughout ontogeny; in early neanic stage, septa in contact distally, drawing apart through ontogeny to leave open axial area in ephebic stage which contains only a few septal lobes; tabulae often incomplete.

Description. — These are solitary corals, of trochoid or cylindrical, and sometimes scolecoid form. In some species (though not the type species) basal-attachment or supportive structures may be present. The theca is a septal stereozone.

The cardinal septum is on the convex side of curved corallites, and the cardinal fossula, when present, is weakly developed.

In the earliest stages, the septa are in axial contact. From at least the early neanic phase onward, the septocoels are open. As the corallite grows, the axial ends of the septa draw apart, as each septum comes to occupy a proportionally smaller part of the radius than in earlier stages. At this point, the axis of the corallite contains only a few septal lobes winding irregularly. No axial structure is developed. The minor septa are short, less than half the length of the radius.

There is no dissepimentarium. The tabulae are often incomplete and are usually convex (through sometimes concave) in the axis.

Discussion. — In his original description of this genus, Hall (1847, p. 17) used the name *Streptoplasma,* designating thereby the axial twisting of the septa as one of the most distinctive features of the form. In this same work ("Corrections", facing p. 338), Hall corrected the generic name, with apparent propriety, to *Streptelasma,* the name currently used. In so modifying the name, he noted that the error had occurred "in changing the generic name from *Streptophyllum,* which was the original name in MS."

In this work, Hall described several species of *Streptelasma* (one of which, *S. expansa* [*sic*], is a parablastoid plate, according to its catalogue card at the American Museum of Natural History). *S. corniculum,* from the original species suite, was designated the type species by C. F. Roemer (1861, p. 19). Hall and Simpson (1887, p. xi) belatedly, and so, invalidly, selected *S. expansum* as the type. Thus, Roemer's timeliness preserved for *Streptelasma* a cnidarian, rather than echinodermal appellation.

Rominger (1876) treated *Streptelasma* as a subgenus of *Zaph-renthis* Rafinesque and Clifford, 1820. He noted (p. 141) that undue attention had been paid by Milne-Edwards to the cardinal fossula in assigning species to *Zaphrenthis*. Rominger felt that *Streptelasma* should be accommodated (as a subgenus) under *Zaphrenthis*, considering it to differ from the latter solely in having a less prominent fossula and in the axial twisting and "entanglement" of its septa, giving rise to a "spongiose-cellulose pseudo-columella".

Rominger did not consider the presence of a dissepimentarium in *Zaphrenthis*, as distinct from its absence in *Streptelasma*, in determining the identity of these two genera. Also, his characterization of the axial structure as spongy suggests that he was including species of *Grewingkia* Dybowski (1873) in his suite of *Streptelasma* species. Indeed, his plates seem to support this idea. The axial mutual-involvement of the major septa of *Streptelasma* is much less developed than in *Grewingkia*, where it is complex and "spongy". This is clearly seen in Neuman's excellent treatment of Scandinavian streptelasmatids of the Upper Ordovician (1969), in which he figured and described the internal features of the holotype of *S. corniculum*, together with numerous other species of *Streptelasma* and *Grewingkia*. *Streptelasma*'s axial structure consists of only a few elongate septal lobes, whereas *Grewingkia*'s is far more intricate and includes many more lobes. In addition, while *Grewingkia*'s neanic septal dilation tends to be great, often excluding the septo-coels, that of *Streptelasma* is characteristically weak. *Streptelasma rusticum* (Billings), the famous Richmondian solitary coral of the Cincinnati Arch area, is in reality a *Grewingkia*.

Wedekind (1927) introduced the genus *Dybowskia* from the Upper Ordovician of Norway selecting his new species, *D. prima*, as the type species. Lang, Smith, and Thomas (1940, p. 55) later noted that the generic name was preoccupied and renamed the genus *Brachyelasma*. Wedekind's figures consist of a transverse view (apparently of the ephebic stage) and a longitudinal view. Taken alone, these sections make this species (and thus, the genus) a candidate for *Streptelasma* Hall. Neuman (1969, pp. 11-17) discussed *D. prima* and illustrated internal views of several specimens from Norway's Upper Ordovician, providing convincing evidence that *Brachyelasma* is indeed identical to *Streptelasma*.

Streptelasma scoleciforme, sp. nov.   PI. 2, figs. 29-36; PI. 12, figs. 1-8

Type material. — Holotype: UCGM 41942 (Pl. 2, figs. 29-31; Pl. 12, figs. 1-4); Paratype: UCGM 41953 (PI. 12, figs. 5-8).

Type locality and horizon. — Locality 1, the top of the Noland Member, Brassfield Formation, near Panola, Kentucky.

Number of specimens examined. — Twenty-two.

Diagnosis. — Solitary, characteristically scolecoid *Streptelasma* with contratingency of septa.

Description. — This species is characteristically scolecoid in form, with many changes in calice orientation and sometimes diameter with growth, making it reminiscent of *Calostylis lindstroemi* (*q.v.*). Some corallites appear less scolecoid than trochoid, but it is rare that either irregular expansions of the diameter or changes of calice orientation do not occur. Specimens were found as long as 23 mm, but most are smaller. Often, a specimen is somewhat compressed laterally, such that its diameter is noticeably greater in one direction than at right angles to it.

The internal structure is similar to that found in most species of *Streptelasma* (*see* Neuman, 1969 for examples). The theca is a septal stereozone, about 0.5 mm thick. The septa in earlier stages are fairly straight and reach to the axis where their tips may become fused. These septa have an axially-tapering central core which is less dense optically than the rest of the septum. At this stage, the septa have a maximum width of 0.3 mm at the wall and taper axially. The cardinal septum may be recognized in some cases because the septa immediately adjacent to it on both sides maintain a fairly constant lateral distance from it. As the corallite grows, the diameter increases at a greater rate than does the length of the septa, such that they seem to withdraw from the axial area. Their proximal ends retain about the same thickness as in earlier stages, but the distal ends are much more attenuated. At no stage does a complex axial structure of intertwined septal ends (as in *Grewingkia*) develop. The minor septa are extremely short, usually no more than 1/5 the length of the major septa. In one specimen they are absent at a diameter of about 5 mm, but present at 7 mm (Pl. 12, figs. 6, 7).

Perhaps the most distinctive feature of this species is the presence of contratingency among the major septa. In general this occurs between a shorter septum and the major septum on its counter side. The appearance is similar to that in *Schizophaulactis densiseptatus* (Foerste) (*q.v.*) with which it is associated in the field. It is unclear if the major septa of *S. scoleciforme* form in the same manner as do those of the other species. The minor septa are not known to be contratingent.

There is no dissepimentarium within the corallite; only a complex pattern of complete and incomplete tabulae (Pl. 12, figs. 3, 4). The former may be horizontal, concave or convex. The curved topography of the tabular surfaces often intersects the plane of a transverse section to give the appearance of dissepiments between the septa.

Distribution. — Kentucky: mid-Llandovery (Brassfield Formation, Noland Member).

Brassfield occurrence. — Localities 1 (near Panola) and 46 (near College Hill) in central Kentucky.

Genus DINOPHYLLUM Lindström, 1882

1882b. *Dinophyllum* Lindström, p. 21.
1900. *Scenophyllum* Simpson, p. 210.

Type species. — *Dinophyllum involutum* Lindström (1882b, p. 21), by monotypy (= *Clisiophyllum hisingeri* Milne-Edwards & Haime, 1851). A detailed description, with figures, is given by Lindström (1896, p. 38).

Type locality and horizon. — Silurian of Olenek, Siberian Platform.

Diagnosis. — Solitary corals lacking a dissepimentarium; theca a septal stereozone; cardinal side convex (when corallite is curved); minor septa short; in ephebic stage, major septa reaching axis with counter-clockwise whorl; tabulae axially convex, sometimes steeply conical.

Description. — This genus consists of solitary corals which vary in form from the curved, rapidly expanding corallite of *D. stokesi* (Milne-Edwards and Haime) (Pl. 1, figs. 1-3) to a more conical, straight corallite of slimmer proportions [*D. hoskinsoni* (Foerste), Pl. 2, figs. 2, 4]. Some species have a cylindrical form (*D. involutum* Lindström, as first illustrated by Lindström, 1896, p. 39). The theca is a septal stereozone, and its outer surface is marked with a series of longitudinal grooves corresponding to the positions of the septa.

In the early stages of ontogeny, the septa show varying degrees of thickening by stereomal deposits, but at calice level, the septa are unthickened. Also in later stages the septa undergo counter-clockwise torsion, typical of many streptelasmatids. This torsion usually affects the cardinal fossula as well, which is typically present in later stages. At this point, the fossula is well developed and may be straight-sided (*D. semilunum*, sp. nov.), or have distinctly bowed lateral borders, giving it an elliptical shape (*D. stokesi* [Milne-Edwards and Haime]).

The major septa reach the axis in the ephebic stage, either singly or in distally-joined groups. The minor septa, in contrast, are short, less than half the length of the major septa.

The tabulae are mostly incomplete and are oriented in such a manner as to slope fairly steeply from the axis down toward the periphery of the calice. The resulting axial convexity of the tabulae, and the twisting of the distal ends of the major septa around this convexity (Lindström, 1896, figs. 88, 95) is regarded as characteristic of this genus. A similar pattern is seen in *Clisiophyllum* Dana (1846), a genus often confused with *Dinophyllum*, but possessing a dissepimentarium.

Discussion. — *Scenophyllum* Simpson (1900) agrees with Lindström's concept of *Dinophyllum* and is generally regarded as a junior synonym of the latter. Simpson based *Scenophyllum* upon *Zaphrenthis conigera* Rominger (1876), a Devonian form from Michigan.

Dinophyllum involutum Lindström is treated in the discussion of *D. hoskinsoni* (Foerste) [*q.v.*].

Dinophyllum hoskinsoni (Foerste)   Pl. 2, figs. 1-4; Pl. 12, figs. 9, 10; Pl. 13, figs. 4-6

Cf. 1882b. *Dinophyllum involutum* Lindström, pp. 21-22.
    1890. *Streptelasma hoskinsoni* Foerste, pp. 344-345, pl. 9, figs. 1-4; perhaps figs. 5, 6.
Cf. 1896. *Dinophyllum involutum* Lindström, pp. 38-42, pl. 7, figs. 87-98.

Type material. — Lectotype (here designated): USNM 84783, the original of Foerste's plate 9, fig. 2 (see Pl. 2, fig. 2; Pl. 12, figs. 9, 10). Paralectotype: USNM 84783, the original of Foerste's plate 9, fig. 1 (see Pl. 2, fig. 1).

Type locality and horizon. — "Brown's Quarry", west of New

Carlisle, Ohio, probably from the Brassfield Formation (mid-Llandovery).

Number of specimens examined. — Three (including type suite).

Diagnosis. — *Dinophyllum* with little or no curvature of the corallite axis; rugose swellings common; no talon or other attachment structure at base; cardinal fossula prominent; tabulae present distinctive bilaterally symmetrical aspect in longitudinal section, like dishes and bowls of different sizes and concavities stacked upside-down; septa not sufficiently dilated to constrict septocoels.

Description. — The paralectotype of this species (USNM 84783) shows only slight curvature of the axis. Vertically it is 3.3 cm high, and the length of its convex side is 3.6 cm. The specimen is longitudinally striated with narrow grooves, separated by much broader ridges. There are three distinct growth discontinuities (though not demonstrably rejuvenescence) occurring at distances of 1.5, 2.25, and 2.95 mm above the base (measured along the convex side). The diameters at these points are 7.0 mm, 15.0 mm, and 18.5 mm.

The lectotype also shows little curvature. Its vertical height is 2.0 cm, and its diameter at the top is 1.3 cm. It is rugose, and the surface pattern of its theca is as in the paralectotype.

In a transverse section through the lectotype taken at a level with a diameter of 8 mm (Pl. 12, fig. 10), the theca is a septal stereozone, about 1 mm thick. The septa are of two orders. There are 20 major septa (including the cardinal septum), of which all but the cardinal septum reach the axis, where they are affected by a counter-clockwise whorl. The cardinal septum is about 1/4 the length of the major septa and is distinguished only by its shortness. In the axis, a sort of fusion of the septal ends appears to be brought about by sparse patches of stereom. The minor septa of this level are almost restricted to the theca.

In a more proximal transverse section of the lectotype, the diameter is 5 mm. There are 19 major septa, reaching to the axis but with no obvious whorl. The theca is 0.3 mm thick at this level.

A longitudinal section of the lectotype (Pl. 12, fig. 9) shows one of the most characteristic features of this species. The tabulae are convex, with flattened tops and down-bowed sides. They are arranged like dishes and bowls of differing depth and diameter piled upside-down, one atop the other. The result is a remarkable bilateral symmetry of the interior of the corallite. In this specimen there is one particularly high-domed, narrow-based tabula whose edges rest upon the underlying tabula. The lower tabulae are spaced about 3 in 1.5 mm vertically. The fuller one on top stands 1.5 mm high. In transverse section (Pl. 12, fig. 10) these tabulae appear as lines in the septocoels, generally concave towards the theca.

One specimen of this species was found in the field, at locality 46 (near College Hill, Kentucky). The corallite is fairly straight, about 2.5 cm long (Pl. 2, fig. 4), and is broadest at the top (2.3 cm at the level of the calice floor). There are localized rugose swellings that give the corallite the appearance of a stick of bamboo.

In a transverse section made near the top (at a diameter of 1.6 cm) the theca is a septal stereozone, about 1 mm thick. The 29 major septa appear to reach the vicinity of the axis (the central portion of the section is destroyed), where they are affected by a counter-clockwise whorl. The minor septa are about 1/6 to 1/7 the length of the major septa at this level. The cardinal fossula is oblong, with the cardinal septum running its full length (about 5 mm) and not reaching the axis. It runs directly along the midline of the fossula and is stopped by the juncture of the bounding septa and the rise of a tabula. A darkish core runs along the centers of the major and minor septa, tapering to a sharp point where they enter the stereozone. It should be noted that in the floor of the calice of the whole specimen, the cardinal fossula is more elliptical, and nearly pointed at its axial end (Pl. 13, fig. 5).

A lower transverse section has a diameter of 9.5 mm in the greatest direction and 7.0 mm in the shortest (Pl. 13, fig 6). Here the theca is 0.2 to 0.25 mm thick, and it is clear that the longitudinal grooves on the theca reflect the positions of the septa, and that the ridges mark the septocoels. The length proportions between the major and minor septa are the same at this level as in the upper section, but the major septa extend only about halfway to the axis. The abundance of sediment in the cavity suggests this may be due to breakage.

In longitudinal section (Pl. 13, fig. 4) the lower, cylindrical portion of the corallite (seen in the cardinal-counter plane) displays the striking bilateral symmetry of the tabulae discussed above. In one case, a tabula appears to have been depressed axially so that its upper surface corresponds to that of the tabula on which it rests. Where the rim of a narrower tabula abuts the surface of the tabula just beneath it, the lamellae of the former merge smoothly with those of the latter, so that the zone of lamellar merging is no thicker than either of the tabulae individually. The spacing of each of the five main complete tabulae in the axis is as follows: the bottom two are 2.0 mm apart, the third from the bottom is 0.5 mm above the second, the fourth is 0.7 mm above the third, and the fifth is 2.0 mm above the fourth. At the top of the section, where the diameter of the corallite swells, the axial tabulae grow higher and narrower while maintaining bilateral symmetry. In the space between their sides and the theca are found large, incomplete tabulae that do not contribute to the symmetrical appearance. In transverse section, as noted in the lectotype, the tabulae appear as lines in the septocoels that are concave toward the theca.

Discussion. — Foerste (1890) noted the characteristic transverse annulations of this species, and that usually only the lower portion of the corallite is curved. The calices, he reported, are at least 8 mm deep, with steep sides. (This latter feature he must have obtained from specimens other than the two in the type suite, as the calices are not exposed in them.) Foerste was unable to provide information on the internal structure, except that the septa vary in number from 40 to 55.

To better understand this species, it would be well to compare it with *Dinophyllum involutum* Lindström (1882b), the type species of this genus. Lindström's specimens came from the Silurian of Gotland (probably the Högklint beds of Early Wenlock age), though he also reported the species from the Silurian of the Olenek River Basin in northern Siberia. In his original description, Lindström described the coral as solitary and either straight or slightly curved. The epithecal ridges (referred to by Lindström as "pseudocostae") are distinct. The major septa are 55 to 60 in number. The cardinal septum extends to the center where it joins with the counter septum. The other major septa also reach the axial region, where they undergo torsion (apparently counter-clockwise), and form an axially-convex "columella" of axially upraised incomplete tabulae. Between the theca and the incomplete tabulae occupying the septocoels, is a "long strip of a loose, spongy mass, which sometimes fills the entire septocoel". No figures accompany this description, but Lindström gave the dimensions of the specimen as 41 mm long and 31 mm wide.

In 1896, Lindström gave a more detailed, and illustrated description of this species. To his earlier description he added that the corallite is distinctly rugose. In the shallow cup there are two orders of septa, with up to 60 major septa. (The figures show the minor septa as virtually restricted to the stereozone.) A cardinal fossula(?) contains three to four small septa. The corallite growth is not accompanied by particularly rapid increase of septal number: In a diameter of 25 mm (about 10 mm above the base) there are 40 major septa; higher, at a diameter of 34 mm, there are found 50 major septa, and this number remains constant until a diameter of 37 mm in reached, at which point the maximum number, 54 septa, is found. The entire length of this sectioned specimen is 58 mm.

In this article, Lindström delved into the early ontogeny of the species. The stereomal filling of the septocoels on the cardinal side during the neanic stage (as seen in *D. semilunum,* sp. nov. [*q.v.*]) does not appear to occur in *D. involutum.*

The major septa approach the axis, where they undergo counterclockwise torsion as they rise up the axial slope of the "columella". The tabulae present a confused pattern, but appear generally to be incomplete, rising from the theca toward the elevated axis. There is no dissepimentarium. The septa seemed to Lindström to have perforations, which he was at a loss to explain, suggesting that they may have been parasitic in origin. No mention is made in this description of the spongy deposits in the septocoels, noted by Lindström in his earlier paper (1882b). The largest specimen he measured was 12.0 cm long and 4.5 cm in diameter.

Lindström figured several variations of his form, possibly representing more than one species. For example, figure 87 shows the septa undergoing torsion as they ascend the axial prominence, while figure 89 shows them going straight to the axis. Figures 94-98 suggest an axial boss elongate in the cardinal-counter plane (as also described by Ryder, 1926, pp. 394-395, pl. 12, figs. 7-9). Apparently no specimen has been designated the lectotype of *D. involutum,* so that it is difficult to judge which of these forms should be regarded as typical.

D. hoskinsoni (Foerste), differs from *D. involutum* Lindström in the bilateral symmetry of its tabulae and the prominence of its cardinal fossula.

Distribution. — Ohio, Kentucky: mid-Llandovery (Brassfield Formation).

Brassfield occurrence. — Type suite from near New Carlisle, southwestern Ohio, probably Brassfield Formation (based on lithology and Foerste's data); also found at locality 46, Noland Member, Brassfield Formation near College Hill, central Kentucky.

Dinophyllum stokesi (Milne-Edwards & Haime)   Pl. 1, figs. 1-7;
                                              Pl. 12, figs. 11, 12; Pl. 13, figs. 1-3

    1851. *Zaphrentis stokesi* Milne-Edwards & Haime, p. 330, pl. 3, fig. 9.
    1876. *Zaphrentis stockesii* Milne-Edwards & Haime, Rominger, p. 145, pl. 51, lower tier.
    1876. *Zaphrentis umbonata* Rominger, p. 146, pl. 51, lower tier.
Cf. 1882b. *Dinophyllum involutum* Lindström, pp. 21-22.
Cf. 1896. *Dinophyllum involutum* Lindström, Lindström, pp. 38-42, pl. 7, figs. 87-98.
    1906. *Zaphrentis intertexta* Foerste, pp. 307-308, pl. 7, figs. 1A, B.
? 1906. *Zaphrentis intertexta,* varieties or young, Foerste, pp. 308-310, pl. 7, figs. 5A-E.
    1919. *Zaphrentis stokesi* Milne-Edwards & Haime, Williams, p. 65, pl. 14, figs. 1, 2.
    1928. *Zaphrentis stokesi* Milne-Edwards & Haime, Twenhofel, p. 116.
? 1960. *Dinophyllum breviseptatum* Ivanovskiy, pp. 92-93, pl. 9, fig. 1.
    1966. *Dinophyllum* (?) *umbonata* Rominger, Bolton, p. 16, pl. 5, fig. 9.
    1968. *Dinophyllum* (?) *umbonata* Rominger, Bolton, p. 48, pl. 13, fig. 30.

Type material. — Possibly in Museum of Natural History, Paris.

Type locality and horizon. — Silurian (probably Llandovery or Wenlock) of Drummond Island, Lake Huron.

Number of specimens examined. — Nineteen.

Diagnosis. — Curved, trochoid *Dinophyllum* with rapidly increasing diameter, widest perpendicular to the cardinal-counter plane; cardinal fossula prominent, elliptical in outline, and located on convex side of corallite; septa in mature stages thick, resulting in narrow septocoels; tabulae without the bilateral symmetry found in *D. hoskinsoni* (Foerste); basal attachment scars and stereomal deposits on the cardinal side, as in *D. semilunum,* sp. nov., are absent.

Description. — This species includes some of the largest solitary corallites found in the Brassfield Formation. The largest specimen studied (Pl. 1, figs. 1-3) has the following dimensions:

length (straight line from base to highest point) $= 6.5$ cm
length along convex side (curvature) $= 8.5$ cm
diameter in cardial-counter plane (at level of calice floor) $= 4.4$ cm
diameter in plane perpendicular to latter $= 4.9$ cm

The proportions remain fairly constant throughout the growth of an individual and show little variation from one corallite to another. All specimens are curved in the cardinal-counter plane with the cardinal side convex, and in all specimens studied the elliptical cross-section is narrowest in the cardinal-counter plane. Corallites increase in diameter rapidly, maintaining an apical angle of about 65° (though occasionally somewhat less). The corallite is weakly rugose, the surficial wrinkles never being so prominent as to detract from the regularity of the corallite's expansion with age. The theca is invested with narrow longitudinal grooves, separated by much broader ridges of low convexity. These ridges near the top of the largest specimen attain a width of about 1.3 mm, while the grooves are no wider than a pencil-line. No attachment scars are evident on the surface.

The calice is typical of *Dinophyllum*, with the distal ends of the septa affected by a counter-clockwise whorl and raised to form a low axial boss. The minor septa do not participate in forming the boss and extend only about 1/3 the distance to the axis. The upper surface of the highest tabula is visible in the septocoels and is raised in the axis along with the septal ends.

The cardinal fossula is similar to that of *D. hoskinsoni* (Foerste), being prominent and elliptical in outline due to the lateral flexing of the two major septa forming its boundaries. Often it is nearly as wide as long, to the extent that it reaches the axis, and is caught up in the axial whorl. The fossula is deepest in the middle, or toward the proximal end (closest to the theca), shallowing distally against the axially rising tabula as the bounding septa come together. The cardinal septum is depressed in the fossula, with a wide space between it and the neighboring septa.

All specimens studied were recrystallized; in only one was the internal structure sufficiently preserved to allow meaningful investigation (Pl. 1, figs. 4-7; Pl. 12, fig. 12; Pl. 13, figs. 1-3). The theca is a septal stereozone. In a transverse section measuring 1.6 cm $\times$ 1.9 cm (Pl. 13, fig. 2) its thickness is 1.5 mm.

The septa in levels ranging in diameter from 1.8 × 1.5 cm to 2.5 × 2.2 cm were thick, leaving septocoels consistently narrower than the septa. An exception was on the counter side of the highest section (Pl. 13, fig. 1), where the septocoel width about equals that of the septa.

In these sections the minor septa are almost totally restricted to the stereozone, and the majors appear to be split along their axial planes. The significance of this is unclear. Possibly it reflects the dissolution of a central zone of different structure from the outer layers of the septum. This idea is supported by the presence of a zone of optically less dense material in the central portions of well-preserved septa, with a darker zone to either side possibly representing secondary deposits. In some instances it appeared that the darker deposits covered only the more proximal portions of the septum, the distal portions consisting of the lighter core-material. Axially the major septa flex generally to their right, each contacting the distal end of its neighbor. Those on the cardinal side attenuate distally somewhat more than the other septa and approach the axis more closely, though they are somewhat sinuous.

So far as can be determined, the tabulae are complete, closely-spaced (often less than 1 mm apart), and distinctly convex upward. Some are smoothly-rounded, while others appear to be more sharply peaked axially.

Discussion. — *D. stokesi* was first reported by Milne-Edwards and Haime (1851, p. 330) from the Silurian of Drummond Island, Lake Huron. The specimen they described is trochoid, without prominent rugae. Its straight longitudinal length is given as 8 cm, the calice diameter (maximum?, minimum?) as 4 cm, and the calice depth as nearly 2 cm, indicating a corallite slightly longer and narrower than my largest specimen.

These authors described the corallite as ". . . légèrement comprimé dans le sens opposé à la courbure." This is translated as "gently compressed in a direction opposed to the curvature", a phrase of uncertain meaning. Their figure (pl. 3, fig. 9) suggests the coral's cross-sectional diameter is slightly greater in the cardinal-counter plane than in any other direction. My specimens, where elliptical, were widest perpendicular to this direction. The cardinal fossula is located on the convex side of the corallite. There is no indication in the description or figure that it is elliptical in plan view. The illustration suggests that the sides are fairly parallel.

Milne-Edwards and Haime's specimen has 64 major septa (compared with about 44 in my largest specimen). These are slightly thickened at the wall-juncture, and reach the axis in a flexuous manner, where they join in an elevation of the calice floor. The proximal portions of these septa are described as composed of a pair of lamellae, possibly corresponding to the paired pattern noted in my thin-sections (in which case they would be secondary deposits remaining after solution of the original septal core). Short secondary septa, apparently almost totally restricted to the thecal region, occur between the major septa.

The major differences between Milne-Edwards and Haime's specimen and those of the Brassfield, are the lack of ellipticity in the fossula of the former and the apparent absence of compression in this specimen. In this sense, the Drummond Island coral would seem to have more in common with the specimen figured by Bolton (1966, 1968) and identified as *Dinophyllum*(?) *umbonata* (Rominger). This specimen (GSC 20521) is from the Fossil Hill Formation (Late Llandovery through Middle Wenlock) of Manitoulin Island, and thus corresponds closely with the Drummond Island specimen in both location and age.

Nonetheless, both the Drummond Island specimen and the Manitoulin Island specimen do show some degree of lateral flexing of the septa bounding the fossula, such that this character, and the apparent lack of compression might be regarded as of subspecific importance. Consequently, my specimens and Bolton's, are regarded as conspecific with the species described by Milne-Edwards and Haime's.

Rominger (1876) described specimens of the present species from the Manistique Formation (Middle Llandovery through Middle Wenlock) of Point Detour, northern Michigan. These are trochoid corallites with broad, elliptical calices, apparently narrowest in the cardinal-counter plane. The size-range corresponds well with that of the Brassfield form. There are reportedly up to 60 to 65 major septa, attaining the axis where they are affected by a counter-clockwise whorl. The whorl is slightly elevated above the more peripheral portions of the calice floor. (The whorl in

Rominger's figure appears to be clockwise because his figures were all reversed in the printing process). Between each pair of major septa is a short minor septum, reaching about half-way to the axis. A large, elliptical fossula occurs on the convex side of the corallite. There seems little doubt that the specimens figured by Rominger (UMMP 8615) are conspecific with mine.

In the same work, Rominger described as a new species, *Zaphrenthis umbonata*. These corals (UMMP 8614) come from the same region and unit as "*Zaphrentis stockesii*", and Rominger noted the close similarity of the two forms. It is unclear why he chose to set these corals apart from the species named by Milne-Edwards and Haime, and in fact Rominger stated that their relationships were open to question. *Z. umbonata* has all the morphic features characteristic of *D. stokesi*: the sizes and shapes are the same, and both have an elliptical fossula on the convex side of the coral-lite, similar septal arrangement, and an axial prominence where the distal ends of the septa join. The septa of the larger figured specimen of *Z. umbonata* go directly to the axis without whorling but this seems acceptable in light of the variation shown by the specimens studied.

Foerste (1906) reported the present species, under the new name of "*Zaphrentis intertexta*", from the Brassfield Formation on the eastern side of the Cincinnati Arch in central Kentucky. His specimens are of the same general size as my larger ones, with the characteristic stout trochoid form, and compressed cross-section (narrowest in the cardinal-counter plane).

The peripheral portion of the calicular area is better-preserved than in my specimens. Foerste's plate 7, figure 1A shows a broad marginal area accounting for about one-half of the radius, sloping down to the calice floor. The point of contact between these two regions is the deepest portion of the calice, forming a sort of peripheral moat around the central floor. Axially, the floor rises into a broad boss. A prominent cardinal fossula which appears to be elliptical is located on the convex side of the corallite and reaches to the axis. The major septa attain the axis where they are affected by a counter-clockwise axial whorl. The minor septa appear to be restricted to the high peripheral zone of the calice. Foreste stressed that the distal ends of the major septa cross one-another, producing a reticulate pattern in the axis. This is seen in some of my specimens but is not considered a character of specific rank.

Foerste regarded his specimens as differing from that of Milne-Edwards and Haime by their more rapid expansion from the base, fewer septa and reticulate pattern in the axis. My assemblage, however, indicates that none of these characters are sufficient grounds for a separate species.

In the same paper, Foerste described and figured specimens which he believed (probably correctly) to be young individuals of the species, *Z. intertexta (D. stokesi)*. In some of these, the cardinal septum appears to fuse axially with what may be the counter septum, dividing the calice into two halves. This condition is approached in some of my specimens and appears to result in a decrease of the axial whorl. It is possible that this state, noted by Foerste, is a normal phase of the ontogeny, prior to development of the axial whorl, but the material at hand is insufficient to provide the necessary evidence.

Williams reported this species in 1919 from the "Lockport" (actually Fossil Hill) Formation of Manitoulin Island and Cabot Head, Lake Huron (Late Llandovery through Middle Wenlock). His figures show two specimens, the smaller (GSC 4688) with a large cardinal fossula and little or no axial whorl, and the larger (GSC 4694) also with a prominent fossula and a well-developed counter-clockwise axial whorl. These specimens agree well in all respects with my concept of this species.

Twenhofel (1928) reported this species from the Jupiter and Chicotte Formations (Late Llandovery to Early Wenlock) of Anticosti Island. The description given, except for Twenhofel's remark that the tabulae are commonly depressed in the center, agrees well with my material. It seems likely that the Anticosti material can be regarded as conspecific with the Brassfield specimens.

Ivanovskiy (1960) reported a new species, *Dinophyllum breviseptatum*, from the Upper Llandovery beds 66 km upstream from the mouth of the Borgiyachin River in the Siberian Platform. The corallite of this species is trochoid, with a maximum diameter of 40 mm and a length (to the rim) of 25 mm on the concave, and 55 mm on the convex side.

The figures indicate a form much like Brassfield specimens in general proportions. Ivanovskiy's diameter measurement was made at the calice rim, while mine was made at the level of the calice floor. The cardinal septum lies on the convex side of the corallite, and occupies an elliptical fossula characterized as indistinct by Ivanovskiy (p. 92), but plate 9, figure 1a of Ivanovskiy appears comparable to mine in this respect. Proximally, stereome thickens the ends of the septa to form a stereotheca.

About half-way to the axis, the septa twist to one side. Many do not reach the axis but unite distally with their neighbors. Ivanovskiy's transverse section suggests that the septa on the counter side are less thickened than those of the cardinal side, a situation opposite that seen in my specimens. The cardinal septum is long and bends before reaching the axis, a feature also found in the Brassfield specimens. The counter and lateral septa both of the Russian form and mine are indistinct.

Ivanovskiy was unable to detect clear alternation between major and minor septa in his specimens. This was the case in my thin-sections (Pl. 12, fig. 11; Pl. 13, figs. 1-3) and often in whole specimens from the Brassfield. In other specimens (Pl. 1, fig. 2) the two orders are more distinct. The discrepancies seem related to the waviness of the septa and to preservation.

The tabulae in the Russian form are complete and rise axially from the stereozone to assume a "cupola-like" aspect. They are spaced about 0.75 mm apart, a figure consistent with my specimens. There is no dissepimentarium.

Ivanovskiy's figure 1a shows the septa affected by a clockwise axial whorl. If this is a faithful illustration, then there is a significant difference between the Russian coral and my form, not to mention all other specimens of the genus *Dinophyllum* encountered. Illustrations of clockwise whorls in this genus have been results of reversal of figures in printing (as in the case of Rominger, 1876) or of mounting transverse sections upside-down. If this is not the case with *Dinophyllum breviseptatum*, then this form should be closely studied in an effort to understand axial torsion.

D. breviseptatum appears to differ from my form primarily in the septal thickening pattern, which appears greatest and seems to extend farther, along the septa of the cardinal side. This is opposite to the pattern found in the Brassfield specimens of *D. stokesi*. Ivanovskiy's section is of greater diameter than that obtained from my specimen. As he indicated (p. 93), he had no information on the ontogeny of the species. As only one of my specimens produced useful sections, it is possible that this difference will ultimately prove less important than it is now considered and the Brassfield form may be conspecific with the Russian species. Until further study is possible, it seems best to leave the relationship between these forms open to question.

D. *involutum* Lindström, the type species of *Dinophyllum* Lindström, was discussed in connection with D. *hoskinsoni* (Foerste) of the Brassfield Formation. It differs from the Brassfield form of D. *stokesi* primarily in being more elongate and thin with a round cross-section (Lindström, 1896, fig. p. 39), in having incomplete tabulae, and in apparently lacking a well-developed fossula. These differences do not preclude the possibility that these two forms represent subspecies of a single species, but present information makes it best to regard them as distinct species.

Distribution. — Drummond Island: Silurian; Manitoulin Island: Late Llandovery through Middle Wenlock; Point Detour, northern Michigan: Middle Llandovery through Middle Wenlock; Kentucky: mid-Llandovery (Brassfield Formation); Anticosti Island: Late Llandovery to early Wenlock. (Possible occurrence — Siberian Platform: Late Llandovery).

Brassfield occurrence. — Localities 1 (near Panola) and 46 (near College Hill) in central Kentucky. Large numbers of specimens have been collected by other workers in this vicinity in the southwest corner of the Panola Quadrangle.

Dinophyllum semilunum, sp. nov.　　Pl. 1, figs. 8-11; Pl. 13, fig. 7; Pl. 14, figs. 1-7

Cf. 1928. *Zaphrentis hannah* Twenhofel, p. 115, pl. 2, figs. 8, 9.
Cf. 1949. *Ptychophyllum* (?) *cliftonense* Amsden, pp. 112-113, pl. 30, figs. 1-7.

Type material. — Holotype: UCGM 41924 (Pl. 1, figs. 10, 11; Pl. 14, figs.1-7).

Type locality and horizon. — Locality 7a, near Fairborn, Ohio, in the clay-coral bed (described under "Brassfield Lithosome") in the Brassfield Formation (Middle Llandovery).

Number of specimens examined. — Three, all found in the field.

Diagnosis. — *Dinophyllum* with the septocoels of the cardinal side filled by stereome in the neanic stage, gradually diminishing into the ephebic stage; cardinal fossula becoming discernible as stereome decreases, with somewhat bulbous end distally and parallel sides proximally; major septa and fossula affected by counter-clockwise axial whorl; carinae absent.

Description. — The solitary corallite is usually trumpet-shaped, fairly narrow through most of its length, and flaring strongly at the calice (Pl. 1, figs. 8, 11). In most specimens examined, the original length appears to have been about 3 to 4 cm. The outer surface is usually worn and has a pattern of longitudinal alternating light and dark lines, the latter representing the septa, and the former representing the septocoels. These lines are of about the same breadth. In one specimen with a well-preserved calice, the depth of the calice was at least 1.3 cm, its lip-to-lip diameter at least 3.5 cm, and the width of the floor about 2 cm. It should be noted that while the calice is strongly flared, it is not reflexed, as in *Ptychophyllum* Milne-Edwards and Haime.

The corallite cross-section tends to be elliptical, with greatest diameter in the cardinal-counter plane. No strong curvature was noted, and while there is no clear evidence of talons or rootlets, the basal region seems to have attachment scars (Pl. 1, fig. 8), suggestive of an erect growth-habit.

The theca is a septal stereozone, its thickness about 1 mm near mid-height. The septa may be traced proximally into this stereozone and emerge from it distally as a result of the sharp diminution of the interseptal deposits.

The septa are of two distinct orders, the minor attaining no more than 1/4 to 1/3 the length of the major. The latter extend to the center from the neanic stage into the ephebic stage. In the axis, their ends fuse with those of adjacent septa, or even with septa entering from 90° or 180° away.

A distinctive cardinal fossula is present from the late neanic stage, continuing into the ephebic stage. Characteristically, its sides are parallel, except for the distal end which tends to be slightly bulbous. The cardinal septum runs through the entire length of the fossula and commonly bends to the right at its distal end, to abut the side of the septum forming the border of the fossula. From here it fuses with that septum and continues with it to the axis (Pl. 14, figs. 1, 2).

The axis of the corallite is affected by a counter-clockwise whorl, involving the cardinal fossula as well as the major septa. The distal bending of the cardinal septum is in the same sense as the axial whorl.

One of the most distinctive features of this species is the stereome filling the septocoels of the cardinal side. Slightly less than half of all the septocoels are thus affected, and the diminishing of these deposits marks the gradual transition from neanic to ephebic stage. The cardinal fossula usually first becomes evident as a split in this in-filled zone (Pl. 14, figs. 1-7).

In longitudinal sections (Pl. 13, fig. 7) the absence of a dissepimentarium is noted. The interior is replete with incomplete tabulae forming a convexity of the axis. It is when they down-bow peripherally, passing through the plane of a transverse section, that they may be mistaken for vesicular dissepiments of a dissepimentarium in the septocoels.

Within the septocoels, there seems to be a pattern to the orientation of the incomplete tabulae: As the calice floor is encountered in transverse section, in-filling sediment first appears in the deeper, peripheral zones of the septocoels. These pockets of sediment are bounded distally by what must be the surface of a down-bowed incomplete tabula, which links the sides of the two septa forming the lateral boundaries of the pocket. The distal boundaries of the pockets lie not at right angles to the bounding septa but rather in an oblique orientation. They always seem to contact the septum on the cardinal side of the pocket more distally than they do the septum on the counter side. The result is that the cardinal-counter plane forms the axis of symmetry in which the shape of these sediment pockets on one side is the mirror-image of those on the other side. This has been clearly seen in one specimen, and is suggested in other, less-complete specimens.

Discussion. — There are a number of species which, though clearly distinct from *D. semilunum,* share similarities with it.

Twenhofel (1928) described *Zaphrentis* [*sic*] *hannah* from the Becscie Formation and Gun River Formation (Middle Llandovery) of Anticosti Island. The holotype (from the latter unit) has a septal pattern in its later stages similar to *D. semilunum.* The cardinal fossula is of the same type, the minor septa are short, and the major septa reach the axis where they are affected by a weak counterclockwise whorl. The theca is a septal stereozone, and no dissepimentarium appears to be present. The earliest stage studied (diameter = 6 mm) shows the septa dilated and completely closing the septocoels. This was not observed in *Dinophyllum semilunum*, but the lowest level studied in *D. semilunum* is about 1 cm in diameter, and the equivalent stage may not have been reached. The absence of stereomal filling of the septocoels on the cardinal side during the earlier stages of *Z. hannah*, indicates a different species from (though one possibly congeneric with) *D. semilunum*.

Ptychophyllum(?) *cliftonense* Amsden from the Brownsport Formation (Early Ludlow) of western Tennessee superficially resembles *D. semilunum*, but the presence of a well-developed dissepimentarium indicates that it is a distinct form, probably belonging to *Ptychophyllum* Milne-Edwards and Haime. The presence of some stereomal in-filling of the septocoels on one side of the paratype (YPM 17691) seems related to the presence of basal support structures in *P. cliftonense*, which is probably not the case in *D. semilunum*. Also, the spinose carinae on the sides of the septa in *P. cliftonense* are absent from *D. semilunum*. These carinae are similar to those seen in many of Sutherland's (1965) corals from the Henryhouse Formation of Oklahoma, of equivalent age to the Brownsport.

Distribution. — Ohio: mid-Llandovery (Brassfield Formation).

Brassfield occurrence. — Locality 7a, near Fairborn, Ohio. The species occurs in the clay-coral bed, described under "the Brassfield Lithosome".

Genus **DALMANOPHYLLUM** Lang and Smith, 1939

1933. *Tyria* Scheffen, p. 33 (*non Huebner*, 1819, p. 166, a lepidopteran; *nec* Fitzinger, 1826, pp. 29, 60, a reptile).
1939. *Dalmanophyllum* Lang & Smith, p. 153 (*pro Tyria* Scheffen, 1933).

Type species. — *Cyathaxonia dalmani* Milne-Edwards and Haime (1851, p. 322).

Type locality and horizon. — Upper Silurian of Gotland.

Diagnosis. — Solitary and compound corals lacking dissepimentarium; theca a septal stereozone; cardinal side convex; minor septa very short; major septa reaching close to axis, sometimes contacting linguoid axial structure formed in ephebic stage by juncture of cardinal and presumably counter septa; interior mostly filled with stereome.

Description. — This genus consists of solitary corals (with a possible exception mentioned in the "Discussion"), which are trochoid, and are convex on the cardinal side. The corallites appear to have been unattached to the substrate, except for *D. minimum* (Ryder), 1926, which often has an apparent attachment scar at its apex (Ryder, 1926, p. 396, pl. 12, fig. 2). The theca, a septal stereozone, bears grooves on its external surface, marking the positions of the internal septa.

The septa are of two orders, the major reaching the axis, but the secondaries barely protruding from the stereozone. The most characteristic feature of this genus is its columella, a linguoid process that protrudes sharply from the base of the deep calice. This structure is elliptical in cross-section, and smoothly curved in the vertical section along its plane of symmetry. It is elongate in the cardinal-counter plane. The columella differentiates in the later ontogenetic stages, and is the result of axial juncture of the cardinal septum with one of the septa on the opposite side of the calice, presumably the counter septum.

In the earliest ontogenetic stages observed in this study, the septa all meet in the axis, and the septocoels are filled by the stereome thickening the septa. As the corallite grows, the septa withdraw from the axis, in typical streptelasmatid fashion, and the septocoels gradually become vacant. At this stage the major septa often join in groups of two or three by uniting their distal ends. At this time, the axial juncture between the cardinal and counter septa becomes thickened by stereome, so that in transverse section it has the shape of a football in profile (Pl. 15, fig. 5).

The stereomal thickening of the septa usually restricts the development of tabulae, so that they are virtually always absent. There is no dissepimentarium.

Discussion. — Lang and Smith (1939) erected *Dalmanophyllum* for *Cyathaxonia dalmani* Milne-Edwards and Haime (1851) from the Silurian of Gotland, when they determined that this species was not congeneric with *C. cornu* Michelin, the type species of *Cyathaxonia* Michelin (1847). Lang and Smith suggested that *Dalmanophyllum dalmani* was related generically to *Tyria insertum*

Scheffen (1933), type species of *Tyria* Scheffen (by subsequent designation of Lang and Smith, 1939). As *Tyria* was a pre-occupied name Lang and Smith proposed the new generic name, *Dalmanophyllum*.

Tyria insertum is from the 5b Zone (Porkuny, variously regarded as Late Ordovician and basal Llandovery — see Hill, 1959a, p. 151; Manten, 1971, p. 17) in the vicinity of Oslo, Norway, and appears to be a compound coral. If so, this would be the only such form known in the genus *Dalmanophyllum*.

Dalmanophyllum (?) obliquior (Foerste)   PI. 2, figs. 5, 10, 11; PI. 15, figs. 1, 2

1890. *Streptelasma obliquior* Foerste, p. 345, pl. 9, figs. 14, 15.
1893. *Streptelasma obliquins* [*sic*] Foerste, p. 601, pl. 34, figs. 14, 15 (figures labelled *Streptelasma? obliquius*).

Type material. — Holotype: USNM 84794 (fig. 14 of Foerste, 1890), specimen shown on Plates 2, 15; Paratype: USNM 84785 (fig. 14 of Foerste, 1890).

Type locality and horizon. — Hanover, Indiana, from the Clinton beds (Brassfield Formation).

Number of specimens examined. — Only holotype and paratype examined. No specimens found in field.

Diagnosis. — The specimens studied provide too little information on this species to allow the formulation of a meaningful diagnosis. My species description is based on the holotype and paratype specimens. More material is needed to give a fuller picture of this form.

Description. — The two specimens of the type suite show little structure on the outside. The paratype shows weak longitudinal striation and rugosity. The only external features shared by these specimens are that they are solitary corals, apparently without attachment structures, and their apertures seem more oblique than is usual in horn corals.

As shown in Foerste's figures, they are of different sizes: The length of the paratype, measured in a straight line from the base to the higher calice margin is about 28 mm, with a diameter of 16 mm, and the length of the holotype is about 19 mm, with a diameter of about 7 mm. Both corallites are gently curved.

The holotype was sectioned transversely at two levels. The lower level (Pl. 15, fig. 2) has a diameter of 5 mm, with a circular outline. The theca is a septal stereozone, just under 1 mm in thickness. The minor septa are almost entirely restricted to the stereozone at this level, but 22 major septa are present. One of these, probably the cardinal, extends through the axis and contacts what may be the counter septum. The cardinal septum is of about constant width, and has a dark, discontinuous midline, as have the other septa. The septa immediately to either side of the cardinal are short, and diverted away from it, such that they terminate against the sides of the septa to their respective counter sides. The rest of the septa, all major, extend axially, without reaching the center of the section. Most join distally with a neighboring septum and terminate without contacting the cardinal septum.

In the higher section (Pl. 15, fig. 1) the outline is elliptical, 7 × 6 mm in diameter. The cardinal septum extends along the shorter diameter. As in the lower section, the cardinal septum reaches into the axis to contact the counter septum, but now the axial portion of these united septa is somewhat swollen. The two septa to either side of the cardinal are bent outward, contacting their neighbors to the counter side. At this stage there are 24 major septa, and the minor septa protrude but slightly from the stereozone in places. The septa are mostly shorter, not coming as close to the axis as in the lower section, and not fusing their ends together. Possibly this is due to recrystallization of the specimen. At this level, the theca is 1 mm thick.

There do not appear to be vesicular dissepiments in the septocoels, and nothing is known of the tabulae or other features ordinarily seen in longitudinal section.

Discussion. — Foerste's original description unfortunately adds nothing to our knowledge of this species, while his later (1893) reference merely recorded its presence in the "Clinton beds" of Hanover, Indiana, and repeated his earlier figures.

In general form, this species seems to fit most readily in the genus *Dalmanophyllum*, in view of the presence of a swollen, elongate axial structure formed by the extended cardinal septum, the short secondary septa, the presence of a septal stereozone, and the apparent absence of a dissepimentarium. Its relationship to other Brassfield species of this genus, predominantly found in the southern region of the Cincinnati Arch in Kentucky, cannot be determined at this time.

Distribution. — Indiana: mid-Llandovery (Brassfield Formation).

Brassfield occurrence. — Hanover, Indiana.

Dalmanophyllum linguliferum (Foerste)   Pl. 2, figs. 6-9, 12-17, 20-23; Pl. 15, figs. 3-12

? 1878. *Lindströmia laevis* Nicholson & Etheridge, pp. 90-92, pl. 6, figs. 4, 4a-e.
? 1887. *Cyathophyllum gainesi* Davis, pl. 104, figs. 1-6.
  1906. *Lindstroemia lingulifera* Foerste, pp. 311-312, pl. 5, figs. 2A-F.
Cf. 1926. *Dinophyllum minimum* Ryder, pp. 395-397, pl. 12, figs. 2-6; text-fig. 3.
? 1933. *Lindströmia laevis* Nicholson and Etheridge, Scheffen, p. 31, pl. 5, fig. 1.
? 1933. *Tyria insertum* Scheffen, p. 33, pl. 5, figs. 2, 3.
Cf. 1961. *Dalmanophyllum minimum* (Ryder), Minato, pp. 81-85, pl. 11, figs. 1a, 2a, 3a, 5, 7-9, 12, 13, 16; text-fig. 20-22.
? 1964. *Dalmanophyllum gainesi* Davis, Stumm, p. 17, pl. 6, figs. 31, 32.

Type material. — Lectotype (here chosen) — USNM 87175, the original of Foerste's (1906) plate 5, figure 2C (shown on Pl. 2, figs. 6, 7, 12-14); paralectotypes are originals of Foerste's (1906) figures 2D, 2E and 2F, under the same catalogue number.

Type locality and horizon. — "Along the road north of Estill Springs, north of Irvine" in central Kentucky (locality 53-53a).

Number of specimens examined. — Ten (the type suite plus seven collected specimens).

Diagnosis. — *Dalmanophyllum* with diameter about equal to length, and cardinal fossula present on convex side in later stages; major septa bearing elongate carinae; in earliest stage, septocoels entirely closed by stereome; no attachment scars present.

Description. — Seven specimens of *Dalmanophyllum*, all of which seem to be conspecific, were found at locality 1 (near Panola, Kentucky). These are curved, solitary corals with the cardinal side convex. The diameter in the cardinal-counter plane is generally (but not universally) somewhat shorter than that in other directions. This, combined with the sense of axial curvature, results in a cuneate, calceoloid corallite. It seems likely that these specimens lay on their cardinal (convex) side on the sediment, unattached, with the oral disc directed obliquely.

The theca is rugose, and fine longitudinal grooves reflect the positions of the major and minor septa. In most specimens there is little evidence of wear, an unusual condition in Brassfield specimens.

Several have virtually perfect basal points.

The following measurements (in mm), taken from the largest and smallest of the specimens found in the field, give some indication of the size-range involved:

	smallest specimen	largest specimen
perpendicular height from base to center of calice floor	7	14
perpendicular height from base to convex calice lip	17	22
maximum diameter across calice floor	7	7
maximum diameter across calice lip	12	17

The external and internal features of three of the field specimens were studied in thin section. These are described in turn below:

The first specimen (Pl. 2, figs. 15-17; Pl. 15, figs. 9, 10) has an extremely well-preserved calice, free of sediment. The diameter across the rim of the calice is 14.5 mm, and at the level of the floor, about 10.0 mm. In this specimen the cross-section is more circular than in the others. The upper edges of the major septa (27 in number) are flat and parallel to the calice floor, extending about 1.5 to 2.0 mm from the theca. From here they drop nearly vertically to the calice floor, a distance of about 4.5 mm. At the floor they curve smoothly through 90° to produce a flat, horizontal surface. They extend axially, their ends uniting with some twisting at the base of the axial structure. This structure is a tongue-like columella raised 2.0 mm above the calice floor, with a horizontal length near its base of 3.5 mm and a basal thickness of about 0.4 mm, tapering upward. It is formed by the fusion of the counter and cardinal septa. The connection with the cardinal septum occurs below the level of the calice floor in a cardinal fossula, so that the vertical edges of the columella and the cardinal septum are separated at the level of the calice floor by more than 1 mm. The minor septa are short, extending just over 1/3 as far from the theca as the major septa. They do not have an upper horizontal shelf, as do the major septa. The sides of the major septa bear weak ribs (elongate carinae) rising axially at a slight angle to the horizontal. It cannot be determined if they occur on the minor septa. They are located and oriented consistently with the position of septal trabeculae, by analogy with modern scleractinian corals.

Two transverse, and one longitudinal section were made of this specimen. Unfortunately, the lower cross-section, with a diameter

of about 4 mm, was partly broken in the process of cutting, and
shows only dilated septa, without open septocoels, that reach to the
axis.

The higher transverse section is circular with a diameter of 11
mm (Pl. 15, fig. 10). The theca is a septal stereozone, slightly over
1 mm thick. The minor septa are almost entirely restricted to the
stereozone, with only the smallest point occasionally projecting into the septocoels. The major septa are thickened proximally by
stereome just distal to the stereozone, and number 26 at this level.
The cardinal septum lies in a fossula, and the two major septa to
either side of it join their axial ends to the sides of the major septum
on the counter side of each. The other major septa join their axial
ends together in groups of two and three, and these connections are
cemented by secondary deposits of stereome. One septum is somewhat thinner than the others, and doesn't reach as far axially as
most, but bends toward the counter side, contacting its neighbor.
The cardinal and presumed counter septa are connected across the
axis to form a diametric line. In the axis, for a length of 3.5 mm, this
line is thickened by secondary stereome, in the midst of which may
be seen the thinner material of the septa proper. This is the mode
of formation of the linguoid columella. Only a few septa are in contact with the columella at this stage.

In longitudinal section (Pl. 15, fig. 9), no tabulae or dissepiments were discerned; only twisted masses of stereome. This presumably reflects the section plane intersecting septa and the
stereome of the axis. The deposits of the more axial region are
perforate in places.

The calice of the second specimen (UCGM 41946) is not so
well preserved as that of the first. The diameter across the rim of
the calice is 11 × 12 mm (the shorter direction being in the plane
of the cardinal-counter line). The base of the calice is about 6.5
mm in diameter, and the length of the axial columella is about 2.5
mm. There are 25 major septa, their axial ends approaching and
sometimes contacting the base of the columella. The major septa do
not have the distinct upper shelf seen in the previous specimen, possibly a result of the poorer state of preservation. There appears to be
a cardinal fossula, though it is largely obscured by encrustations in
this region. The minor septa are not as prominent as in the previous
specimen. Two possible minor septa appear on either side of the

cardinal septum, though these could be weakly developed major septa. The septa bear faint ribs, similar to those described in the first specimen (UCGM 41952, above).

. One transverse section (Pl. 15, fig. 11) was taken just below the calice floor, and a longitudinal section (Pl. 15, fig. 12) was also taken. The former has a circular cross-section, 7.5 mm in diameter. On one side, the perimeter of the circle has been slightly depressed as a natural consequence during growth. Twenty-one major septa appear in the section, though no minor septa are present. The septa are secondarily thickened by stereome, and this thickening increases peripherally, the stereome extending into the stereotheca to give the portions of the septa in the stereotheca a bulbous appearance. This pattern was seen in the previous specimen, though not so clearly. The septa join axially in groups of two, three, and (in one case) four. The presumed cardinal and counter septa join across the axis (the counter septum somewhat indirectly approaching the columella), to form a diametric lamella which is only slightly thickened in the axis, the thickened portion measuring about 1.5 mm in length. There is no clear symmetry of the septa around the presumed cardinal septum as seen in the previous specimen, but this may be due to the smaller diameter (earlier growth stage). The theca is about 0.7 mm thick.

The longitudinal section is identical to that of the previous specimen, with copious stereomal deposits, particularly in the axis. No tabulae or dissepiments appear to be present. Perforations occur in some areas of the axial stereome, and in the base of the corallite are apparent axial ends of the early septa.

A third specimen (Pl. 2, figs. 20-23; Pl. 15, figs. 5-8) was studied in transverse section only. The diameter at the calice rim is 17 × 15 mm (the latter in the cardinal-counter plane). The calice, unfortunately, was filled in.

Four transverse sections were taken. The lowest (Pl. 15, fig. 8) is 4 × 3.5 mm in diameter, slightly broader in the cardinal-counter plane. The septocoels are completely closed by stereome. There appear to be about 21 septa present, with no clear distinction of orders. The cardinal septum lies on the convex side, and to either side of it the neighboring septa parallel the cardinal so that they abut the side of the septa to their counter side. This forms a sort of fossula, clearly marking the cardinal. The cardinal septum extends into the axial area and is the longest septum present. It does not appear to be in direct contact with the counter septum at this stage. In the peripheral zone, wedges appear between the septa, possibly representing incipient septa.

The next-highest section (Pl. 15, fig. 7) has a diameter of 6.5 mm. Here there are 24 major septa, two of which are inserted about 90° from the cardinal septum. The two septa neighboring the cardinal to either side are distinctly curved axially toward their respective counter sides, contacting the sides of their neighboring septa. No minor septa are present. Some of the septa have bulbous peripheral ends where they enter the stereozone. The septocoels are becoming open. The axial area is obscured by dolomitization.

The third section, higher still in the corallite (Pl. 15, fig. 6) has a diameter of 9 mm. There are 24 septa present now, all major. One of them is only a large stub in a septocoel. I first thought that this was a septum in the process of being inserted. As, however, it corresponds in position with a fully-developed septum of the lower transverse section, it seems more likely that this is a septum which was aborted by crowding. Despite the nearly 50% increase in diameter of this stage over its predecessor, this septum is only slightly shorter than the one corresponding to its position in the preceding section. The septa of this level are proportionately shorter than those of the previous level. The latter reached nearly to the axis, while the former leave an axial space fully 1/3 the total diameter. From a septum on the counter side (but not directly opposite the cardinal) there extends a thin band of skeletal material connecting it with a small feature that appears to be the remnant of a nearly destroyed axial structure.

In the fourth and highest section (Pl. 15, fig. 5) the diameter is 7 × 6 mm, shorter in the cardinal-counter plane. There are 27 major septa, with dilated proximal ends, and minor septa represented by small bumps in the septocoels. The theca is 1 mm thick. The septa at this level are more slender than at lower levels. The cardinal septum is not so clearly marked as below. A columella extends along the cardinal-counter plane, presumably connecting these two septa, and measuring 4 mm in length, and 1.5 mm at its greatest breadth. The septa are joined axially in twos and threes, but only in a few cases do the axial ends contact the columella.

Discussion. — Nicholson and Etheridge (1878) described *Lind-stroemia laevis* from the greenish mudstones of Penkill, near Girvan, Scotland. They regarded these beds as late Llandovery in age, and the latest estimates (Ziegler, Richards, and McKerrow, 1974) agree.

These solitary corals are curved, with the diameter equal to or slightly greater than the length. The figures give no impression of foreshortening of the cardinal-counter diameter, a feature commonly seen in the Brassfield form. The calice is oblique and was filled with sediment in all specimens studied by these authors. All that could be determined is that the calice is deep, expanding abruptly, and the septa are quite short near its lip. At various levels, from 15 to 23 major septa were found, which join axially into a columella projecting up into the calice. Minor septa were not clearly observed. Presumed arched tabulae appear in the base (Nicholson and Etheridge, pl. 6, fig. 4b), but dissepiments are absent.

The columella is elongate in the horizontal plane, just as in my specimens, and presumably lies in the cardinal-counter plane. Considerable stereome is involved in its formation. The proximal ends of the septa expand where they enter the stereozone. In a transverse section, supposed to be 6 mm in diameter, the columella is 2.4 mm long with a maximum width of about 0.6 mm. In this stage (Nicholson and Etheridge, pl. 6, fig. 4e), the septa have nearly all lost contact with the columella, but have not joined distally in groups.

The corallites are apparently smaller than mine, averaging "three lines" (about 6.5 mm) in length, and "3.5 lines" in greatest diameter. Large specimens reach a length of "6 lines" (about 13 mm) and a diameter of 6 to 7 "lines". The largest Penkill specimens are comparable in size to my smallest. As the authors indicate that they found up to 23 major septa (presumably in their larger specimens), it is likely that they were dealing with small individuals of a species that could attain larger dimensions.

No mention is made of a cardinal fossula in these specimens (though in a previous description, of *L. subduplicata,* these authors specifically noted the absence of such a feature), and no mention is made of ridges on the septa (as the septa were observed only in thin-section). The two transverse sections presented (of different specimens) do not clearly show the symmetry about the cardinal septum so commonly seen in the Brassfield specimens.

The over-all similarities between the Girvan form and my specimens are so strong as to make it possible they are conspecific. In the absence of definitive information on the presence of a fossula in the Girvan form, however, their conspecificity must be viewed as questionable, and Nicholson and Etheridge's trivial name is not presently applied to the Brassfield species.

Davis (1887) figured the species *Cyathophyllum gainesi* from the Louisville Formation (Late Wenlock and Earliest Ludlow) near Louisville, Kentucky. Later, Stumm (1964) refigured two of the specimens and described the species. He gave the collecting locality as Brunerstown, 12 miles east of Louisville.

The corals are curved, trochoid, with a length of 1.0 to 1.5 cm and a maximum diameter of 0.5 to 1.5 cm (comparable in size with the Brassfield specimens). The calices are funnel- or cup-shaped, and similar in depth to ours. Some specimens have a peripheral platform on top of the septa, as seen in the first of the three Brassfield specimens described above. Stumm reported about 50 septa (apparently referring to the total of both orders), and a bladelike columella formed by the junction of the cardinal and counter septa. The major septa extend to the base of this columella. The minor septa are very short, largely restricted to the periphery, as in my specimens. The figures and description do not make it clear if there is a fossula, or if the septa bear ridges.

The syntypes of *C. gainesi* (MCZ 7551-7556) were not available for study, so that it is best to treat this form as probably conspecific with my material due to the many features they share, but to withhold a final decision pending further study.

The earliest reference in the literature to a form unquestionably conspecific with my specimen is that of Foerste's (1906) *Lindstroemia lingulifera* from the Waco bed, Noland Member of the Brassfield Formation near Panola, Kentucky (locality 1 of this study), and from the same horizon north of Estill Springs, Kentucky (vicinity of locality 53-53a). The type assemblage consists of three specimens (USNM 87175), which are the originals of Foerste's plate 5, figures 2C-F.

The calice areas of these specimens are poorly preserved: the walls are destroyed, leaving only the calice floor. The exterior of the theca is as found in my specimens. The calice shows a cardinal fossula on the convex side, and a linguiform columella of the same form and orientation as in the Brassfield specimens. In a diameter of 11.5 mm, the specimen of fig. 2D shows 28 septa (major). The minor septa are apparently restricted to the stereozone. This specimen presents what is essentially a cross-section, as the calice walls are gone, and the ratio of septa-to-diameter is in good agreement with that in my specimens.

The specimen shown in Foerste's fig. 2E and F was sectioned (Pl. 15, figs. 3, 4). This specimen was about 17 mm in length (measured straight, from the base to the highest part of the convex side), and about 11.5 mm in diameter. Note that these dimensions are of the specimen without the calice wall.

The transverse section is 9.5 mm in diameter. There appear to be about 25 major septa, all peripherally thickened (secondary alteration somewhat obscures the exact number). The columella connects the presumed cardinal and counter septa and is 3.5 mm long, with a maximum breadth just under 0.5 mm. Clearly, this structure is the result of stereomal thickening of the cardinal-counter lamella in the axis. The septum presumed to be the cardinal is curiously bent and slightly out of line with the columella. The septum to one side of it aborts against the side of its neighbor toward the counter side, but this does not happen in the case of the septum on the other side of the presumed cardinal. The stereotheca is slightly over 1 mm thick, and no minor septa appear in it. The major septa join axially in groups of two and three, and in some cases these groups seem to be joined by stereomal deposits to the side of the columella.

The longitudinal section is similar to those of the field specimens. The axial area is filled with a perforate mass of stereome. Between this mass and the theca, there are a few axially convex tabulae, separated from one another by about 0.4 to 0.6 mm.

Ryder (1926) reported a species from the Wenlock of Gotland, *Dinophyllum minimum*, which clearly belongs to the genus *Dalmanophyllum* Lang and Smith. It is a trochoid coral, reaching a maximum length of 26 mm and diameter of 12 mm. The calice is fairly deep, with a small linguoid columella projecting from the floor often united with the cardinal septum. The ratio of coral diameter to number of major septa appears to be comparable to that of my specimens. A depression or flattening, interpreted as an attachment-scar, often appears on the lower convex side (Pl. 2, fig. 19)*. This

---

*For other views of this specimen, see Pl. 2, figs. 18, 24 and Pl. 15, figs. 13, 14.

feature is not found in the Brassfield specimens. Ryder reported two forms of this coral, one "short, squat, and turbinate" and the other "elongated, this latter with the calyx [*sic*] usually set oblique to the axis of growth". It seems likely that he was dealing with two separate species. The cardinal and counter septa are united by the columella, and the axial ends of the other major septa unite in groups of from two to four. The septa appear to be thickened slightly peripherally, and development of minor septa is feeble.

In the earliest stages (diameter = 2 mm) the axial lamella is present, with the ends of the other septa united with it. This structure thickens in later stages. The transverse sections suggest a fossula or pseudofossula containing the cardinal septum, but no specific mention is made of this in the text. (Nonetheless, Minato [1961] in an ontogenetic survey of specimens from Gotland which he considered to be *D. minimum,* reported a cardinal fossula.) The figures also indicate that in the earliest stages (diameter = 2 mm) the septocoels are open. Minato's figure 22 substantiates this. This fact, and the presence of basal attachment-scars makes it best to regard the species described by Ryder as distinct from my form.

Scheffen (1933) described two forms that resemble *D. linguliferum,* from the Ringerike area near Oslo, Norway. These come from Zone 5b, which Neuman (1969, p. 2) assigned to the Porkuny Stage. Neuman considered this stage to be Upper Ordovician, while Manten (1971) regarded it as lowermost Llandovery in age. In either case, it is pre-Brassfield in age.

The first of these species is *Lindstroemia laevis* Nicholson and Etheridge, 1878, which is discussed above. This has a simple columella arising from the union of the septa. It is free-standing in the ephebic stage and joins the cardinal and counter septa. The major septa are single or united axially into groups, and are peripherally thickened. Minor septa are weakly developed. The corallite diameter is from 5 to 10 mm. There are 21 major septa in a diameter of 6.5 mm, similar to the ratio found in the Brassfield specimens. The stereotheca is about 0.5 mm thick. There is no sign of tabulae.

The information on this species given by Scheffen is too incomplete for any conclusions to be safely drawn. The figure shows only a portion of a transverse section with a free-standing columella about 2 mm long and 0.7 mm at its greatest breadth. It is not possible to determine if a fossula is present. Further information on the external features, especially the length and nature of the base, would make it easier to determine the relationship between the Norwegian and Brassfield forms. From what may be seen, it is likely that they would prove to be conspecific.

The second form described by Scheffen, *Tyria insertum,* has a simple columella, easily recognized even before separation from most of the septa. In the stereotheca, the proximal ends of the septa stand out clearly. Stereome fills the interior of the corallite in early stages. The septa reportedly withdraw evenly from the axial columella. Though the characters of this species described by Scheffen agree with those of *D. linguliferum* (Foerste), there is too little information available to draw conclusions on the relationships between the two forms.

Distribution. — Kentucky: mid-Llandovery (Brassfield Formation).

There are forms similar to Brassfield specimens occurring in northwest Europe. In no instance can they be confidently said to be conspecific with the present species. In some instances, such as *D. minimum* (Ryder) morphologic features preclude conspecificity. In others, such as *L. laevis* Nicholson and Etheridge, *L. laevis* N. and E. of Scheffen, *C. gainesi* Davis, *D. gainesi* (Davis) of Stumm, and *T. insertum* Scheffen, details are lacking in the descriptions that might make conspecificity conclusive. It seems likely that the ancestry of *D. linguliferum* (Foerste) may be in the Late Ordovician or Early Llandovery of Scandinavia or western Europe.

Brassfield occurrence. — Locality 1 (near Panola, central Kentucky); Foerste's specimens included some from north of Estill Springs, central Kentucky (vicinity of locality 53-53a).

Genus **RHEGMAPHYLLUM** Wedekind, 1927

1927. *Rhegmaphyllum* Wedekind, p. 14.
1927. *Regmaphyllum* Wedekind, p. 74.
1937. *Regmaphyllum* Soshkina, p. 85.
1940. *Rhegmatophyllum* Lang, Smith & Thomas, p. 114. (*Nom. van. pro Rhegmaphyllum* Wedekind, 1927).
? 1960. *Briantelasma* Oliver, p. 89.

Type species. — (SD Soshkina, 1937, p. 85) *Turbinolia turbinata* Hisinger (1831); Lang, Smith, and Thomas (1940, p. 114) mistakenly implied that the type species is *R. turbinatum* (Hisinger) of Wedekind (1927, p. 74).

Type locality and horizon. — Silurian of Gotland. Lang, Smith, and Thomas said the species is "very abundant at the Slite Marls at Slite and at Visby. Widely distributed in Gotland and ranges from Wenlock to Ludlow."

Diagnosis. — Solitary corals lacking dissepimentarium; theca a septal stereozone; minor septa very short; major septa reaching axis, sometimes with counter-clockwise whorl; well-marked cardinal fossula on convex side of corallite, with parallel sides; interior mostly filled with stereome.

Description. — These are solitary, non-dissepimentariate rugose corals. Their forms vary from straight-conical to ceratoid. No basal attachment structures are known in this genus, the lower portion of the cone simply coming to a smooth point. The theca, a septal stereozone, is thick.

The major septa reach to the axis. The minor septa are characteristically short, often barely protruding from the theca.

A straight-sided cardinal fossula appears to be ubiquitous. Within it is a cardinal septum depressed below the level of the calice floor but reaching the axis. The septa forming the walls of the fossula often distally truncate the septa between them and the cardinal septum.

The axis of *Rhegmaphyllum* species is commonly affected by a counter-clockwise axial whorl (Pl. 16, figs. 1, 12).

The interior of the corallite is mostly filled with stereome. This serves to thicken the septa in the earlier stages of growth, filling the septocoels and leaving only sutures betwen the septa. In later stages, incomplete tabulae are sometimes present, sloping from the axis downward toward the theca (Pl. 15, fig. 19).

The septal trabeculae may thicken to form elongate carinae, projecting up toward the axis from the periphery. On worn specimens, these may be manifest on the surface as a pattern of diamond shapes in the planes of the septa (Pl. 3, fig. 8), a pattern referred to by Foerste (1906, p. 310) as "quincuncial".

Discussion. — Soshkina (1937, p. 85) designated *Turbinolia turbinata* Hisinger, 1837 [*sic*] as type species of this genus. Hisinger's earliest use of this specific name was in 1831 (p. 128), in which he seems to have attributed it to Lamarck. In 1837 Hisinger briefly described *T. turbinata,* and provided an illustration (pl. 28, fig. 6) for the first time. Lang, Smith and Thomas (1940, p. 114) incorrectly believed Soshkina's selection to have been Wedekind's (1927) *R. turbinatum* (Hisinger).

Lang *et al.* (1940) considered *R. turbinatum* (Hisinger) of Wedekind, and *T. turbinata* Hisinger (*in part*) to be identical to *Zaphrentis*(?) [*sic*] *conulus* Lindström, 1868, and opined that the correct name of the type species is *R. conulus* (Lindström). Apparently these authors felt that *R. turbinatum* (Hisinger) of Wedekind was not identical to the typical *T. turbinata* Hisinger (though I am unaware of designation of a lectotype or neotype for the latter), but rather was conspecific with *Z. conulus* Lindström. In his unillustrated description of *R. turbinatum* (Hisinger), Wedekind reported the absence (or lack of observation) of a cardinal fossula, an element which is distinct in Lindström's figures of *Z. conulus.* As noted above, Lang *et al.* were in error in giving the type species of *Rhegmaphyllum* Wedekind as *R. turbinatum* (Hisinger) of Wedekind. The correct type species is *Turbinolia turbinata* Hisinger, and while Lindström's species is probably a *Rhegmaphyllum,* its features cannot be regarded as definitive of the genus.

Further, if Hisinger's specimens of *R. turbinatum* prove to lack cardinal fossulae it will necessitate a re-evaluation of the importance of this feature as a taxobasis for *Rhegmaphyllum,* as stressed here. This emphasis is largely based on the common opinion that *Z. conulus* Lindström, a form better documented than *T. turbinata* Hisinger, is a *Rhegmaphyllum.*

Another possible explanation for Lang *et al.*'s view that *Z. conulus* Lindström is the type species of *Rhegmaphyllum* Wedekind, would be that these authors felt that *T. turbinata* had originally been described by Lamarck, as Hisinger (1831) seems to imply. If they felt that Hisinger's specimen did not belong to Lamarck's species, but rather to Lindström's, their opinion on the correct name of the type species of *Rhegmaphyllum* would be understandable. No mention, however, is made of Lamarck in Lang *et al.*'s discussion of this genus, so this possibility seems unlikely.

Before describing *Rhegmaphyllum,* Wedekind (1927, p. 14) listed the genus in a systematic array. In this list, the spelling was "*Rhegmaphyllum*". In the subsequent description (1927, p. 74), Wedekind (whether by design or typographical error) spelled the name as *"Regmaphyllum"*, without further comment. This latter spelling was used by Soshkina (1937). Because the first spelling of the generic name is *"Rhegmaphyllum"*, a name apparently properly in keeping with the rules of nomenclature, and because Wedekind gave no indication that he had intentionally changed his mind in altering the spelling on the later page, *"Rhegmaphyllum"* must be regarded as the correct name. Lang, Smith, and Thomas (1940, p. 114) emended the spelling to *Rhegmatophyllum*, a modification also regarded as invalid.

Briantelasma Oliver (1960) is a solitary rugose genus from the Silurian and Lower Devonian of North America, whose characters mark it as closely related to *Rhegmaphyllum*. The two forms share the general radiality of the septa, which in the case of the majors, extend to the axis without forming an axial structure. The septocoels are largely filled with the stereome of dilated septa in both genera. In addition, both *Briantelasma* and *Rhegmaphyllum* have a cardinal fossula. On the other hand, the minor septa of *Briantelasma* tend to be longer (1/3 to 2/3 the length of the major septa) than is usual in *Rhegmaphyllum*, and its tabulae are reportedly complete. More study is necessary on the type material before the relationships between *Briantelasma* and the present genus can be ascertained.

Rhegmaphyllum daytonensis (Foerste)   Pl. 2, figs. 25-28;
PI. 3, figs. 1-17; PI. 15, figs. 15-19; PI. 16, figs. 1-12

- Cf. 1868. *Zaphrentis conulus* Lindström, text-fig. p. 428; pl. 6, fig. 8.
- Cf. 1882b. *Zaphrentis conulus* Lindstrom, Lindström, pp. 16, 20-21.
- Cf. 1885. *Zaphrentis vortex* Lindström, p. 19 (*nom. nud.*).
-      1890. *Cyathophyllum celator daytonensis* Foerste, pp. 339-340, pl. 9, figs. 9-11 (*non Zaphrentis celator Hall*, 1877).
-      1890. *Cyathophyllum facetus* Foerste, p. 341, pl. 9, fig. 8.
-      1893. *Cyathophyllum celator* (*Hall*) var. *Daytonensa* Foerste, Foerste, p. 601, pl. 34, figs. 9-11.
- ?   1894. *Zaphrentis vortex* Lindström, Weissermel, p. 630, pl. 50, figs. 3, 4.
- Cf. 1896. *Zaphrentis conulus* Lindström, Lindström, pp. 32-34, pl. 6, figs. 65-68.
- ?   1896. *Zaphrenthis vortex* Lindström, Lindström, pp. 34-35, pl. 6, figs. 69-73.
-      1906. *Zaphrentis charaxata* Foerste, pp. 310-311, pl. 7, figs. 4A-E.
- ?   1933. *Lindströmia torsa* Scheffen, pp. 31-32, pl. 5, figs. 6, 7.
- ?   1962b. *Briantelasma* sp. Oliver, p. 27, pl. 14, figs. 8-12.
- ?   1971. *Leolasma kaljoi* Scrutton, pp. 211-212, pl. 2, fig. 13; pl. 3, figs. 1-8; text-fig. 7.

Type material. — Lectotype: USNM 84796, the original of Foerste's (1893) plate 34, fig. 10 (identical to Foerste's (1890) pl.

9, fig. 10), figured on Plate 2, fig. 27, and Plate 16, figures 1, 2 of the present work. Paralectotypes: 2 additional specimens under USNM 84796, possibly the originals of Foerste's (1893) plate 34, figures 9, 11 (identical to Foerste's [1890] pl. 9, figs. 9, 11), shown on Pl. 2, figs. 28, 26 respectively; and USNM 93518 (shown on Pl. 2, fig. 25).

In the museum box with USNM 93518 is a card bearing no catalogue number, designating as type of the species the original of Foerste's (1893) plate 34, figure 10. This figure bears no resemblance to the specimen in the box. In the box with specimens catalogued as USNM 84796, is another card with the number 84796 on it, again designating figure 10 as the type. Both labels appear to be in Foerste's handwriting. In neither his 1890 description, nor that of 1893 did Foerste indicate which figure represented the type of the species, though he did refer to a type specimen. Consequently, the museum label is the only source of information as to which specimen he regarded as the type. It appears that the card in the box with USNM 93518 was placed there by mistake, and that the true type, as intended by Foerste, is the one of three specimens included under USNM 84796 that best agrees with Foerste's figure 10, here designated the lectotype. The other two specimens, labelled as cotypes on their museum cards, thus become paralectotypes.

Type locality and horizon. — Dayton, Ohio, site of the Veterans' Hospital (called Soldiers' Home by Foerste), from the Brassfield Formation (mid-Llandovery).

Number of specimens examined. — Over 100, including the type suite of *C. celator daytonensis* Foerste, 1890 (4 specimens), and 29 specimens of *Zaphrenthis charaxata* Foerste, 1906 under USNM 87178 (including the type suite of 4 specimens).

Diagnosis. — Trochoid or turbinate *Rhegmaphyllum* with counter septum lengthening through ontogeny as cardinal shortens; cardinal and counter septa in contact, or nearly so, at all levels below floor of the fossula; axial structure solid (no axial tube), often whorled counter-clockwise; septa ridged in later growth-stages; major septa rhopaloid.

Description. — This is one of the most remarkable species found in the Brassfield. It is by far the most widespread (Table 3, Text-fig. 5e), occurring at 18 localities, and possibly at 9 others. Locally its abundance can be striking, as at localities 5, 37, and others. More significant is the fact that it is often abundant where other species are almost totally lacking (locality 5). Its plasticity of form makes it difficult to formulate a description of this species that is universally applicable. Certain distinct features found in the *R. daytonensis* group seem to occur together in various combinations, but none are so closely-correlated as to allow confident division of this form into more than one species.

The corallite may be trochoid or turbinate, varying widely in the width-to-length ratio. It is always curved with a convex cardinal side. The outer surface may be smooth (perhaps due to wear) or distinctly striate longitudinally. Rugosity occurs but is not particularly pronounced. Foerste (1906, p. 310) noted that the longitudinal striae, especially on weathered specimens, appear as "zig-zag lines dividing the surface into small polygonal facets arranged in more or less quincuncial order." This pattern reflects peculiarities of the septal structure (see below) brought to the surface by wearing of the theca.

The calice is distinctive, dominated by a distinct, deep cardinal fossula, parallel-sided or narrowly wedge-shaped, and reaching the axis. The fossula borders are formed by two major septa. The cardinal septum lies in the axis of the fossula, dipping deeply below the level of the calice floor before reaching the axis, but reappearing distally to rise closer to the axis. It is flanked by a pair of major septa that are contratingent to the pair of major septa that actually form the walls of the fossula (Pl. 3, figs. 9, 12). This is due to the fact that those abutting lie parallel to the cardinal septum, while the bounding septa (like all the other septa) are radially-disposed relative to the axis. This pattern of the cardinal region is sometimes encountered in *Dalmanophyllum*.

In many specimens, the septa bear ridges (elongate carinae) sub-parallel to the calice floor, and slightly higher axially than peripherally. These are the same features as were recognized in *Dalmanophyllum linguiferum* (Foerste). They occur on both sides of a septum, and are arranged both oppositely and alternately, even on a single septum (Pl. 3, fig. 14). The ridges appear to be absent in early stages. In three specimens in the collection of Ohio State University, they appear at the following elevations above the base: 12 mm (OSU 22598), 12 mm (OSU 22603), and 17 mm (OSU 22597).

It is presumably the intersection of these carinae with the weathered corallite surface that produces the zigzag lineations mentioned above (Pl. 3, fig. 8). This relationship is shown especially well in Plate 3, figure 11. The carinae are presumed to be due to unusual thickening of trabeculae at intervals along the septum.

The minor septa do not extend so far axially as do the major septa, but they bear ridges when these features are present on the major septa.

Axially the major septa unite, and their juncture may or may not be affected by a whorl. Invariably, the sense of the whorl is counter-clockwise. In the absence of a whorl, the counter septum extends to, and apparently across the axis. From the depths of the cardinal fossula, a septum arises in the axis and joins with the counter septum. It was not determined if it is the axial end of the cardinal septum rising from the fossula, or the axial end of the counter septum extending into the fossula. When the axial ends of the septa are whorled, juncture between the counter and cardinal septa is ruptured.

Usually, a broad, flat-topped convexity occupies the calice, separated by a fairly narrow moat from the calice wall. This convexity appears to be due to whorling of the septal ends, accompanied by stereomal deposits between them. The fossula extends into the boss and is usually affected by the whorl. The moat contains a series of radially-disposed marginal pits in most cases, which are the septocoels restricted to the periphery by axial deposits of stereome (Pl. 3, fig. 15).

Occasionally, there is abortion among the major septa, where one or more appear to be crowded out from the axis by their neighbors, such that they reach only mid-way between the ends of the major and minor septa. In at least one case (Pl. 3, fig. 9) there is a certain symmetry to this: Two septa are aborted, one being the seventh to one side of the cardinal septum, the other being the eighth to the other side.

There is considerable variation in the configuration of the calice floor, depending upon whether or not there is an axial whorl or boss. In one specimen with a boss (Pl. 3, figs. 15-17), the calice diameter (at the rim) is about 2.0 cm, with a depth (to the top of the boss) of about 7 mm. The boss diameter is 8.5 mm. The specimen itself measures 2.1 cm along its conxex side, and 1.2 cm along its concave side.

A second specimen, without a whorl and boss (Pl. 3, figs. 9, 10), has a rim diameter of 1.7 cm with an approximate depth of 3.5 mm. The vertical profile of the calice is a segment of a circle. The length, measured along the convex side, is 2.1 cm, and along the concave side, 1.1 cm.

A third specimen (Pl. 3, figs. 12-14) has a slight axial prominence but no whorl. The rim diameter is 2.1 cm, about equal to that of the previous specimen, but its convex and concave sides measure, respectively, 3.5 cm and 1.8 cm, making it more elongate than either of the two specimens described above. Its septal pattern is identical to that of the previous specimen, being without a whorl, with a continuous septum running from the counter across to the cardinal side. On the other hand, this specimen has prominent carinae, while these are absent in the previous specimen.

A series of nine transverse, oriented sections were taken from a single specimen (Pl. 16, figs. 3-11):

The lowest section is 2 mm in diameter. All that can be determined is that the approximately 10 septa present fill the entire cross-section.

The next level is 5 mm in diameter. Here the entire cross-section is still filled by fewer than 20 septa, all of one order.

The third section is 7 mm in diameter. About 21 septa are present, with well-developed incipient septa wedged between them. Again, the entire cross-section is filled.

The fourth section is 9 mm in diameter with approximately 27 septa. It becomes difficult to count the septa now, for while some are fully inserted, reaching the axis, others are not yet of full length, and reach only about half-way to the axis. So far, all are of the major order. The marginal pits (discussed above) are beginning to form. The cardinal septum is one of the longest, reaching directly to the axis with the other septa abutting it. From the opposite direction, a second septum (presumably the counter) does the same, but does not contact the cardinal. At this level the counter-clockwise axial whorl first becomes apparent. The fossula opens to either side of the cardinal septum as two short spaces, midway between periphery and axis.

The fifth section has a diameter of about 11.5 mm with about 31 or 32 septa. The axial space, which was a pin-point in the previous section, is broader now, with only a few septal ends intruding (notably that of the counter, and possibly, to a lesser degree, the cardinal). This trend of the shortening of the cardinal septum, concurrent with lengthening of the counter, is one of the distinctive features of this species. The marginal pits are more elongate, approximating full septocoels. The axial whorl is less obvious than in the previous section. The cardinal region appears to have suffered somewhat from diagenesis, but it seems that the cardinal septum touches the end of the counter septum, though these two are not colinear, nor are they readily differentiated.

In the sixth section, the outline is more elliptical than below, measuring about 13.5 × 11.5 mm in diameter. The shortest diameter lies at about 45° to the cardinal-counter plane. The counter septum reaches and crosses the axial area, but a small gap appears to separate it from the cardinal. The cardinal fossula is larger here, but alteration obscures its true nature. There are about 31 or 32 major septa present. What may be incipient minor septa appear in the stereozone, but do not yet protrude into the septocoels. The thickness of the stereotheca is about 2 mm. Stereome deposits give the axial ends of the septocoels a rounded aspect, and restrict the marginal pits peripherally, as well as adding to the solidity of the axial structure. A gap remains between one side of the counter septum in the axis, and the axial ends of the approaching septa. Otherwise, the axial structure is solid.

The seventh section measures 15.0 × 13.5 mm across, compressed as the last section. The cardinal fossula nearly reaches the axis. About 36 major septa are present. The marginal pits are longer than at lower levels, shaped more like ordinary septocoels. Stereome is still present in the axis, but the axial structure is now more loose. The axial whorl is well-defined, and there is no connection between the counter and cardinal septa. The minor septa do not yet protrude from the stereozone, which has about the same thickness as in the above section.

The eighth section measures 16.0 × 15.0 mm across, compressed as below. There are 35 major septa present. The axial whorl, though recognizable, is weak. The presumed counter septum crosses the axis, but does not contact the cardinal septum. The distal ends of the major septa in the axis are still slightly thickened by stereome, but not to the degree encountered in lower sections. The axial structure has become more open. The minor septa do not yet protrude from the stereozone. The cardinal fossula reaches the axis and is slightly affected by the whorl. The septocoels are largely free of stereome, and in them may be seen the first signs of tabulae.

The ninth, and highest section is incomplete. It appears compressed as below, with a maximum diameter of 17.5 mm. The stereozone is 2 mm thick. The septa do not reach the axis, as this section is just above the calice floor. As a consequence, the cardinal fossula is parallel-sided (whereas below, the sides bowed outward slightly), and opens axially. The cardinal septum is considerably shorter than in lower sections. The theca at the cardinal area is about half the thickness found elsewhere at this level, and the cardinal septum hardly extends beyond the inner edge of the adjacent stereozone. The minor septa protrude as short points in the septocoels, though the base of the protruding portion is much narrower than the segment of wall from which it emerges.

Other specimens show in transverse section the same features as this specimen. In some cases, the sutures between the septa in the stereozone are zigzagged, reflecting the intersection of the septal carinae with the plane of the transverse section. The axial whorl may occur at one level, but not at another. In one specimen (Pl. 3, figs. 12-14), for example, the calice floor does not have an axial whorl, and the cardinal and counter septa appear to be continuous, and in contact. But a cross-section just below the calice level (Pl. 15, fig. 18), shows a small whorl, with one of the septal ends blocking contact between the cardinal and counter septa.

Longitudinal sections of this species (Pl. 15, fig. 19; Pl. 16, fig. 2) show a corallite filled basally with stereome, but becoming more vacuous distally. Strongly convex incomplete tabulae may be seen between the theca and the axial structure. The section on Plate 15 shows the septal carinae, visible as short bars nearly perpendicular to the corallite's outer surface, such that they are slightly higher axially than peripherally.

Discussion. — In a series of papers (1868, 1882b, 1896), Lindström discussed *Zaphrenthis conulus,* from the Baltic area and the Siberian Platform. The first reference (1868) consists of two transverse views of the corallite, showing 25 major septa (including the cardinal) with an equal number of minor septa, these at most 1/3 the length of the major septa. The cardinal septum, along with two other short septa on either side of it, lies in a parallel-sided fossula, which extends to the axis. The axis is vacant, the void occupying about 1/7 the total diameter of the corallite. This axial space, connected with the fossula, gives the appearance of a key-hole. The septa and fossula are affected by a counter-clockwise axial whorl. The specimen is from Gotland, but neither its size nor its horizon are given.

In his 1882b paper, Lindström referred to occurrences of this species in the Silurian (probably Early or Middle) of the Middle (Podkamennaya?) Tunguska River Basin of the Siberian Platform. These are specimens considerably larger than the average, reaching 40 mm long and 23 mm broad at the calice rim. The largest examples are more expansive distally and not so elongatedly conical. Also, they are somewhat curved. Lindström asserted that, in thin-section, they are in agreement with the Gotland specimens.

In the same paper, he referred to specimens of this species of Silurian age from near the Olenek River. These are conical and straight. The cross-section near the calice is elliptical to circular. The calice is funnel-shaped, deepened at the innermost (axial) end of the cardinal fossula. The axis is whorled. There are 34 to 35 major septa, and an equal number of minor septa. The cardinal fossula is large, and keyhole-shaped. The tabulae are frequently incomplete and arched against the axis.

Lindström's 1896 reference is more detailed, and deals with the Gotland form. The corallite is simple, straight, regularly conical, and seldom curved (and then only slightly). The theca is weakly rugose. There are no signs of attachment devices. The calice is circular to slightly elliptical, its depth extending for 1/5 the total corallite length. A corallite 55 mm long has a calice 12 mm deep, and about 30 major septa, alternating with as many minor septa which project only slightly from the wall. Two or three septa, including the cardinal, occur in the fossula. The major septa extend to the axis and meet along a circular rim which bounds the axial space. The septa bear ridges and grooves as in the Brassfield specimens. In thin-section, the major septa appear as twisted cords rimmed by dark stereome. The axial edges of the major septa ar irregular, so that when they meet and grow together, a spongy mass forms in the axial area. The sparse tabulae are strongly convex axially.

There is general resemblance between Lindström's material and the Brassfield specimens. The form of the cardinal fossula is similar, as is the ridging of the septa. In details, however, differences are evident. The two septa neighboring the cardinal in the fossula abut the sides of their neighbors, the major septa forming the fossula walls. This character, found in the Brassfield form, does not appear in Z. conulus. Also, while something approaching an axial void is on rare occasions seen in my specimens (e.g., Pl. 3, fig. 15), it is never so well-developed as in Lindström's form. Finally, R. daytonensis is invariably curved, while Z. conulus is generally straight. The conclusion drawn is that, while Z. conulus is surely congeneric with R. daytonensis, the two forms are distinct species.

Foerste (1890) introduced a new form, Cyathophyllum celator daytonensis from the Brassfield Formation at Dayton, Ohio. He regarded it as a new subspecies of Zaphrenthis celator Hall (1877). Examination of the type of Hall's species (AMNH 1888) shows that these two forms are not conspecific. Foerste's type suite consists of the lectotype (USNM 84796) and three paralectotypes (two catalogued as USNM 84796 and one as USNM 93518). The lectotype is shown on Plate 2, figure 27, and Plate 16, figures 1, 2 and the paralectotypes are shown on Plate 2, figures 25, 26, 28. All the specimens are trochoid, and bear rugosities and longitudinal striae on their outer surfaces. On the surface of the lectotype, the pattern of septal insertion makes it clear that the convex side (see figure) is the location of the cardinal septum.

The lectotype was studied in transverse and longitudinal sections (Pl. 16, figs. 1, 2, respectively). The transverse section has a diameter of 2 cm, and shows about 38 major septa, including the cardinal. The minor septa are represented by extremely short points barely protruding from the stereozone (which is 3 mm thick, with zigzagged sutures between the proximal edges of the septa). The structure of the cardinal fossula is somewhat obscured by fragmentation. The major septa and the axial end of the fossula are affected by a counter-clockwise axial whorl, and the axial ends of the major septa have stereomal deposits between them, making the axial structure more solid. There is no appreciable axial void.

The longitudinal section shows the meaning of the zigzagged
sutures in the stereozone: The septa bear carinal ridges on their
sides, which rise axially. The ridges are spaced about three per mm,
and are more strongly developed in the stereozone than on the sides
of the free septa. Very steep tabulae, some incomplete, occur in the
space between the stereozone and the axial complex. This form is
undoubtedly a species of *Rhegmaphyllum* Wedekind, and is identified with the Brassfield form. (In 1893, Foerste recorded the
presence of this species at Dayton, Ohio, in the "Clinton", and repeated the figures from the 1890 paper, without a description.)

    Cyathophyllum facetus was described by Foerste (1890) from
Todd Fork near Wilmington, Ohio (locality 13) (probably Brassfield). He remarked that it was similar to *Cyathophyllum celator
daytonensis* Foerste, described as a new "variety" in the same work,
differing only in being of considerably narrower diameter at a
given level. The specimens of *C. facetus* in Foerste's collection rarely
exceed 25 mm in length, and in all, the fossula is on the convex side
of the corallite. The outer surface is longitudinally striate, and may
be faintly rugose. The calice contains fifty to sixty-five septa, and
has a depth of 4 or 5 mm. The septa are of two orders, the minor
septa failing to reach the base of the calice. Each septum is crenulated (carinate?) on both sides, 5 crenulations occurring in 1.8 mm,
and the axis of the corallite exhibits torsion of the septa. Two or
three septa are "intercalated" towards the border of the fossula
(possibly alluding to the condition shown in Pl. 3, figs. 9, 12, where
a pair of septa flanking the cardinal septum terminate against the
sides of the septa forming the fossula walls). While the description
and figure do not provide as much detail as might be desired, the
likelihood is strong that *C. facetus* is conspecific with *Rhegmaphyllum daytonensis* (Foerste).

    In 1896, Lindström presented a detailed description and figures
of *Zaphrenthis vortex*, for which he had in two earlier publications of
the 1880's given only the name, with no description or figures. The
specimens are from Gotland, in the oldest marls in the area of Visby
[Upper Llandovery or Lower Wenlock, according to Manten (1971,
p. 43 and geologic map)]. This coral is conical, generally curved,
short, and expands rapidly distally. The rugae are coarse. The
sutures of the septa in the stereozone are undulose as in the Brassfield form. The calice is quite deep, about 17 mm in a corallite 25

mm long. The septa bear sub-horizontal ridges, as in the Brassfield specimens. The cardinal fossula is prominent, deepest axially, and has a keyhole-like form. It is often affected by an axial whorl whose direction cannot be determined from the figures. The axial region is thickened by stereome. Where not obscured by stereomal deposits, a few tabulae may be seen apparently arching upward axially.

Lindström's figure 71 suggests the counter septum goes into the axial end of the cardinal fossula, a feature distinctive of my form. If this does in fact occur, then it is likely that the two are conspecific, possibly representing two varieties of the same species. Pending further study, we must regard the Brassfield and Gotland forms as only possibly conspecific.

Weissermel (1894) described what he regarded as specimens of *Zaphrenthis vortex* from Königsberg, Trömpau, and Bergenthal in Prussia. He gave a general description of what were apparently fragmentary examples.

The corallite is short, conical, rapidly expanding distally, straight or weakly-curved. The thecal striae are pinnately arranged against three principal striae. The cardinal fossula is wide and long, reaching the axis, and remaining about constant in width. It contains the short cardinal septum, and what Weissermel regarded as two minor septa. The cardinal fossula is bounded by two major septa, parallel to the cardinal. In the rest of the calice, the septa are radially arranged. In addition to the cardinal septum, there are 25 to 43 major septa, alternating with an equal number of minor septa. When the corallite is curved, the fossula lies on the convex side.

Where the interior is not filled with stereome, tabulae may be seen, horizontal in the axis, and sloping steeply peripherally. There is no dissepimentarium.

Weissermel's description is too sketchy to allow confident identification of his form with the Brassfield specimens, though there are clear similarities. No mention is made of ridged septa, nor of the fate of the counter septum on reaching the axis. Also, the lack of strong curvature is more reminiscent of *Rhegmaphyllum conulus* (Lindström) than of *R. daytonensis* (Foerste).

Foerste (1906) described what he regarded as a new species, *Zaphrenthis charaxata* from the Waco bed of the Noland Member, Brassfield Formation, north of Irvine, Kentucky (locality 53-53a of this study). The type suite (USNM 87178, of which the lectotype here chosen is shown on Pl. 3, figs. 1-5) consists of four specimens, labelled as cotypes, from among the total of twenty-nine specimens included under this catalogue number. In all specimens, the calice walls have been damaged, leaving only the calice floor, with occasional fragments of the wall and the lower portions of the corallites. Most of the specimens are around 1 cm in diameter, trochoid, ranging from 1.83 to 0.82 cm in length (measured in a straight line from the base of the corallite to the calice floor on the convex side). In all but one instance the corallites are curved, and in all but two instances (or possibly three) the cardinal fossula is on the convex side of the corallite. Of the thirteen specimens in which the axis was clearly visible, all but two have a counter-clockwise axial whorl. The number of major septa range from 27 to 32. In some cases the axial ends of the major septa where they meet in the axial complex, are dilated (Pl. 3, fig. 1), almost forming small separate knobs. The peripheral ends of the septa are also thickened. The sides of the cardinal fossula are generally parallel, and the fossula reaches the axis. A quincunx pattern could clearly be seen on the surfaces of twenty-two of the specimens. In two of the largest specimens (including the original of Foerste's fig. 4A) the sides of the septa are ridged, though this is not always obvious due to wear on the specimen. In transverse section (Pl. 15, figs. 15, 16) the proximal thickening and the parallelism of the sides of the fossula are most obvious. Without question, this form is conspecific with *R. daytonensis* (Foerste).

Lindstroemia torsa Scheffen (1933) from Stage 6 (probably Llandovery) in the Oslo area of Norway, bears some resemblance to the Brassfield form. The major septa are fused together axially, where they and the cardinal fossula are affected by a counter-clockwise whorl. This torsion, visible even in the more compact younger stages, is considered typical of the species by Scheffen. The whorl rises as an axial convexity in the floor of the calice. The minor septa are weakly developed. Tabulae cross the plane of section in the septocoels, sloping downward from the axis. The corallite diameter is about 15 mm, and the septa seem to have lateral spacing similar to that in the Brassfield specimens. From the little that can be seen in Scheffen's figures and discerned from his description, it seems possible that this form will prove conspecific with mine, judging from the septal arrangement, features of the fossula, and the stereome-thickened axial structure. Lack of information on the development of the counter septum, however, and on the presence or absence of septal carinae, make it wisest to reserve judgement.

From the Sayabec Formation (Wenlock in age) in the Lake Matapedia region of Quebec, Oliver (1962b) described *Briantelasma* sp., a form similar to the present species. As Oliver's description is readily available, only the similarities and differences between these species will be considered here.

Oliver's measurements of 60 mm and about 30 mm as the lengths of his two specimens, and 13 mm and 12 mm as the corresponding maximum diameters, indicate corallites more elongate than mine. The number of major septa in a given diameter (29 in 12 mm, 20 in 7 mm, for example) indicate ratios similar to those in the Brassfield specimens. The septa are often rhopaloid (dilated distally), as is common in my specimens, and the axis is affected by a counter-clockwise whorl. There does not appear to be a void in the axial structure. The stereozone is about 1 mm thick at the level of a corallite diameter of 1.2 cm.

The minor septa project only a very short distance from the stereozone, reaching no more than 1/4 the length of the major septa. The fossula is not clearly shown in the figures, but generally seems similar to my specimens. Marginal pits appear in one cross-section that is just over 1 cm in diameter, and so far as can be determined, the interior is filled with stereome.

There is no indication from lower sections of the progressive shortening of the cardinal septum, and coincident lengthening of the counter septum found in my form, nor is there mention of septal carinae. These two factors are sufficiently important to require a suspension of judgement on the affinities of the Quebec and Brassfield forms pending further study, but the available information indicates the likelihood of a close relationship.

Scrutton (1971) described a new species, *Leolasma kaljoi*, from Rio Caparo near the Paso Caparo in the Mérida Andes of Venezuela. Scrutton's is the only reference encountered during this study of Silurian corals from South America. Scrutton assigned an Early Llandovery age to his material.

These are straight, conical corallites, reaching at least 30 mm in length, with steep-sided calices about 10 mm deep. The early stages are free of septocoels, the septa occupying the entire cross-section. Incipient septa appear as wedges between the proximal ends of the septa. In some instances, an axial vortex twists the septal ends in a counter-clockwise manner. A cardinal fossula appears to be the first space to form in the cross-section, followed by an axial thinning of the major septa. With this axial thinning, the dilated axial ends of the major septa remain to fill out the axial complex. From the time of axial separation of the major septa, the cardinal septum shortens rapidly, and the rhopaloid ends of the septa in the axial complex break apart, causing disintegration of the complex. As noted in some of my specimens (e.g., Pl. 15, fig. 18), the cardinal fossula extends proximally into the stereozone, making the thecal area thinner at this point than elsewhere. The corallite diameter ranges up to 17 mm, with 38 major septa present. Tabulae were observed in only one longitudinal section, and seem similar to those in the Brassfield specimens.

It is in its straight, conical form that *L. kaljoi* most clearly differs from the Brassfield form. This, plus the absence of evidence that the counter septum extends axially at the expense of the cardinal septum, and the lack of mention in the description of septal carinae make it best not to conclude that these forms are conspecific. Further study may, however, show that they are conspecific, as there remain many similarities.

Distribution. — Ohio and Kentucky: mid-Llandovery (Brassfield Formation).

Possibly conspecific forms occur in the following localities: Gotland, Upper Llandovery or Lower Wenlock; Norway, Llandovery?; Quebec, Wenlock; Venezuela, Early Llandovery.

Brassfield occurrence. — Localities 1, 5?, 7a, 11, 12?, 13, 23, 26?, 30?, 31, 32?, 34, 36, 37, 38?, 40?, 41, 42?, 46, 47, 48, 49, 50, 51, 52?.

Specimens collected by Foerste (reported 1890) in Dayton, Ohio; (reported 1906) north of Irvine, Kentucky (vicinity of localities 53-53a of this study).

Genus **SCHLOTHEIMOPHYLLUM** Smith, 1945

1945. *Schlotheimophyllum* Smith, p. 18.

Type species. — *Fungites patellatus* Schlotheim, 1820, p. 347, *partim.*

Type locality and horizon. — Silurian of Gotland.

Diagnosis. — Solitary or compound corals, lacking dissepimentarium; theca a septal stereozone; corallite patellate, with reflexed calice walls; septa of both orders dilated, often in contact laterally along sutures, with no septocoels remaining; minor septa indistinguishable from major septa in outer portion of calice wall, both orders of nearly equal length; major septa reaching axis, sometimes forming axial structure; calice wall with fibrous microstructure, lacking dissepiments.

Description. — This genus contains corals of solitary and compound form, which often grow to unusually large size, and beautiful countenance. The calices are broad and explanate, so that the corallite often appears like the modern scleractinian coral *Fungia*, a possible ecologic analogue. The calice walls commonly form a broad horizontal surface which appears to have encrusted the sediment surface, and the more central portion of the corallite is similar in position to the stalk of a mushroom. The septa of both orders are dilated throughout almost their entire length, so that the septocoels are reduced to sutures over the entire horizontal platform formed by the outer calice walls. It is only in the central region that they are differentiated visibly into major and minor septa. Here, the minor septa terminate abruptly, their axial ends describing an ellipse or circle around the corallite axis. The major septa continue axially, often first having their upper edges depressed below the level of the outer parts of the calice to form a calicular pit. In the axis, the ends of the major septa of some species intertwine to form a ropy axial structure which may protrude to form an axial boss.

The internal structure of this genus is striking. There are no vesicular dissepiments in the marginarium. Rather this portion of the corallite consists of layers of trabecular sclerenchyme (representing the septal microstructure) with no interlayered dissepiments. The surfaces of these layers slope downward toward the corallite axis in the same direction as would layers of dissepiments in a dissepimentarium.

Tabulae are restricted to the narrow calicular pit and tend to be incomplete.

Discussion. — *Schlotheimophyllum, Mucophyllum, Chonophyllum,* and *Craterophyllum* are four rugosan genera all sharing the basic characters of a wide, reflexed calice wall with dilated dissepiments meeting along sutures, and a patellate form. For this reason, they have often been confused. Their basic differences lie in combinations of two characters: calice walls with a trabecular or dissepimental structure, and axial ends of the major septa either entering or not entering the calicular pit area. Table 9 shows the relationships of these genera, all of which are Silurian forms.

Table 9. Wall structure and septal characteristics of four genera of patellate rugosan corals.

	STRUCTURE OF CALICE WALL	
	Trabecular	Dissepimental
Septa enter calice pit	*Schlotheimophyllum*	*Chonophyllum*
Septa do not enter calice pit	*Mucophyllum*	*Craterophyllum*

Schlotheimophyllum patellatum (Schlotheim)   Pl. 5, fig. 1;
                                                                    Pl. 20, figs. 1, 2

    1820. *Fungites patellatus* Schlotheim, p. 347 (*partim*).
    1837. *Fungites patellatus* Schlotheim, Hisinger, p. 99, pl. 28, fig. 3.
Non 1854. *Ptychophyllum patellatum* (Schlotheim), Milne-Edwards & Haime,
          p. 291, pl. 67, figs. 4, 4a.
  Cf. 1865a. *Ptychophyllum canadense* Billings, p. 107.
  Cf. 1901. *Chonophyllum canadense* (Billings), Lambe, pp. 185-186, pl. 17,
          figs. 1, 1a-c, 2, 3, 3a, 3b, 4.
  Cf. 1901. *Chonophyllum belli* Lambe (*partim*), pp. 186-187, pl. 16, fig. 6 (*non*
          fig. 5; *non* Billings, 1865b).
    1926. *Chonophyllum patellatum* (Schlotheim), Lang, pl. 30, figs. 4-6.
    1927. *Chonophyllum planum* Wedekind, pp. 41-42, pl. 7, figs. 2, 3.
    1927. *Chonophyllum patellatum* (Schlotheim), Wedekind, p. 42, pl. 7, fig.
          1.
    1945. *Schlotheimophyllum patellatum* (Schlotheim), Smith, pp. 18-19, pl.
          32, fig. 1.
    1970. *Schlotheimophyllum patelloides* Flügel & Saleh, p. 276, pl. 4, fig. 4.
    1971b. *Schlotheimophyllum patellatum* (Schlotheim), Lavrusevich, pp. 66-
          67, text-fig. 15.

Type material. — Lectotype of Smith (1945, p. 19): "a specimen from Gotland bearing Schlotheim's original label, in the Schlotheim Collection, Geologisch-paläontologisches Institut und Museum (Naturkunde Museum) der Friedrich-Wilhelms Universität, Berlin."

Type locality and horizon. — Silurian of Gotland.

Number of specimens studied. — One.

Diagnosis. — *Schlotheimophyllum* with trochoid early stage, but developing a broad, reflexed calice wall, forming a horizontal platform in later stages; septa of platform broad, flattened, and laterally-contiguous along sutures; calicular pit in axis, containing convex axial structure that is circular or elliptical in horizontal section, formed by axially-contorted major septa and tabulae.

Description. — The specimen is a solitary corallite, of which about 1/3 is missing. Assuming the corallite was originally symmetrical, it was elliptical in the horizontal plane, measuring about 5 × 4 cm in diameter and 1.25 cm high. In its earlier stages it was trochoid, or perhaps turbinate in form. In later stages, its width increased at a far faster rate than its height, and the calice walls became reflexed, such that they formed horizontal extensions of a broad upper platform. In general appearance, then, it was similar to *Fungia* (the "mushroom coral"), a Recent scleractinian. In the center of this platform is the elliptical calicular pit, 2.2 × 1.1 cm in diameter, and elongate in the same direction as the corallite's upper surface.

The septa are wedge-shaped in surface aspect, dilating proximally to a maximum width of 1.25 mm at the periphery of the horizontal calice platform. Throughout most of this horizontal region, the minor septa are not morphologically distinct from the major septa, and all septa are laterally contiguous along straight suturelines. Distally (about 1.4 cm from the periphery of the corallite) the septa all thin to a blade-like edge, and plunge toward the floor of the calicular pit. The minor septa extend no further towards the axis than the rim of this pit. The major septa, however, continue across the floor of the depression, and their axial ends twist and fuse into an elevated, elliptical axial structure. This structure has a relief of about 0.25 cm and is elongate, with basal diameters of 1.5 × 0.5 cm. The texture of this structure is ropy, and its elongation is in the same direction as that of the calicular pit.

In vertical section, the corallite's broad platform has a fibrous structure (Pl. 20, fig. 2) characteristic of the genus. These fibers represent the trabecular elements of the septa. The septocoels are closed, leaving no room for dissepiments. The calicular pit (Pl. 20, fig. 1) contains the contorted elements of the axial structure, consisting of twisted septal ends and incomplete tabulae. Between this structure and the walls of the pit arch several incomplete tabulae, separated vertically from one another by about 1 mm.

At one point, during its reflexed stage (Pl. 20, fig. 1) the corallite decreased in radius at one edge by about 0.5 cm, and then expanded by about 0.35 cm, just prior to final cessation of growth. A similar feature may be seen in *Chonophyllum planum* Wedekind (1927, pl. 7, fig. 3).

Discussion. — In examining descriptions and figures of specimens assigned to *S. patellatum,* it was tempting to consider the Brassfield specimen, with its elongate calicular pit and axial structure, as distinct from other specimens of the species in which these features were more circular in outline. It appears preferable, however, to mold the concept of the species described by Schlotheim to include this range of variation. The length:width ratio of the Brassfield specimen's calicular pit is 2. In comparison, a specimen from Gotland, consisting of four laterally-fused corallites of this species (UCGM 4814) has three corallites in which this ratio is 1.5, and a fourth with a partially broken pit region that appears to have been nearly circular.

Smith (1945, p. 19) designated as lectotype of this species the single specimen remaining of Schlotheim's six syntypes. Unfortunately, he did not figure the specimen, and gave a rather general description: "Although a somewhat worn specimen it admirably typifies the species, as interpreted by Hisinger, Edwards [*sic*] and Haime, and others. It is a flat, oval specimen of the usual mushroom shape 20 mm high and 70 mm wide through its longer diameter."

Hisinger (1837) mentioned the occurrence of *S. patellatum* on Gotland, and his figure (pl. 28, fig. 3), though showing nothing of the interior of the calicular pit, agrees with my concept of the species. He noted its lamellar structure, round outline, and flattened surface.

Milne-Edwards and Haime (1854) described and figured a specimen identified as *Ptychophyllum patellatum.* They thereby synonymized with the present species one described by Lonsdale (1839), *Strombodes plicatum* (see synonymy and discussion under *Schlotheimophyllum benedicti*). In fact, Lang and Smith (1927, p. 482) noted that Milne-Edwards and Haime's figures are of the same specimen (GS 6569) as illustrated by Londsdale (fig. 4b, c). This specimen is not here considered as conspecific with the species described by Schlotheim, but is assigned to *S. benedicti* (*q.v.*).

Lambe's (1901) illustration of the type of *Ptychophyllum canadense* Billings from Anticosti Island appears to be a different form from *S. patellatum*. The broad, reflexed calice is present, and the dilated septa are contiguous laterally along sutures. The central pit, however, is enclosed in a steep-sloped, relatively narrow mound, more narrowly conical than is usual in this species. More important, while Billings mentioned the septa entering the calicular pit, the major septa meeting in the axis with a whorl, the type figured by Lambe (pl. 17, figs. 1, 1a-c) appears to be without septa in the central pit. It appears more like *Craterophyllum* Foerste, with septa extending only to the margin of the pit, and only tabulae present within the axis. Further, Lambe's figure gives no evidence of a convex axial structure, here considered typical of *S. patellatum*. In fig. 3b, however, Lambe showed the calice of another specimen in which the septa do clearly reach the axis of the calicular pit (though no convex axial structure is indicated). It will be necessary to study the type material of *P. canadense* first-hand before its relationships can be understood.

In the same publication, Lambe redescribed *Chonophyllum belli* of Billings (1865b). He figured two specimens, one of which (pl. 16, fig. 6; *non* fig. 5, the lectotype) bears a strong resemblance to *S. patellatum* from the Brassfield. However, examination of this specimen (GSC 2624a) revealed what appear to be vesicular dissepiments in the reflexed calice wall, suggested that this does indeed belong to *Chonophyllum*, as does the lectotype. These specimens are from the Silurian of Grand Manitoulin Island, Lake Huron.

Lang (1926) figured sections of what appear to be *S. patellatum*. One specimen (pl. 30, fig. 4) is from "horizon *c* of Lindström, Häftings Klint (North Gotland)", probably of early Ludlow age. The second specimen (figs. 5, 6) is reportedly from beds of Wenlock age on Gotland. These figures show the trabecular structure of the corallite, and the nature of the axial structure (which is somewhat elongate in fig. 4).

· Wedekind (1927) described and figured a new species, *Chonophyllum planum*, from Skovengen, Tyrifjord, Norway. No age is given for the specimens on which the description was based. In the same paper, Wedekind described and figured *Chonophyllum patellatum* from the coast near Visby, Gotland, in the *Dinophyllum-Chonophyllum* horizon (Högklint beds, of Early Wenlock age). Both forms are figured as silhouettes, but it is clear, through comparison with his other figures, that Wedekind intended to show that these are not vesiculose. The difference between these two species, according to the author, is the presence of rejuvenescence, and apparently the sharper break at the calice rim between the near-horizontal upper surface and the descending wall. Examination of specimens and published figures indicates that these differences are all accommodated within *S. patellatum*.

From the Silurian Niur Formation of eastern Iran, Flügel and Saleh (1970) described *Schlotheimophyllum patelloides*, which appears in no significant way different from *S. patellatum*. The form is patellate, with reflexed calice walls, and a central circular calicular pit containing a convex axial structure of twisting, fused major septal ends and tabulae. The trabecular nature of the corallite deposits, and absence of vesicular dissepiments is clear from the figure. The septa dilate peripherally and are in lateral contact along sutures.

Finally, Lavrusevich (1971b) reported *Schlotheimophyllum patellatum* from the Late Llandovery of the Zeravshan-Gissar Mountain region of Uzbekstan. The specimen is patellate, reflexed, 16 mm high and 50 mm wide. The central pit has a width of 7 mm, and a depth of 5 mm. The apical angle, typically obtuse in this species, is about 100°. Within the calicular pit is a convex axial structure formed by septal ends and domed tabulae. The septa dilate peripherally, leaving no septocoels. The reflexed calice is thickest around the calicular pit (10 mm), gradually thinning to 2 mm near the periphery.

Distribution. — Gotland, Early Wenlock — Early Ludlow; Norway, Silurian(?); Iran, Silurian (probably post-Middle Llandovery); Uzbekstan, Late Llandovery; Ohio, mid-Llandovery (Brassfield Formation); Anticosti Island(?), Silurian.

Brassfield occurrence. — Locality 7a, in southwest Ohio (Fairborn).

Schlotheimophyllum benedicti (Greene)   Pl. 3, figs. 18-20;
                                            Pl. 4, figs. 1-4; Pl. 17, figs. 1-5

1839. *Strombodes plicatum* Lonsdale, pp. 691-692, pl. 16 *bis*, figs. 4b, c (*non* figs. 4, 4a; *non Cyathophyllum plicatum* Goldfuss, 1826, p. 59).
1854. *Ptychophyllum patellatum* (Schlotheim), Milne-Edwards & Haime, p. 291, pl. 67, figs. 4, 4a (*non Fungites patellatus* Schlotheim, 1820, p. 247).
1887. *Omphyma verrucosa* Davis, pl. 105, fig. 11 (*non* Rafinesque & Clifford, 1820, p. 235; *nec* Milne-Edwards & Haime, 1851, p. 403; *nec* Rominger, 1876, p. 118).
1900. *Ptychophyllum benedicti* Greene, p. 28, pl. 12, figs. 4-6.
1927. *Chonophyllum*(?) *patellatum* (Schlotheim), Lang & Smith, p. 482 (*partim*); (*non Fungites patellatus* Schlotheim, 1820, p. 247).
1964. *Schlotheimophyllum fulcratum* (Hall), Stumm (*partim*), p. 25, pl. 25, figs. 4, 6 (*non* figs. 3, 5, 7, 8; *non Ptychophyllum fulcratum* Hall, 1884).

Type material. — Lectotype (herein designated): AMNH 23505, original of Greene's fig. 5 (Pl. 4, figs. 1-4, Pl. 17, figs. 3-5); Paralectotypes: AMNH 23504, 23506.

Type locality and horizon. — Louisville Formation (Late Wenlock and Early Ludlow) near Louisville, Kentucky (Beargrass Creek quarries).

Number of specimens examined. — 1 specimen found in field, plus the type suite.

Diagnosis. — *Schlotheimophyllum* in which the septa are not fused along sutures in the more horizontal outer portion of calice, but retain a narrow valley between themselves; septa "feathery" in texture; axial boss absent; septa meeting in axis.

Description. — The specimen collected from the Brassfield is broadly turbinate, but the configuration of what remains of the base suggests that the corallite was trochoid in its earlier stages. The walls display several episodes of sharp change in their profile, shifting from near-horizontal to near-vertical (Pl. 3, fig. 19). This produces a "stepped" or "shelved" profile, apparently common in this species. The theca has broad ribs (Pl. 3, fig. 20), separated by somewhat narrower troughs. Near the base of the corallite, these ribs are about 0.3 mm across (measuring from trough axis to trough axis) and broaden upward to about 1.3 mm near the top.

The specimen's axial length is 2.2 cm to the level of the calice rim, and the calice is 0.9 cm deep. In addition, an estimated 0.4 cm has been lost from the base of the corallite. The calice reaches its maximum diameter of 4.7 cm at an estimated height of 2.3 cm above the original base (the minimum diameter at this level is 3.5 cm). At this point, rejuvenescence occurred, the calice contracting, and the growing area shifting to a segment of the calice constituting about 40% of the septocoels. From this region, a new calice developed, while the rest of the old calice filled in with sediment. The new calice grew with one side contingent with the nearest wall of the old calice, and at the time that growth ceased, had attained a lip-to-lip diameter of 3.5 × 2.5 cm.

The calice wall consists of two zones: the upper half lies at an angle of about 25° to the horizontal, while the lower half is steeper, about 35°. The result is a central steep-walled area with an outer more nearly horizontal wall. The boundary between these two regions is gentle, but distinct, and the hollow it encloses is about 2.0 × 1.3 cm in diameter at its upper limit. The calice floor itself is concave.

The upper, rejuvenated calice has approximately 30 major septa, with what appear to be minor septa between some, but not all of them. All the septa are broad and flattened in the outer, more horizontal portion of the calice, and separated by grooves less than half as wide as the septa themselves. The septa taper downward and, in the deeper portion of the calice, are as narrow as, or narrower than the grooves separating them. It seems that only the major septa cross the calice floor to meet in the axis, but as the major and minor septa are not readily distinguishable, it is safest to say that some of the septa stop short of the axis, going no farther than the lower portion of the steep central wall. The rest of the septa cross the calice floor and meet in the axis, though many join their ends with those of their neighbors just before the axis, closing the distal ends of their septocoels. The axis is affected by a weak, narrow, counter-clockwise whorl.

A transverse section of the specimen just below the calice floor (Pl. 17, fig. 2), where the diameter is 1.8 cm, reveals clearly the major and minor septa. The minor septa lie between each pair of major septa, and extend from the periphery about halfway to the axis.

Perhaps the most distinctive feature of this specimen is the "feathery" texture of the septa. This may be seen both in superficial examination of the specimen, and in thin-section, and is most obvious in the outer portion of the calice where the septa are broadest. It appears to be due to pitting of the septal surfaces and the presence of very short splinter-like branches, diverging from the axis of each septum towards the periphery of the calice.

In vertical section the calice walls appear to consist of distinct layers, each having trabeculae projecting about 0.5 mm normal to its surface (Pl. 17, fig. 1). The septa, as seen in this section, are undulatory. No tabulae or vesicular dissepiments were noted in vertical section, though what may be thin-walled dissepiments appear in the septocoels as seen in transverse section.

There is no evidence of basal attachment-structures on this specimen.

Discussion. — The first appearance of this species in the literature is in Lonsdale's section of Murchison's *Silurian System* (1839). In his brief description, the only feature Lonsdale firmly referred to was the axial spiral contortion of the septa. His plate 16 *bis,* figure 4b shows the whorling as clockwise, which, if accurate, would be the only coral with its axis whorled in this direction encountered in this study from the Lower Paleozoic. It seems more likely that the figure has been somehow reversed, a not uncommon occurrence (e.g., Rominger's 1876 coral plates, see Laub, 1978, p. 739). Lonsdale (pl. 16 *bis,* fig. 4b), seems to have tried to communicate to the reader the idea of a roughened surficial texture on the septa, interpreted as the "featheriness" noted in the Brassfield specimen. Also, in his figure 4b, the septa in the more horizontal outer portion of the calice, while flattened, remain separated by spaces too broad to be termed "sutures". Figure 4c shows the broad ribbing and stepped nature of the theca. Lonsdale's figures 4 and 4a may represent worn examples of this species, but they cannot confidently be identified as S. *benedicti* at this time. The specimens described by Lonsdale are from the Wenlock Limestone (Late Wenlock in age, according to Gignoux, 1955) on the western slopes of the Malvern Hills, England.

Lang and Smith (1927), in their critical revision of Lonsdale's rugose corals, gave the following information on the specimen (GS 6569) on which Lonsdale's figs. 4b and c are based: The diameter is 4.5 cm, and the height less than 2 cm (with part of the base missing). The calice has a wide, moderately deep central pit, and a reflexed margin, which succeeds an earlier, broader calice. The axial whorl of the septal ends rises as a boss from the calice floor. The lower surface of the corallite is slightly concave, with a stout central cone representing the early, pre-expansion portion of the specimen. Lonsdale's figures suggest that the axial whorl does not rise in a particularly pronounced boss, though there does appear to be a small convexity. This feature is insufficient to preclude the conspecificity of Lonsdale's specimen and mine.

Lonsdale identified his specimens as *Strombodes plicatum*, referring the species to Goldfuss (1826, p. 59). Goldfuss (pl. 18, fig. 5) however, showed the septa of the outer theca as broad and flat, bordering each other laterally along sharp sutures. This is distinct from my specimen and from Lonsdale's figures. Lonsdale's synonymy also includes *Strombodes plicatum* of Ehrenberg (1832, Abhandl. König. Akad. Berlin (for 1831?), p. 312), a reference to which I have not had access, so that it is not possible at this time to determine the relationship between Ehrenberg's coral and mine. Lonsdale's final entry in his synonymy is Milne-Edwards' description of *Cyathophyllum plicatum* in the second edition (1836) of Lamarck's *Histoire Naturelle des Animaux sans Vertèbres* (vol. II, p. 431). Milne-Edwards' reference is clearly to Goldfuss' description and figure of a specimen from the limestones of Sweden, and so is no more related to the present species than was Goldfuss' material.

Milne-Edwards and Haime (1854) described and figured a specimen under the name *Ptychophyllum patellatum*. According to Lang and Smith (1927, p. 482) this is the same specimen as was shown by Lonsdale (1839) in his figs. 4b, c, and is thus regarded as conspecific with the Brassfield form.

Davis (1887, pl. 105, fig. 11) figured a calice-view of a specimen from the "Upper Niagara white clay, near Louisville" in Kentucky, probably of Late Wenlock or Early Ludlow age, and either from, or associated with the Louisville Limestone. While the general form of this solitary coral cannot be discerned due to the limited view provided by the figure, the calice is clearly similar to that of my specimen. The septa are broad in the outer, more horizontal portion of the calice, and taper towards the axis, which lies in the central, steeper-walled region. Here the major septa (the only ones to cross the calice floor) are affected by a weak counter-clockwise axial whorl. The septa in the outer portion of the calice are not sutured laterally, but are separated by valleys, apparently broader than are found in the Brassfield specimen, relative to the septal width. In places the septa appear undulose, perhaps feathery.

Davis identified his specimen as *Omphyma verrucosa*. Stumm, however (1964, p. 47), noted that this name, originally used by Rafinesque and Clifford (1820, p. 235) apparently referred to specimens of Early Mississippian age from Kentucky, and considered the description too indefinite. Milne-Edwards and Haime (1851, p. 403) applied this name to a specimen from the Silurian of Drummond Island in Lake Huron, figured by Bigsby (1824, pl. 29, figs. 1, 2a, b) which is clearly not comparable to my specimen, nor apparently to Davis'. Rominger's *Omphyma verrucosa* has rootlet talons and a prominently-convex calice floor, separated from the wall by a distinct moat. Davis' specimen is thus distinct from *Omphyma verrucosa*, probably as this name had originally been used, and certainly as it has subsequently been used, so that this name cannot be applied to the Brassfield specimen.

Ptychophyllum benedicti Greene (1900) from the Beargrass Creek quarries near Louisville, Kentucky (Louisville Limestone, Late Wenlock and Early Ludlow in age) seems to agree with the Brassfield specimen in most features (Pl. 4, figs. 1-4; Pl. 17, figs. 3-5). The outer surface is "stepped" and broadly ribbed. The general dimensions of the lectotype (figured here, and under the name of *Ptychophyllum fulcratum* Hall by Stumm, 1964, pl. 25, figs. 4, 6) are close to those of my specimen: axial height is 2.3 cm (with one edge of the upper surface slightly higher); diameter about 3.9 cm; the calice depth is about 8 mm. There are 41 major septa crossing the calicular pit, reaching close to the axis (with a possible void between them), and the minor septa do not extend beyond the calice wall, at the base of the pit. The axial ends of the major septa are affected by a narrow counter-clockwise axial whorl, most clearly seen in transverse section, and appear to contain some patches of stereome. Though the specimen is silicified, the trabecular nature of the walls is preserved, and the septa in the outer walls, though dilated, yet have valleys, not sutures, separating them from one another. In these outer portions of the calice, the septa have a somewhat roughened texture. Whether this is the "featheriness" observed in the Brassfield form or an artifact of silicification, is uncertain.

The differences between these two corals lie mainly in the more narrow and regular calicular pit in Green's specimen, and its greater number of septa in a corallite of about equivalent size to mine. In light of the variation seen in similar forms, however, it seems reasonable to consider the Brassfield specimen conspecific with Greene's.

Incidentally, if Greene's figure 4 is conspecific with his other specimens, this would indicate that this species can occur in compound, as well as solitary form.

Distribution. — Ohio, mid-Llandovery (Brassfield Formation); England, Late Wenlock; Kentucky, Late Wenlock or Early Ludlow.

Brassfield occurrence. — Locality 15, in a clay pocket near the top of the unit, at West Milton, Ohio.

Schlotheimophyllum ipomaea (Davis)   Pl. 4, figs. 5-7

1887. *Ptychophyllum ipomaea* Davis, pl. 104, fig. 7; pl. 105, fig. 9 (*non* fig. 10).
1890. *Ptychophyllum ipomea* [*sic*.] Davis, Foerste, pp. 342-343.
1964. *Schlotheimophyllum ipomaea* (Davis), Stumm, pp. 25-26, pl. 25, figs. 1, 2.

Type material. — Lectotype (chosen by Stumm, 1964, p. 25): MCZ 8164, the specimen figured by Davis (1887) in his plate 104, figure 7 and plate 105, figure 9; shown in present work on Plate 4, figures 5-7.

Type locality and horizon. — Louisville Formation (Late Wenlock and Early Ludlow) at Louisville, Kentucky.

Number of specimens examined. — One (the lectotype).

Description. — The only indication of this species in the Brassfield Formation is Foerste's (1890) description of the best of several specimens he collected from the former quarry in the Brassfield Formation east of Centreville, Ohio. Unfortunately, no illustration accompanies this description. Foerste's account of the specimen, however, is in such good agreement with the lectotype of *S. ipomaea* (Davis) that this species is included in the list of Brassfield corals. Foerste's description is here given *in toto*:

> The specimens here described are from the upper layers of the Clinton Group at Allen's Quarry, east of Centreville, Ohio. The calyx of the best preserved specimen has a diameter of 56 mm. and is composed of ninety-seven septa, which have an average width of 2 mm. near the border of the calyx, where they are contiguous; they gradually narrow towards the centre of the calyx, becoming separated; at a distance of 6 mm. from the centre they are separated to the extent of about .5 to .8 mm. At the centre the lamellar septa of our specimens are not visible but are presumed to be more or less twisted

as is common in the genus. At the border of the calyx the tops of the septa are broad and flattish, very moderately convex, and maintain this character for a greater or less degree for about 10 or 12 mm. where they gradually assume the lamellar form. The calyx is broad and flat or but moderately concave for about three-fifths the distance from the border to the centre of the calyx, after which the septa rapidly descend forming a central cup about two-fifths of the width of the calyx in diameter, and *at least* 5 mm. deep, although on account of the partial filling up of this central depression exact measurements cannot be given. Towards the border the margin of the calyx is usually slightly elevated, thus adding to the suggestiveness of the specific name. The septa as a rule are almost straight in the flattened portion of the calyx, although sometimes slightly curved, yet never as much so as in the Niagara specimen figured from Louisville, Kentucky. The vertical height of the specimen is about 33 mm. The external surface of the specimen is vertically marked by the septal furrows, which are in many places rather indistinct, and by concentric quite strong wrinkles with very few concentric striae; the preservation of the exterior of the specimen is such as to suggest that fine striae, even if originally present, have probably not been preserved. Various pits and markings on one side of the specimen may indicate the bases of former rootlets, but of this no certainty is felt. A section at the base of the specimen shows transverse dissepiments when the septa are not contiguous.

Discussion. — Foerste referred to this species in the heading to the foregoing description as *"Ptychophyllum ipomea,* sp. nov.,*"* a puzzling fact, since he did mention the "Niagara specimen figured from Louisville, Kentucky". Clearly, this must be an error in the text, and I do not feel that Foerste regarded his specimens as representing a species for which he was providing the initial description.

Stumm (1964) gave a description of the specimen figured (but not described) by Davis. This specimen is shown in the present work (Pl. 4, figs. 6, 7) from side, top, and bottom. Although Foerste included no figure, his description is sufficiently detailed to allow meaningful comparison between his specimen from the Brassfield and the lectotype.

The lectotype is broader than Foerste's specimen (64 mm vs. 56), but has fewer septa (70 vs. 97). In both specimens, the septa reach a maximum breadth at the periphery of similar proportions (2 to 2.5 mm in the lectotype, 2 mm in Foerste's). The septa in Foerste's specimen behave the same as in the lectotype on approaching the calice pit, becoming separate and narrower. Unfortunately, the axial configuration of the septa is obscured by partial in-filling of the pit in the Brassfield specimen. The depth of this pit is similar in the two specimens. Foerste referred to a slight elevation of the marginal platform of the calice near the border (of the pit?) which is also found in the lectotype. The height of the lectotype (excluding a strange fold of one portion of the periphery) is about 28 mm, compared with 33 mm in Foerste's specimen. Some features possibly identical with the "pits and markings" that Foerste saw on the exterior of his specimen, and which he felt might mark the bases of attachment rootlets, appear on the lectotype. Foerste's final remark, concerning "transverse dissepiments" seen in a basal section where the septa are not contiguous, is difficult to interpret. "Transverse dissepiments" usually refer to tabulae, and he may have been referring to tabulae in the axial area. Unfortunately, no sections were made of the lectotype, so that this aspect of the two specimens cannot be compared.

A feature noted in the lectotype, and not mentioned by Foerste, is that in the peripheral portions of the calice platform, where the septa are most strongly-dilated, there is an alternate widening and narrowing of the septa. When the septa are traced axially, the minor septa are almost invariably broader, and alternate with the narrower but slightly-longer major septa.

S. *ipomaea* differs from S. *patellatum*, one of the two other species of this genus in the Brassfield, in lacking a convex axial structure, and in the differential breadth of the dilated major and minor septa.

From the other Brassfield species of this genus, S. *benedicti*, it differs in lacking the "feathery" texture of the septa, and in having sutures, rather than narrow valleys between the peripheral dilated septa.

Family **PALIPHYLLIDAE** Soshkina, 1955

Solitary rugosans with dissepimentarium, typically with well-developed talon for basal support; septa thin, with no dilation; tabulae generally incomplete, forming flattened domes; tabularium may contain septal lobes, lamellae, and pali, but these do not intertwine to form a dense axial structure.

Soshkina and Kabakovich (1962) included both *Paliphyllum* and *Cyathactis* in the family Cyathactidae. In her original description of these two genera, however, Soshkina (1955) placed *Paliphyllum* in her new family Paliphyllidae (p. 121), and *Cyathactis* in her new family Cyathactidae (p. 122). Thus, if these two genera are to be considered co-familial (and in this it seems that Soshkina and Kabakovich are justified), Paliphyllidae clearly is the name to be used, by virtue of its priority over Cyathactidae.

Genus **PALIPHYLLUM** Soshkina, 1955

1955. *Paliphyllum* Soshkina, pp. 121-122.
? 1956. *Sclerophyllum* Reiman, pp. 37-38.

Type species. — *Paliphyllum primarium* Soshkina, 1955, p. 122.

Type locality and horizon. — Siberian Platform, basin of the Podkamennaya Tunguska River, from the upper level of the Stolbov Suite (Upper Ordovician: Upper Caradocian)

Diagnosis. — Solitary rugose corals with dissepimentarium; basal talon commonly present; septocoels open in all stages observed; minor septa long (1/3 to 1/2 the length of major septa); septa in contact axially in early stages, drawing apart as growth progresses, but leaving septal lamellae and pali in axial space; median septum may be located in axial space during neanic and ephebic stages; tabulae generally convex, often incomplete; theca a septal stereozone.

Original description. —

Corals solitary, conical, weakly-curved. On outer surface are visible fine longitudinal striae and attachment scars. On the calice floor is a small cupola-like boss. Septa of two orders of length, weakly thickened peripherally, gradually thinning axially. The cardinal and counter septa are joined, and together with disjunct axial ends (pali) of the rest of the septa, form the axial elevation. Tabulae convex marginally, and a broad zone of dissepiments is developed peripherally.

Discussion. — Neuman (1968, p. 231) expressed the belief that the differences between *Paliphyllum* Soshkina and *Sclerophyllum* Reiman (Porkuny Stage, Upper Ordovician or Lower Llandovery of Estonia) are not significant at the generic level. In this he shared the opinion of Kal'o (1958b) and Ivanovskiy (1963) that intermediate species or varieties link these two genera forming a single genus. However, the differences in structure of the axial complexes are distinct between *Paliphyllum* and *Sclerophyllum*. (In the latter a transverse section reveals a reticulate pattern in the axial complex, apparently without a median lamella or pali.)

These authors did not present material that convincingly shows synonymy of these genera, so that it is best to reserve judgement on the matter.

Paliphyllum primarium Soshkina   Pl. 18, figs. 1-3

1955. *Paliphyllum primarium* Soshkina, p. 122, pl. 10, figs. 3a, b.
Cf. 1956b. *Sclerophyllum sokolovi* Reiman (*in* Kal'o), pp. 38-39, pl. 10, figs. 5-9; text-fig. 4.
Cf. 1958b. *Paliphyllum soshkinae soshkinae* Kal'o, pp. 109-110, pl. 2, figs. 10-14.
Cf. 1958b. *Paliphyllum soshkinae karinuense* Kal'o, pp. 110-111, pl. 3, figs. 1-9.
1961. *Paliphyllum primarium* Ivanovskiy, pp. 204-205, pl. 3, figs. 2a, b.
Cf. 1968. *Paliphyllum suecicum* Neuman, pp. 231-237, text-figs. 1-3.

Type material. — Holotype: PIN 587/3057 (Paleontological Institute of Novosibirsk), the specimen figured by Soshkina (1955).

Type locality and horizon. — Upper Caradocian (upper levels of the Stolbov Group), at the lower course of the Podkamennaya Tunguska River, above the Stolbov River, on the Siberian Platform.

Number of specimens examined. — One.

Diagnosis. — *Paliphyllum* with long, straight major septa, and relatively narrow axial space (about 5 mm across in a corallite diameter of 2.75 cm); septa relatively numerous proportional to diameter (e.g., 71 of each order in a diameter of 2.75 cm); septa not markedly thickened peripherally.

Description. — A single specimen was collected, encased entirely in rock, so that its external features and dimensions are unknown. Two transverse, and one longitudinal section were made. Apparently the coral was curved, and the absence of the theca, as well as irregularities of the margins, indicate that the specimen had been subject to considerable wear prior to final burial.

The lower transverse section (Pl. 18, fig. 3) was originally about 2.5 cm in diameter, though chunks had been broken from the periphery. The septa are quite abundant, considering the diameter: there are 64 major septa, and nearly as many minor septa (not all of the latter are present at this stage). The major septa are fairly straight, extending for about 11 mm of the 12.5 mm radius, so that there is a narrow space in the axis, free of septa. The major septa are broadest at their proximal ends, reaching a width of about 0.6 to 0.7 mm. About 4 mm toward the axis they are thinnest and beyond this they broaden to about 0.2 to 0.3 mm, with only the slightest tapering as the axis is approached. None of the septa appear to have a midline.

The minor septa at this level are perhaps 2/3 the length of the major septa. There is an imaginary line through the axis that presumably represents the cardinal-counter plane. On one side of this line the minor septa are contratingent on the side of the major septum to their left, while those on the other side of the line are contratingent to the right. This orientation carries through the entire periphery of the coral. Of the two septa within this imaginary line, the one from which the axial ends of the minor septa diverge is here considered the cardinal, by comparison with the pattern in other corals. This presumed cardinal septum is not clearly seen, but one which is in about that position is elongate, its end entering the axial space as a sinuous, thickened line. This is accompanied by other such septal elements in the axial space which, however, are disjoined from the septa from which they arise. These form the pali so characteristic of *Paliphyllum*. There appears to be a median lamella at this stage, but it is not in the presumed cardinal-counter plane, and it is unclear if it connects these two septa.

Rejuvenescence at this stage is suggested by discontinuity of the minor septa (but not the major septa) in one area, and an accompanying infilling of sediment.

The surfaces of the tabulae and the dissepiments intersect the plane of the section in the septocoels. Those nearest the axis are concave toward the periphery, while those closer to the periphery are concave toward the axis.

The higher transverse section (Pl. 18, fig. 2) has a diameter of 2.75 cm, with 71 major septa and an equal number of minor septa. The latter are only about one-half the length of the former and are contratingent in the same manner as in the earlier phase. The major septa extend fairly straight towards (but not quite to) the axis, leaving a central gap of about 5 mm. Several septa do continue for a greater distance into the space. Judging by the disposition of the minor septa, by their position relative to the septa of the earlier stage, and by the fact that they are at opposite sides of the coral, two are believed to be the cardinal and counter septa. These extensions are somewhat thicker than the more proximal parts of the septa, and usually are not quite colinear with them.

The major septa are thickest at the periphery, about 4 mm wide at most, and appear to be narrowest at a point just proximal to the distal end of the adjacent minor septa. Beyond this point they widen almost imperceptibly and continue with little change in width to the axis. In some specimens, the septal structures of the axis are apparently disjunct from the rest of the septa forming pali. Most, however, appear to be conjunct.

As in the lower section, the tabulae and dissepimental projections in the septocoels are concave outward in the axial area, and concave inward in the peripheral area. There is a zone between these two sets of plates in which few such plates appear, and this is believed to signify a moat-like depression between the tabularium and the dissepimentarium.

The longitudinal section (Pl. 18, fig. 1) was not well-oriented and missed the narrow axial area. It shows the straight distal edges of the major septa with plates (probably of the tabularium) lying generally at right angles to the septa in the septocoels. These plates and those of the dissepimentarium, are 1/3 to 1/2 the thickness of the septa. The dissepiments of the dissepimentarium are about 1.5˚ to 2.5 mm long (in the axial plane of the coral) and slope down axially, at about a 60° angle from the horizontal, to the corallite axis. The tabularium is about 1.5 cm across near the highest preserved portion of the specimen, and probably consists of incomplete, peripherally-inclined tabulae (as found in most species of this genus). Rejuvenescence is suggested by two sharp outward-directed lips of the dissepimentarium spaced 12 mm apart.

The close-spacing of the septa near the axis, especially in the upper section, results in some septa being crowded out from the center, so that they remain shorter than the others.

Discussion. — Soshkina described *Paliphyllum primarium* as solitary corals, adhering to one-another, broadly conical, weakly curved, bearing fine longitudinal ribbing on the outer surface, and well-developed attachment scars. The cup is shallow, with a rounded edge and a broadly raised axial structure. The major septa are thick at the periphery and not fused in this area, but rather, fully isolated (presumably meaning they form a stereozone), and thin gradually toward the axis. Their axial ends again thicken, and small oval pali form a rather dense axial structure. The cardinal and counter septa unite in the axis, crossing the axial structure. The cardinal septum lies in a long, narrow, open fossula, located on the convex side of the corallite. The minor septa are about 1/2 the radius in length, somewhat thickened, and gradually taper toward the axis. At the confluence of their axial ends and the thickened major septa a sort of inner wall is formed (i.e., the region of the axis within this confluence is visually separate). In a diameter of 24 mm are found 56 septa of each order. The tabulae are convex peripherally, while in the center they intersect the axial structure. It is unclear from Soshkina's figures if the tabulae are complete.

The Brassfield specimen agrees sufficiently with *P. primarium* to be regarded as conspecific. The general features of Soshkina's transverse section are seen in mine, namely the rather narrow axial space, with numerous septa (56 major septa in the Siberian form, as compared with 64 in the portion of the Brassfield form with similar diameter). The median septum (axial lamella) in my form is rather less regular (not so straight or symmetrical in the axial space) than that of Soshkina's specimen, but considering the variability of corals, this might be regarded more as a sub-specific, than a specific difference.

Reiman (*in* Kal'o, 1956b) described what he considered a new genus and species, *Sclerophyllum sokolovi*, from the Porkuny Stage (Borkholmian) of the Upper Ordovician or Lower Llandovery at Porkuny, Estonia. Neuman (1968) and Kal'o (1958b) regarded *Sclerophyllum* as a junior synonym of *Paliphyllum*, and so it is appropriate to examine this species in comparison with the Brassfield representatives of the genus.

According to Reiman, *S. sokolovi* (based upon one incomplete, but well-preserved specimen) is long (reaching 11.0 cm) with a maximum diameter (at a level of rejuvenescence) of 3.5 cm, and a minimum diameter of only 1.6 cm. The corallite is cylindrical throughout. The distance between rejuvenescent levels, as well as the sharpness of rejuvenescence, is dissimilar. The minimum distance between these levels is 20 mm. Very fine transverse "ornamentation" (rugae?) is present, as well as distinct longitudinal furrows, up to 14 occurring in a width of 1 cm. The relationship between furrows and septa was not determined.

The septa in transverse section are thin, and broadened peripherally, where their ends fuse to form a stereotheca. The major septa reach the axial complex, their ends uniting in groups of two and three. The minor septa are over 1/2 the length of the majors. Peripherally, the septa of both orders are identical. The minor septa bend axially toward an adjacent major septum, as a rule effecting contratingency. Judging from Reiman's transverse sections, the contratingency is of the same pattern as in the Brassfield specimen. Reiman described this bending of the minor septa as "changing", with "groups of united septa bending to one side, but septa of other groups, adjacent along the periphery, to the other side." He regarded this pattern as related to the arrangement of the quadrants with regard to protosepta. Reiman's plate 10, figure 7 shows a transverse section 3.2 cm in diameter, with 65 septa of each order.

The axial complex in transverse view has the appearance of a net with cells of different sizes. The thickness of the separate elements of the complex is not greater than that of the septa. In transverse view, the complex occupies 1/4 the coral's diameter.

The convex tabularium consists of incomplete tabulae, spaced with 12 to 13 disposed vertically in 1 cm, but of varying convexity. There are supplementary plates between the tabularium and dissepimentarium. The dissepimentarium, according to Reiman, is "highly primitive", consisting of large dissepiments of "unequal development" (convexity?). In areas immediately above levels of rejuvenescence, where the diameter of the corallite is minimal, the tabulae nearly reach the wall of the corallite, and dissepiments have very steep, curved plates. In areas of maximal diameter, dissepiments tend to be elongate. The dissepiments average 5 mm in length, and are inclined axially with an angle of less than 40° to the corallite wall. Some dissepiments apparently do not rest upon other dissepiments, but rather, lap onto the tabulae.

The similarities between Reiman's species and the Brassfield specimen are several. In both cases, there are abundant septa, with the minor septa of comparable length and contratingent in apparently the same manner to the major septa. The axial structure, in both cases, is quite narrow and lies in the midst of the fairly straight, radiating major septa. The Estonian specimen's tabularium appears to be typical of those found in *Paliphyllum,* and is therefore of the sort expected in my specimen. The two cannot be compared as to external form, as too little information is obtainable from the Brassfield specimen. Rejuvenescence is a prominent feature of *S. sokolovi,* and it is suggested in my specimen by distinct marginal projections seen in longitudinal section, and by apparent interruption of the minor septa in the lower transverse section. The two lips of rejuvenescence seen in my longitudinal section are 12 mm apart, while Reiman reported the minimum interval in his specimen as 20 mm.

The differences are as great as the similarities between these forms. The septa of the Estonian coral appear to be somewhat thin- ner than in mine. Further, the nature of the axial structure is dif- ferent. The Brassfield specimen has definite pali, and at least in the lower section, a feature that might pass for a median septum. Rei- man mentioned neither of these features in his coral, and his figures bear out his description of the axial structure as being reticulate. It seems unlikely, then, that these two forms are conspecific, though they may be fairly closely related.

Kal'o (1958b) described two subspecies of a new species, *Pali- phyllum soshkinae*, from the eastern Baltic. The first, *P. soshkinae soshkinae*, is from the Llandoverian Juurusian Stage, at the Vakh- trepa Canal in the island of Khiyumaa. The corallites are subcylin- drical, 50 to 60 mm high, with a maximum diameter of 20 to 25 mm. The floor of the shallow calice is convex, and the calice margin is rounded. The epitheca is thin, with its finely furrowed outer sur- face often destroyed.

The major septa are long, gradually-thinning, with thickened axial ends. The septa are weakly sinuous, partially unequal in thick- ness. The axial complex is generally fairly dense, with coarse ele- ments, granular in appearance (pali?), but in some specimens, later stages of growth have thinner elements. The number of septa in a given diameter is as follows: At 19 mm, 33 and 34 septa of each order (two specimens); at 18 mm, 36 septa; at 16 mm, 37 septa; and at 15 mm, 33 and 35 septa.

The minor septa of very young stages are short. The tabulae are convex, and especially in the peripheral parts, are "split" (in- complete?). The dissepimentarium consists of five to six rows, principally of small vesicles. Elongate dissepiments are found among short, highly convex ones.

This form differs from the Brassfield specimen in having ap- parently finer axial elements, in lacking a median septum, in the thickening of the axial ends of the major septa, and in having a nar- rower maximum diameter. In view of these dissimilarities, it seems clear that they are different species.

Kalo's second subspecies, *P. soshkinae karinuense*, is from the Raykkyulaskian Stage (Llandovery) of Karinu. The corallite is cylindrical, reaching a height of 60 mm, with a maximum diameter of 30 mm. It displays parricidal budding. The epitheca is finely-striated, and the calice floor is nearly flat.

The septa of both orders are long (the minor septa are over half the length of the major septa) and generally thin axially. In five specimens studied at different levels, the number of septa of each order ranged from 33 to 43, in diameters ranging from 13 to 29 mm.

In the proximal portion of the corallite, the axial ends of the septa of the first order are thickened, and the axial complex is best developed. In the middle stage, the minor septa are longer, reaching about 3/4 the length of the major septa, and the axial thickening of the major septa has decreased. In late stages, the axial thickening has nearly disappeared, and the axial complex contains separate, oval, coarse fragments (pali).

The tabulae are convex and incomplete. The dissepiments commonly are small and convex, with interspersed larger, elongate ones.

This subspecies, too, is not conspecific with my specimen, having far fewer septa in a given diameter and having thickened septal ends. Also, the Baltic form has proportionally longer minor septa than does the Brassfield form.

Ivanovskiy (1961) described specimens which he assigned to *P. primarium*, from the Upper Ordovician Dolborsk Stage in the basin of the Podkamennaya Tunguska River in the Siberian Platform. These are solitary corallites, broadly-conical, or shaped like curved horns, relatively small (the height of the best-preserved specimen is not over 30 mm, with a maximum diameter of 18 mm).

The major septa are equal, thickened with layers of stereome. They thin gradually as the reach the axis. Their axial ends are slightly thickened and isolated, forming a dense axial structure of pali. The minor septa are thin, and about 1/2 the length of the major septa. Their axial ends are often contratingent on the sides of the adjacent major septa, though whether or not the pattern of contratingency is as found in my specimen cannot be determined from the figures or description. In diameters varying between 16 and 24 mm, the number of septa of each order generally varies within the limits of 37 to 56. This higher figure is in good agreement with Soshkina's specimen, and slightly lower than that found in my lower transverse section (25 mm with 64 septa of each order).

The tabulae are incomplete in Ivanovskiy's figure (though in his description he says they are complete), and are slightly convex axially. The tabulae are close-spaced, with an average vertical spacing of 0.5 mm. The marginarium consists of 5 to 9 rows of rather large, strongly swollen dissepiments.

In the earliest-known stages, the septa are short, and only thickened to a modest degree, with slightly-widened peripheral ends forming a narrow stereozone. Ivanovskiy stated that there are no dissepiments, and offered as evidence his plate 3, fig. 2b, a transverse section about 6 mm in diameter. We cannot, however, be certain that the dissepiments have not been subject to wear. Ivanovskiy's longitudinal section goes down to the level from which his small, transverse section must have been cut. It shows at this stage a narrow, but definite dissepimentarium, which widens distally.

The interiors of different specimens show considerable variation. Variability is seen in the septal thickness, the length of the minor septa (1/3 to 1/2 the corallite radius), the density of the axial complex, and the number of rows of dissepiments (5 to 9), as well as the form of the dissepiments. Not all stages of a single specimen show the same degree of contratingency of minor on major septa.

The data show that Ivanovskiy's specimens are close to mine, if not actually conspecific. The straightness of the septa, the narrowness of the axial structure with its pali, the large number of septa relative to the diameter all suggest *P. primarium*. The absence of a distinct median septum cannot be weighed too heavily if the Brassfield form is considered conspecific with Soshkina's, as the median septum in my specimen is not so distinct as that in hers. It seems that, for the time being, Ivanovskiy's specimens may be regarded as conspecific with mine and Soshkina's.

Neuman (1968) described a new species, *Paliphyllum suecicum*, from the Upper Ordovician Boda Limestone at Osmundsberg, Sweden. This is a ceratoid or cylindrical coral, 5 to 10 cm high, with corresponding diameters of 3.0 and 3.5 cm. The theca is a septal stereozone. A talon is present at the base of the convex side. Neuman noted the presence of very distinct growth-lines, similar to those seen in rejuvenescence, where the direction of growth is altered, but made no definite statement about whether or not rejuvenescence actually occurred. Weak interseptal ridges occur on the theca. The calice is shallow, with a flat bottom. In the early neanic stage, the cardinal and counter septa are joined by a dilated median septum, with the axial ends of the other septa fused with it into an axial complex. With growth, the major septa lose contact with the median septum, and the minor septa grow to about 1/2 the length of the major septa. The septa here are thicker than at later stages. The cardinal septum is in what Neuman regarded as a fairly distinct pseudofossula.

In later neanic stages, the counter septum loses contact with the median septum. The axial complex now consists of the median septum with some septal lobes, some of which are in contact with it. This is at a level where the diameter is about 13 mm, and the talon is still present in the section. The septa of both orders are now broader peripherally and narrower axially. At a higher level, where the diameter is about 2.2 $\times$ 2.0 cm, septal lamellae (pali) develop in the axis. The cardinal septum at this level is detached from the median septum.

In the ephebic stage, the dissepimentarium is lonsdaleoid, with the largest dissepiments in the most peripheral area. In later stages, there are three to four series of dissepiments. The theca remains a stereozone. The major septa are now about 2/3 the radius in length, and the minor septa are about 3/4 the length of the major septa, and contratingent as in the Brassfield form. There may be as many as 50 septa of each order at this stage. The axial structure, which has expanded at the expense of the major septa, at this level consists of round and elongate septal lamellae (pali) without the median lamella.

The tabularium consists of incomplete tabulae forming a floor generally flattened or weakly convex, occasionally becoming weakly concave in the very late stages. Its borders drop steeply, resulting in a marginal moat, often invested with supplementary plates (generally concave, small plates which are added to the moat to decrease its depth).

This form is clearly not conspecific with mine. The degree of septal thickening in all stages, the relatively small number of septa

(46 in a diameter of about 3.0 cm), the waviness of the septa, and the relatively broad axial space are all features pointing to this conclusion.

Distribution. — Ohio, mid-Llandovery (Brassfield Formation); Siberian Platform, Late Ordovician.

Brassfield occurrence. — Locality 11, near West Union, Ohio, in the upper Brassfield.

Paliphyllum regulare, sp. nov.   Pl. 20, figs. 3-5

Cf. 1955. *Paliphyllum primarium* Soshkina, p. 122, pl. 10, figs. 3a, b.
Cf. 1956b. *Sclerophyllum sokolovi* Reiman (*in* Kal'o), pp. 38-39, Pl. 10, figs. 5-9; text-fig. 4.
Cf. 1958b. *Paliphyllum soshkinae soshkinae* Kal'o, pp. 109-110, pl. 2, figs. 10-14.
Cf. 1958b. *Paliphyllum soshkinae karinuense* Kal'o, pp. 110-111, pl. 3, figs. 1-9.
Cf. 1961. *Paliphyllum primarium* Ivanovskiy, pp. 204-205, pl. 3, figs. 1-9.
Cf. 1968. *Paliphyllum suecicum* Neuman, pp. 231-237, text-figs. 1-3.

Type material. — Holotype: UCGM 41969, shown on Plate 20, figures 3-5; paratypes: UCGM 41999 and UCGM 41772.

Type locality and horizon. — From locality 7a, near Fairborn, Ohio, from a limestone lens in the clay-coral layer in the middle Brassfield at this locality, immediately overlying the ferruginous bed.

Number of specimens studied. — Three.

Diagnosis. — *Paliphyllum* with marked radial symmetry in ephebic stage; septa dilated proximally, tapering smoothly distally; minor septa contratingent in earlier stages (diameter = 1.0 cm), but only infrequently in later stages (diameter = 1.9 cm); axial structure relatively broad (c. 6 mm across in total corallite diameter of 1.9 cm), containing long median septum and pali.

Description. — Of the three *Paliphyllum* species found in the Brassfield, this is the most regularly-radial in transverse section. The three specimens collected were too poorly preserved to provide much information on the external features, such as presence or absence of talons.

The best specimen (UCGM 41969; Pl. 20, figs. 3-5) has a diameter of 1.9 cm near its distal end, and a length of about 3 cm along its curved axis. About 1 cm more is missing from the base. Two transverse and one longitudinal section were made.

The lower transverse section (fig. 5) was taken just over 1 cm above the lowest preserved portion. The cut was made somewhat oblique to the axis of the corallite, and the original diameter was slightly over 1 cm. There are approximately 39 major septa at this level and probably an equal number of minor septa. The latter appear to be 1/2 the length of the former and are mostly contratingent upon them. On either side of an imaginary line through the center, the contratingency is to the left and to the right. The presumed cardinal septum (from which the minor septa diverge) is thicker than the others, and appears to lie on the convex side. The presence of a fossula or pseudofossula could not be determined.

The septa of both orders are thickened peripherally and taper smoothly toward the axis. They undergo only the slightest flexing on their way to the axis, and most of the bending occurs at the axial end.

The theca is not preserved at this level. The axial space, with a diameter of about 0.3 cm, contains a loose array of septal lobes and lamellae, all showing considerable curvature. A number of the lobes and lamellae align in the axis, forming a median lamella roughly in the cardinal-counter plane. Connections with the septa are obscure.

The tabular and dissepimental projections in the plane of the section are concave axially in peripheral portions of the septocoels and concave peripherally in the more axial areas.

The regularity of form is most obvious in the higher section (Pl. 20, fig. 4), which is only slightly elongate in the cardinal-counter plane. The diameter is 1.9 cm. Forty-eight major septa, with the same number of minor septa, are present at this level. At this point, however, the minor septa are not contratingent. The minor septa are half or less the width of the major septa (and about 1/3 to 1/2 their length) at a given distance from the periphery. All septa are distinctly thickest peripherally, tapering smoothly toward the axis. Again, the cardinal septum is a bit thicker than the other major septa. The septa of both orders immediately to either side of it flex to either side (toward their counter side), but only at their most axial ends. A septum opposite the presumed cardinal is slightly thicker than the adjacent major septa, and projects further axially. This may be the counter septum.

An axial space about 5.7 × 4.9 mm in diameter and elongate in the cardinal-counter plane is bounded by the axial ends of the major septa. A long median septum extends the length of the space, and while it is definitely colinear with the cardinal and counter septa, it is unclear if it contacts either. If not, the gap is quite narrow. To either side of this median septum and radiating out toward the periphery, are long pali ($\leq$ 1.4 mm long). Of all of the specimens of *Paliphyllum* studied, this has the longest, most symmetrically disposed median lamella.

The projections of the tabulae and dissepiments in the septocoels are as in the lower transverse section. As in the other specimens of this genus studied from the Brassfield, the axial limit of the axially-concave plates is about coincident with the axial ends of the minor septa.

The theca is too poorly preserved to allow analysis.

In longitudinal section (Pl. 20, fig. 3) the tabularium is convex, with a moat separating it from the dissepimentarium. At the distal end of the corallite, the tabularium is about 1.6 mm in diameter, and consists of broadly-convex, incomplete tabulae. In the moat area, the form of the tabulae is less regular. Nearer the proximal end of the corallite, there is a suggestion that some of the tabulae are convex peripherally and depressed axially, but this is unclear. The dissepimentarium is poorly preserved, with only a few vesicles shown in this section. Those most axial appear to lie nearly parallel to the corallite axis, a pattern seen in the other two species of this genus in the Brassfield. The dissepimentarium thins appreciably toward the base of the corallite, but this may simply reflect wear.

The axial edges of septa intersected in the longitudinal section show that they are straight in the vertical plane.

Two other specimens of this species were studied. UCGM 41999, from locality 55 (near Ludlow Falls, Ohio), has a maximum diameter of 2.5 cm, and contains 42 septa of each order at this level. The minor septa are 2/3, or rarely, 3/4 the length of the major septa. The axial area is recrystallized beyond analysis, but the straightness of the septa, their smooth tapering, the absence at this stage of contratingency, and the radial regularity all point to *Paliphyllum regulare*.

In longitudinal section, this specimen is identical to the holotype. The tabularium, which consists of incomplete tabulae forming a flattened or slightly-depressed floor, is separated from the dissepimentarium by a moat, in which short vesicles constitute supplementary plates added to decrease the depth of the moat. The dissepiments are of variable length, but reach a maximum of 3 mm. The theca and outer surface are not preserved.

Another specimen, UCGM 41772, from locality 15 (West Milton, Ohio), was seen only in transverse section. In a diameter of 2 cm are 38 septa of each order, with the minor septa about 1/2 the length of the major septa, and often contratingent upon them. The pattern of contratingency is unclear. The septa are somewhat more crooked than is usual for this species, but still are relatively straight in comparison with other species of the genus. They taper smoothly toward the axis. The axial structure is obscured by the irregular plane of the section.

Contratingency of the minor septa in this specimen, though accomplished by only a few of the minor septa, shows that lack of contratingency cannot be considered a hallmark of this species. Rather, it can be said that, compared with other species of *Paliphyllum*, such as *P. suecicum* Neuman, contratingency is relatively uncommon.

Discussion. — This species is unlike any other described of this genus, and is clearly an entity unto itself. As the other species to which it will be compared have already been described in the discussion of *P. primarium*, only features distinguishing the present species from these others will be given here.

The Brassfield form has less crowded septa than *P. primarium*, with greater dilation of the septa peripherally. A greater proportion of the cross-section is occupied by the axial structure than in the species described by Soshkina.

From *Sclerophyllum sokolovi*, my specimens differ in having greater dilation of the proximal ends of the septa, and in lacking the reticulate axial structure of the Estonian specimens. The septa seem to be less densely-packed in *P. regulare*, but this is uncertain, as levels in the corallite comparable to mine are not shown in Reiman's transverse sections.

From *Paliphyllum soshkinae soshkinae* and *P. soshkinae karinuense*, my specimens differ in having a long, symmetrically disposed median septum.

From Ivanovskiy's specimen of *P. primarium*, my form differs in the same features as from Soshkina's original material of that species.

From *Paliphyllum suecicum*, my form differs principally in having straighter, more radially symmetrical septa, and a well-developed median septum with long, radiating pali. At this diameter in Neuman's specimens (see his figure 3B), the median septum is irregular, and the pali chaotic.

Distribution. — Ohio: mid-Llandovery (Brassfield Formation).

Brassfield occurrence. — Locality 7a, the clay-coral bed near Fairborn, Ohio. (See discussion under "The Brassfield Lithosome"); locality 15, near West Milton, Ohio; locality 55, near Ludlow Falls, Ohio, in the clay bed at the top of the Brassfield.

Paliphyllum suecicum Neuman **brassfieldense,** subsp. nov.
PI. 4, figs. 8, 9; PI. 19, figs. 1-5; PI. 36, figs. 4, 5

Cf. 1955. *Paliphyllum primarium* Soshkina, p. 122, pl. 10, figs. 3a, b.
Cf. 1956b. *Sclerophyllum sokolovi* Reiman (*in* Kal'o), pp. 38-39, pl. 10, figs. 5-9; text-fig. 4.
? 1958b. *Paliphyllum soshkinae soshkinae* Kal'o, pp. 109-110, pl. 2, figs. 10-14.
? 1958b. *Paliphyllum soshkinae karinuense* Kal'o, pp. 110-111, pl. 3, figs. 1-9.
Cf. 1961. *Paliphyllum primarium* Ivanovskiy, pp. 204-205, pl. 3, figs. 2a, b.
1968. *Paliphyllum suecicum* Neuman, pp. 231-237, text-figs. 1-3.

Type material. — Holotype of *P. suecicum* Neuman: UM D 1305, in the collection of the Museum of the Institute of Palaeontology, University of Uppsala. Holotype of *P. suecicum brassfieldense* is UCGM 41970 (Pl. 4, figs. 8, 9; Pl. 19, figs. 1-3); paratype is UCGM 41974 (Pl. 36, fig. 4, 5).

Type locality and horizon. — *P. suecicum* Neuman: Boda Limestone (Upper Ordovician) at Osmundsberg, NE quarry, Siljan District, Sweden. *P. suecicum brassfieldense*: Brassfield Fm. from locality 7a, near Fairborn, Ohio, probably in the clay-coral bed described in section on "Brassfield Lithosome".

Number of specimens examined. — Three.

Diagnosis (of the species). — *Paliphyllum* with sinuous septa, surrounding a relatively broad axial structure (1/4 to 1/3 the total corallite diameter) whose boundaries are ill-defined due to the uneven length of the major septa; from intermediate growth-stages, short median septum (lamella) usually present, attached to cardinal septum; minor septa mostly contratingent on adjacent major septa.

Diagnosis (of the subspecies). — *P. suecicum* without lonsdaleoid outer dissepimentarium.

Description. — Three specimens were studied in thin-section, and confirmed as conspecific. Each will be described in turn:

The largest specimen (Pl. 4, figs. 8, 9; Pl. 19, figs. 1-3) is also the best preserved. It comes from locality 7a, near Fairborn, Ohio. The corallite is 5.3 cm long (from base to calice rim), and attained a diameter of 3 cm. At its base is a talon for attachment. The basal cross-section is elliptical, with greatest diameter in the presumed cardinal-counter plane, while the more distal cross-section is more circular. The diameter across the calice rim is 3.9 $\times$ 2.8 cm.

Two transverse sections were studied, spaced about 17 mm apart. The lower (Pl. 19, fig. 2) is from the most worn portion of the specimen, but appears to have originally measured about 2 $\times$ 1.2 cm. It contains 36 major septa, and an equal number of minor septa. The major septa are about the same thickness throughout, and are rather crooked (as compared with those of *P. primarium*). In many cases their distal ends unite. They are quite variable in length, so there is no sharp border (as there was in *P. primarium*) separating the septal zone from the enclosed axial space. Septal ends intrude into, and occasionally cross the axial space, generally undergoing sharp changes of direction from their more proximal ends. Many of these axial ends are disjunct, forming pali, which appear to radiate from the axis and are mostly 2/3 to 1 mm long.

The minor septa are about 1/2 the length of the major septa, and tend to be contratingent upon them. As in *P. primarium*, an imaginary line (running along the greatest diameter of the corallite) divides those minor septa contratingent to the left from those contratingent to the right. Presumably this is the cardinal-counter plane, along which the corallite is elongate. By comparison with other related species, it is assumed that the minor septa are contratingent in such a way as to diverge axially from the cardinal septum, and converge toward the counter septum. Nothing readily recognizable as a median septum is found in this section. The theca is preserved only in one area, where its thickness is about 0.4 mm. The nature of the theca is unclear. The small portion preserved suggests the slightly bulbous proximal end of a septum embedded in the wall, but possibly this is a stereozone. Thin projections of tabulae and dissepiments lie in the plane of section within the septocoels.

The higher transverse section (Pl. 19, fig. 1) is about 3 cm in diameter and fairly circular. There appear to be about 44 major septa and an equal number of minor septa, though at this level, the greater length of the minor septa (often 2/3 to 3/4 the length of the major septa) makes them hard to differentiate. The major septa are thickened proximally. No theca is preserved at this stage. The minor septa are contratingent in the manner described for the lower section.

As in the lower section, the axial ends of the septa extend for varying distances into the axial space. One in particular extends more directly toward the axis than do the others, forming the closest thing to a median septum (lamella) seen in this specimen. It is uncertain if this is the cardinal or counter septum, though the direction of distal flexing displayed by some of the minor septa suggest it is more likely the former. If so, it would indicate a rotation of this septum through about 120° from its position in the lower section.

In the septocoels are projections of the tabulae and dissepiments. These are concave toward the periphery when nearer the axis, and concave toward the axis when nearer the periphery. Those closest to the periphery are so concave towards the axis that their sides are nearly parallel to the septa, thus enhancing the appearance of thickening in this area.

In longitudinal section (Pl. 19, fig. 3) the tabularium is 2.1 cm broad near the top of the specimen, and 1.4 cm broad 1.2 cm lower. It consists of incomplete tabulae forming an axially convex structure of relatively thin vesicles, sloping peripherally to a moat filled with coarser, more irregularly shaped vesicles that lie against the dissepimentarium. The vesicles of the dissepimentarium slope axially at an angle of 45° to 60° to the horizontal, and are generally 1 to 3 mm long. The septal ends that extend into the plane of the section are straight in a vertical direction.

A second, smaller specimen (Pl. 36, figs. 4, 5) from the same locality and horizon is confidently assigned to this species. As preserved, its length is about 2.1 cm from the base to the calice rim, and 1.8 cm to the base of the calice. About 0.3 cm is missing from the base. Its maximum diameter is about 2 cm across the calice rim.

A single transverse section (Pl. 36, fig. 5) just below the calice floor measures 1.6 × 1.3 cm in diameter. The theca is as described in the previous specimen and reaches a maximum thickness of 0.6 mm. There are 32 major septa with an equal number of minor septa. The latter are 1/2 to 2/3 the length of the major septa, and are mostly contratingent on them. Several of the major septa are in axial contact. Numerous pali are present in the axial space, as is a median septum. This latter is about 2 mm long, an extension of the cardinal septum, and is aligned with (though not touching) the counter septum. Both these septa are well-defined in this section by virtue of the pattern of contratingency, and lie along the shorter diameter of the corallite. The direction of concavity of the tabular and dissepimental projections in the septocoels is similar to the previous specimen.

The longitudinal section (Pl. 36, fig. 4) shows a convex tabularium of low, incomplete tabulae, which is slightly less than 1 cm across the top. A deep moat similar to that in the previous specimen (UCGM 41970) separates this area from the dissepimentarium. The axial ends of the septa, in this section, are straight vertically.

A third, poorly preserved specimen (UCGM 41929), closely resembles this species. It is from locality 21, near Cross Plains, Indiana. As it was entirely encased in rock, little is known of its external features except that the corallite was probably curved.

Two transverse sections were taken, one 1.3 cm and the other 1.7 cm in diameter. The number of septa of each order in the two sections is 36 and 37 respectively. In places the minor septa are very short, barely protruding from the theca, while in other areas they are longer, and appear to be contratingent on the major septa in the same manner as in the other specimens. The axial ends of the major septa are often fused, and a somewhat sinuous median septum, as well as several pali, occur in both sections. The theca is clearly a septal stereozone. The plates of the tabulae and dissepiments are poorly preserved in these sections.

Little is discernible in the longitudinal section, but a tabularium and dissepimentarium of the sort described above seem to be present.

Discussion. — The following species have been described above. Consequently, I will deal here only with the specific similarities and differences between them and the present species.

From *P. primarium* Soshkina, this species differs in having fewer septa in a given diameter, a greater proportion of the transverse area occupied by the axial complex, and more sinuous septa.

From *Sclerophyllum sokolovi* Reiman, it differs in the structure of the axial complex (reticulate in *S. sokolovi*, in contrast to the paliferous Brassfield form), in having a proportionally broader axial structure with more sinuous septa, and in lacking the marked rejuvenescence of the Estonian form.

From *P. soshkinae soshkinae* Kal'o, my form differs in having a median septum, and a more sinuous septal apparatus. In all other important respects, the two forms are identical.

From *P. soshkinae karinuense* Kal'o, it differs in the absence of parricidal budding, and in having more sinuous septa. In other respects it is quite similar.

Ivanovskiy's (1961) *P. primarium* differs from my form in the same ways as do Soshkina's specimens.

The Brassfield form most closely resembles *P. suecicum* Neuman (1968) from the Late Ordovician of Sweden. Our specimens show the presumed cardinal septum still in contact with the median septum at about the same level as in Neuman's specimens, after contact with the counter septum has been lost. The number of septa at any level is similar in both forms, as is their sinuosity, and their length relative to the corallite diameter. In both, the axial area is relatively large compared with *P. primarium* and *S. sokolovi*. The length of the minor septa relative to the major septa is comparable in both. The flatness of the tabularium surface in the Swedish form is approximated in mine, though the axial concavity of some of the tabulae in the former specimens was not seen. Finally, the supplementary plates reported by Neuman in the moat of his specimen are present in mine.

There are, nonetheless, some differences. The proximal dilation of the septa seen in the earlier stages of the Swedish form is not matched in my specimen. Also missing is the lonsdaleoid aspect of the septa in the outer dissepimentarium reportedly present in Neuman's specimens. Neuman noted that in the ephebic stage, the median septum is gone, leaving only pali in the axis. The lamella forming the median septum is still present at the fairly advanced stage shown on Plate 19, figure 2. It is, however, still in contact with the cardinal septum, a stage which Neuman indicated exists prior to the total isolation and ultimate disappearance of the median septum. In a wider stage of the first specimen described (Pl. 19, fig. 1), the median septum has apparently become nonexistent. More sections would be necessary to determine the extent to which the Brassfield form agrees with the Swedish in this respect.

It seems likely that the Brassfield specimens are conspecific with Neuman's material, though differences are such that they are regarded as a distinct subspecies.

Distribution. — Ohio: mid-Llandovery (Brassfield Formation). There are similar forms in the Llandovery of Estonia (Juurusian and Raykkyulaskian Stages), and the Late Ordovician of Sweden.

Brassfield occurrence. — Locality 7a (near Fairborn, Ohio), in the coral-clay bed described under "Brassfield Lithosome"; locality 21 (near Cross Plains, Indiana).

<h3 style="text-align:center">Genus CYATHACTIS Soshkina, 1955</h3>

1955. *Cyathactis* Soshkina, pp. 122-123.
? 1961. *Protocyathactis* Ivanovskiy, p. 205.

Type species. — *Cyathactis typus* Soshkina (1955, p. 123).

Type locality and horizon. — Upper Kochumdekskaya Group (Llandovery) of the Siberian Platform.

Diagnosis. — Solitary rugose corals with dissepimentarium, and well developed basal-attachment structures; minor septa long, about 1/2 the length of major septa; septa thin, in axial contact in earliest stages, drawing apart from axis with upward growth of the corallite; in ephebic stage few septal lobes remaining in axial region; tabulae often axially-depressed, incomplete.

Original description. —

Solitary corals with strong attachment structures. In cup are thin, long septa of both orders. Fossula at cardinal septum located on convex side of coral, cup somewhat displaced laterally. Primary septa numerous, thin, nearly discontinuous, more-or-less reaching axis. Secondary septa long; their length is about half the radius of the transverse section, often their axial ends contact the primary septa. The tabulae are marginally convex, sometimes depressed medially and incomplete. Vesicular zone quite broad, consisting of small swollen dissepiments. In early ontogeny, septa thin.

Discussion. — The most important points in Soshkina's description are that this is a solitary, dissepimentariate coral, in which the major septa reach, or nearly reach, the axis, but do not become mutually involved to form an axial structure. It might be said that its nondissepimentariate counterpart would be *Streptelasma* (rather than *Grewingkia*). Another point of importance stressed by Soshkina is that the septa remain thin (unthickened by stereome) throughout the coral's ontogeny. This thinness of the septa, even in the early growth stages is stressed in contrast to other Silurian dissepimentariates, such as *Phaulactis, Mesactis, Lykophyllum,* and *Semaiophyllum.*

Ivanovskiy (1961, p. 205) created the genus *Protocyathactis* for an Upper Ordovician coral of the Siberian Platform which agrees with *Cyathactis* Soshkina in most respects; but *Protocyathactis* has thickened septa. The possibility that these two constitute one genus (with priority belonging to *Cyathactis*) is strong, but in the absence of sufficient information to allow a sound decision, they are treated in this work as separate genera.

Cyathactis strongly resembles *Paliphyllum* Soshkina except for the presence of a distinctive axial structure in the latter and they are here considered co-familial.

Cyathactis typus Soshkina   Pl. 5, figs. 6, 7; Pl. 36, figs. 1-3

Cf. 1906. *Cyathophyllum sedentarium* Foerste, pp. 315-316, pl. 6, figs. 3A-C.
    1955. *Cyathactis typus* Soshkina, p. 123, pl. 9, fig. 2; pl. 11, figs. 1a, b.
Cf. 1955. *Cyathactis tenuiseptatus* Soshkina, p. 124, pl. 11, figs. 2a, b.

Type material. — PIN 587/759, at Paleontological Institute of Novosibirsk.

Type locality and horizon. — Left bank of Podkamennaya Tunguska River, above Severnoy River in Siberian Platform; from level B-1-5 in the upper part of the Kochumdekskaya Group (G^{k3}), Middle and Upper Llandovery.

Number of specimens examined. — One.

Diagnosis. — *Cyathactis* with no supplementary plates; minor septa about 1/2 the radius in length, often contratingent on major septa; tabularium of incomplete tabulae, broadly concave axially, convex on periphery; dissepiments not small (as in *C. tenuiseptatus*); in early stages at least, cardinal septum short, in fossula that communicates directly with axial space, and counter septum long, extending to or beyond axis; cardinal fossula does not form a round pit at its axial end.

Description. — This species is represented in my collection by a single small corallite, presumably a young individual (Pl. 5, figs.

6, 7). It measures 2.1 cm in vertical height (not along growth axis), with about 0.3 cm more missing from the base. The corallite apparently had first grown to an axial height of about 12 mm with a diameter of about 13 mm, forming a broad cone. At this point, a talon of about 9 mm length extended about 45° down from the calice, in the plane of the corallite's greatest diameter. The corallite grew about 5 mm more when a second, thicker talon, about 10 mm long, grew from the calice on the same side as the first. At this stage, possibly through rejuvenescence, the axial growth changed direction through about 40°, so that the calice was inclined toward the talons, in the plane of greatest diameter. 8 mm more of axial growth is preserved, and at this stage, a talon of about 10 mm length, and thicker than the last, extended in the same direction, as an elongation of the calice, and bearing septa along at least the proximal half of its upper surface (Pl. 5, fig. 7).

The side of the corallite is a wrinkled, rugose surface, with relatively weak longitudinal striae. The calice shows thin septa approaching, but not quite reaching, the axis, so that a space about 3 mm in diameter occupies the axis of the corallite. The calice is fairly round, except for the elongation of the third talon, and is about 13 mm in diameter. It is too poorly preserved to allow distinction of minor septa without thin-sections.

In transverse section, the thinness of the septa is striking, their width being at most 0.1 mm. Two transverse sections were studied, one near the base of the specimen, with a diameter of 8 mm, and one just below the base of the calice, with a diameter of about 11 mm elongated an additional 10 mm by the uppermost talon. The higher section shows possible rejuvenescence.

In the lower section (Pl. 36, fig. 3), 38 septa are present. If these include minor septa, they are not recognizeable as such. No midline is seen in any of the septa, though this may be due to re-crystallization. One septum in the lower section is longer than the others, and is the only one to reach the axis, and slightly beyond. The others come within 4/5 radius of the axis. The distal ends of the septa bend slightly to form four triangles, one with its apex at the axial end of the long septum, and the others at 90°, 180°, and 270° from this point. Opposite the long septum is a very short one, extending through only half the radius to the axis, and lying in a rather broad gap between the adjacent septa. This short septum is thinner than the others, and is presumed to be the cardinal septum, while the long one is probably the counter septum. It could not be determined whether the cardinal gap represents a true cardinal fossula or a pseudofossula. If these septa are correctly identified, then the talons lie on the counter side of the corallite. Dissepiments occur in the peripheral portions of the septocoels, but neither their projections in the plane of section nor those of the tabulae, occur within half a radius of the axis. Those projections closest to the axis are concave peripherally, but those in the more peripheral portion of the septocoels are concave axially. At one part of the periphery is preserved what looks like a zone of broad dissepiments into which the proximal ends of the septa do not extend. This is similar to a lonsdaleoid pattern, but is not fully clear.

In the higher transverse section (Pl. 36, fig. 2) there are 48 major septa, but all the minor septa do not yet seem to be present. Those that are recognizable as minor septa are all on the counter side of the corallite, and are less than 1/2 a radius in length. Most are contratingent (or nearly so) on the side of an adjacent major septum. The radius is about 5.5 mm, and the axial gap has a diameter of nearly 2 mm. The major septa are gently or strongly curved in places, mostly closer to the axis, and are in some instances in mutual contact distally. However, the four triangles of the lower section are now gone and no cardinal septum, counter septum, or fossula are recognizable at this level.

In this section there is the appearance of rejuvenescence, the new skeletal material lying on the side away from the talon. The presumably earlier septa of the talon extension are continuous with the major septa of the rejuvenated corallite on one side, but there appears to be no continuity between the minor septa of the two areas (though they are aligned). Again, the plate projections in the septocoels cloest to the axis are generally concave toward the periphery, while those in the peripheral portion of the septocoels, though irregular, seem generally concave axially.

In longitudinal section (Pl. 36, fig. 1), the presence of the talon somewhat confuses the internal structure. The tabularium attains a maximum diameter of about 6 mm, and consists of mostly incomplete tabulae forming a broadly concave axial region, and a sharply convex periphery. The distance between the peripheral crests reaches 5 mm. The nature of the dissepimentarium is unclear, but it appears to consist of axially sloping dissepiments about 1 mm long in the corallite's axial plane. The theca is poorly preserved, but appears only about twice as thick as the major septa. The talon is difficult to analyze from a structural standpoint, but appears to consist of dissepimental and thecal material.

Discussion. — Soshkina's description of this species gives the following information (paraphrased):

This species consists of solitary corals, barrel-shaped, with sharp contractions of the diameter, and well-developed basal attachment scars. The theca bears fine longitudinal ribbing. The calice is broad, with a somewhat reflexed rim. On the broad, flat base of the calice is a distinct cardinal fossula on the convex side of the corallite, which is also the side of the attachment-scar's occurrence. The major septa are thin and weakly curved. From the periphery they extend not quite to the axis. In places, their axial ends are disjunct. The counter septum projects somewhat into the vacant central space. The cardinal septum is sometimes weakly shortened and lies in a narrow fossula, which widens axially and opens broadly on the vacant axial space. The minor septa are about half a radius in length, and are contratingent on the neighboring major septa. In a diameter of 25 to 26 mm, there are 55 to 56 septa of each order. The tabulae, mostly incomplete, are broad, convex on the periphery, and weakly concave in the center. The dissepimentarium consists of vesicles, strongly convex, and inclined to the axis. In transverse section, the vesicles and the tabulae are convex in opposite directions.

Soshkina considered the distinguishing characters of this species to be the thin septa, narrow opening of the axial end of the fossula, and the axial concavity of the tabulae.

If one makes allowances for the fact that the Brassfield specimen is ontogenetically younger than Soshkina's (based upon overall size and septal number), the similarity between these two specimens is indeed striking. In both specimens the septa are thin, with contratingent minor septa of about equal proportional length. In both the tabulae are incomplete with broadly concave axial portions and convex peripheries. In both the vesicular dissepiments are of about the same size (in contrast with the small dissepiments of *C. tenuiseptatus* Soshkina (1955)). Soshkina spoke of a distinct cardinal fossula in the calice, a feature not obvious in my somewhat recrystallized specimen. The shortened cardinal septum and the elongate counter septum reported by Soshkina are in the lower section of my specimen, but are not distinguishable in the upper section.

The probable cardinal fossula of the Brassfield specimen is of the sort reported by Soshkina, opening into the axial space (though not as broadly as she reported), rather than forming a pit where it enters the axial region, as is the case in *C. tenuiseptatus*. Soshkina noted that the "attachment scars" occur on the cardinal side of the corallite, whereas in the Brassfield specimen they occur on what is taken to be the counter side. This is not regarded as being significant enough to preclude a close relationship between these specimens, and it seems safe to regard them as conspecific.

Cyathactis typus is compared with *C. sedentarius* in the discussion section dealing with the latter.

From *C. tenuiseptatus* (described in the discussion section for *C. sedentarius*), *C. typus* differs primarily in lacking the round pit formed by the axial end of the fossula between the tabularium and dissepimentarium in *C. tenuiseptatus*, and in lacking the more abundant, and finer dissepiments of *C. tenuiseptatus*. Further, the tabularium of *C. tenuiseptatus* is convex, while that of *C. typus* is flat, or even concave, with convexity only at its margins. These are clearly two distinct species.

Distribution. — Kentucky, mid-Llandovery (Brassfield Formation); Siberian Platform, Middle or Upper Llandovery.

Brassfield occurrence. — Locality 1 (near Panola, central Kentucky).

Cyathactis sedentarius (Foerste)   PI. 5, figs. 8-11; PI. 21, figs. 1-3; PI. 38, figs. 1-3

1906. *Cyathophyllum sedentarium* Foerste, pp. 315-316, pl. 6, figs. 3A-C.
Cf. 1955. *Cyathactis typus* Soshkina, p. 123, pl. 9, fig. 2; pl. 11, figs. 1a, b.
Cf. 1955. *Cyathactis tenuiseptatus* Soshkina, p. 124, pl. 11, figs. 2a, b.

Type material. — Lectotype (designated as type by Foerste on museum label): USNM 87177, the original of Foerste's (1906) plate 6, figure 3B. This specimen is figured on Plate 5, figures 9, 11. Several other specimens are included with the lectotype under this number, but the two others figured in Foerste's plate, one of which (fig. 3A) Foerste (1906) had referred to as "the typical form," are not among them, and are apparently lost.

Type locality. — Waco bed, Noland Member of the Brassfield Formation, ". . . from along the road north of Estill Springs, north of Irvine, Kentucky". This is equivalent to my localities 53-53a.

Number of specimens examined. — None found in field. Only the lectotype and associated specimens under USNM 87177 were studied.

Diagnosis. — *Cyathactis* with supplementary plates between tabularium and dissepimentarium; dissepiments relatively large (compared with *C. tenuiseptatus*), being 1.0 to 2.5 mm long in longitudinal section; proximal ends of septa "feathered", perhaps diagenetically.

Description. — As no specimen of this species was found in the field during this study, description and analysis is here based upon a set of specimens (USNM 87177) described and collected by Foerste from a single locality.

This assemblage consists of the lectotype and twelve other specimens. Where the base is preserved, a well developed talon occurs. In two instances the talon attaches to a bryozoan frond, representing apparently the same species in both cases. Rejuvenescence is common.

Most specimens show some degree of wear. Their form is quite variable, ranging from long, slender forms (6.5 cm long and 2.5 cm wide) to low, broad forms (3.5 cm long and 4.5 cm wide). All specimens are elliptical in cross section, the talon lying generally along the greater diameter.

The calices, where well preserved, are flaring (Pl. 5, fig. 9). The peripheral portion, in one example, lies at about 30° to the plane of the mouth, while the inner zone is steeper. There appear to be two orders of septa, the major septa reaching the vicinity of the axis, and the minor septa apparently restricted to the wall of the calice.

Exterior of lectotype (Pl. 5, figs. 9, 11). — The axial height from the base to the calice floor is about 5 cm. The height to the highest calice lip is 7 cm. Maximum diameter across the rim of the calice is 4 cm, and the minimum diameter (reconstructed) is about 3 cm. The outer surface of the corallite is strongly rugose. At an early growth-stage, the corallite apparently rejuvenesced, at the same time changing the direction of growth by about 90°. A second rejuvenescence occurred at about 3 cm from the base. The calice floor has a diameter of 1.75 × 1.25 cm. Judging by relief and width, there are two orders of septa. The minor septa appear to end near the base of the calice wall, but this is uncertain. There are approximately 60 septa of each order present.

Interior of lectotype (Pl. 38, figs. 1-3). — The lower transverse section (fig. 2) is 24 × 17 mm in diameter, and was taken about

1.5 cm above the base of the specimen. The theca, as preserved, is about as thick as the proximal ends of the septa are wide. The major septa approach the axis of the corallite with some counterclockwise torsion, and their distal ends exhibit sharp changes of direction and some waviness, forming septal lobes. The axial region occupied by these septal lobes is elongate in the same direction as the cross-section of the corallite at this level, and measures approximately 5×3 mm. A precise count of the number of septa at this level is difficult because of diagenetic alteration, crushing and the generally confusing appearance of some sectors of the cross-section, but it appears there are about 100 septa. Judging from clearer portions of this section, probably half are major septa, and half are minor septa. The latter, not always clearly distinguishable, seem to reach a bit more than halfway to the border of the axial zone (which is defined by the presence of the septal lobes), and are contratingent against the major septa. The direction of contratingency is bilateral in distribution, the minor septa touching the major septa to their right on one side, and to their left on the other side of an imaginary line that seems to coincide with the shortest diameter in this cross-section. Some of the major septa show a thickening that begins about 2/5 the way from their proximal to their distal ends, and then tapers gradually distally. In no instance, however, does this thickening reach a width equal to the adjacent septocoels. The minor septa, where clearly recognizable, show little if any change in width throughout their length. In the septocoels are the projections of the tabulae and dissepiments. Near the periphery of the cross-section these tend to be concave toward the corallite axis, with the most strongly concave examples occurring near the juncture of the septa with the theca. Where these strongly concave projections are abundant, they give a "feathery" aspect to the proximal ends of the septa with which they are in contact, though this appearance is achieved only in a few areas. The axially concave projections dominate for about the first third of the distance to the axial area of septal lobes. Further toward the axis is a region where the projections are neither predominantly concave nor convex toward the axis, but on average tend to be straight. Projections within the septocoel that are concave toward the periphery occur along about the final quarter of the length, after the minor septa have terminated, and are mixed with straight projections.

The upper transverse section (Pl. 38, fig. 1) is 31 × 18 mm in diameter, and is from just below the calice floor, about 4.5 cm above the base of the corallite, as preserved. It is not clear if the peripheral border of this section is the theca, as in places a long smooth line of skeletal material smoothly truncating numerous septa proximally, as would the theca, and with dolomitic sediment adhering to its entire outer side, is found to have apparent corallite skeletal material projecting from its outer surface. The apparent thecal material may be the walls of broad vesicular dissepiments of the outer dissepimentarium. The outer surface of the specimen itself, however, seems to have most of the theca preserved at this level. The septa approach the axial space of the corallite with a weak counterclockwise whorl. In the axial space itself is a more pronounced whorl that is affected by relatively sharp deflections of septal ends into what might be considered septal lobes (though not so well-defined as in the lower transverse section) and vaguely defines an axial space similar to that of the lower section, measuring about 5 × 3 mm, and elongate in the same sense as the entire cross-section at this level. The longest axes of the upper and lower cross-sections lie at about 45° to one another in the corallite. As in the lower cross section, it is difficult to get an accurate count of the septa, but the total (including both major and minor septa) seems to be about 130. There is contratingency of septa, and while the pattern suggests bilaterality of the direction of contratingency, this is not certain, as the sinuosity of septa makes the differentiation of major and minor septa difficult. Where the minor septa are clearly recognizable (in few instances) they seem to extend a little over half the distance from the periphery of the corallite to the margin of the axial space. The septa appear to thicken over their axial halves, but this may be in part due to the adherence of contratingent minor septa on their neighboring major septa. In the septocoels may be seen the projections of tabulae and dissepiments. These projections, at the peripheral ends of the septocoels, are often concave toward the axis of the corallite, sometimes so deeply concave as to impart a "feathery" appearance to the proximal ends of some of the septa, as seen in the lower cross-section. Peripherally concave projections occur toward the axial ends of the septocoels, mixed with straight projections.

The longitudinal section (Pl. 38, fig. 3) extends between the two transverse sections. It appears broader at the base and narrower at the top because the direction of the greatest diameter of the corallite cross-section rotated sharply along the interval represented in this longitudinal section, so that the upper part of this section parallels the smallest diameter at that level, while the lower part of this section parallels a much greater diameter at a lower level of the corallite. The tabularium consists of tabulae, seemingly both complete and incomplete, spaced vertically from 0.5 to 1.0 mm apart in the axis. The central portions of the tabulae are lower than the peripheral portions, and are either flat or broadly concave upward. The peripheries of the tabulae are strongly convex, forming rims, though it is unclear if this is because of the presence of incomplete tabulae helping to raise the rims as in *Cyathactis typus* Soshkina (Pl. 36, fig. 1), or because of an increase in the convexity of the peripheral portions of complete tabulae. The dissepimentarium consists of dissepiments, some of which are gibbous, and some depressed, but all of which are longer than high. The dissepimental layers form angles from nearly 0° to about 30° with the corallite axis. Between the tabulae and dissepimentarium is a depressed, moat-like sector containing supplementary plates that are mostly flat or concave upward, and that have about the same spacing as the central portions of the tabulae. The tabularium (exclusive of supplementary plates) is 6 mm wide in the upper (distal) portion of the section and 8 mm wide in the lower (proximal) portion. The dissepimentarium in the upper portion of the section is 5 mm thick, and about 7 or 8 mm thick in the lower portion. The region of supplementary plates between tabularium and dissepimentarium is visible only in the upper portion of the section, where its width is 1.0 to 1.5 mm.

Interior of a paralectotype specimen (Pl. 21, figs. 1-3). — Two transverse and one longitudinal section were made of the paralectotype shown on Plate 5, figures 8, 10.

The lower transverse section (Pl. 21, fig. 2) is 17 × 11.5 mm in diameter. Both orders of septa are thickest at a point 1.5 mm from the periphery. The major septa extend to the axial region, where their twisting, anastomosing, sometimes fusing ends form a loose axial structure. The minor septa are about half the length of the major septa, and are not contratingent. The septocoels contain many plates of the tabulae and dissepiments. Most of those near the periphery are concave axially and those further toward the axis are concave peripherally. About 41 septa of each order are present at this level. The theca is not preserved.

The higher transverse section (Pl. 21, fig. 1) measures 29 × 23 mm in diameter and was taken from a level about 1.5 cm above the lower section. As in the lower section the theca was not preserved. The minor septa are nearly as long as the major septa, though the total number of septa appears to be the same. The peripheral (proximal) ends of the septa are "feathery", with what appear to be branches diverging toward the theca. This "featheriness" was not noted in the lower section, but this could be due to the greater degree of alteration in that level. As below, the septa are broadest somewhat inward from the periphery, and the orientation of the incomplete tabular, and dissepimental plates in the septocoels is as in the lower section. The nature of the axial structure remains about the same. There is no evidence of a cardinal fossula. The concavity of the plates in the more peripheral parts of the septocoels (probably representing the dissepimentarium) is much greater than that of the more axially located plates (probably representing the tabularium).

In the longitudinal section (Pl. 21, fig. 3) the dissepimentarium near the top of the section is about 5 mm thick, with rather markedly convex dissepiments. The tabularium at this level is about 1 cm wide, and consists of incomplete tabulae inclined downward toward the periphery, lying among the ends of the major septa. Between the tabularium and dissepimentarium is a zone, about 2 to 3 mm wide, of concave, often incomplete supplementary plates.

Discussion. — Foerste (1906) apparently based his description of this species on unsectioned specimens. In addition to the type locality, he reported specimens from the Waco bed of the Noland Member, Brassfield Formation east of Panola (locality 1 of this study) and from the same horizon "half a mile east of Waco, where the road to Cobb Ferry turns off from the pike to Irvine", in Kentucky.

Soshkina (1955) described *Cyathactis typus,* a new species from the Siberian Platform in the upper Kochumdekskaya Group (Middle and Upper Llandovery). Her description is paraphrased in the discussion of *C. typus* (*q.v.*).

C. sedentarius differs from *C. typus* in having an intermediate narrow zone of supplementary plates between the tabularium and dissepimentarium. The "featheriness" of the peripheral portions of the septa in *C. sedentarius* are not reported in *C. typus*. While these species are here considered distinct, further study may show them to be identical.

Soshkina, in the same paper, described a second species of *Cyathactis, C. tenuiseptatus,* from the right bank of the Podkamennaya Tunguska River, above the mouth of the Stolbov River on the Siberian Platform. Stratigraphically, this species occurs lower than *C. typus,* in layer 2v (Soshkina uses the Russian letter *B*) of the middle Kochumdekskaya Group (G^{k2}). These are solitary corallites, as large as my second sectioned specimen of *C. sedentarius* (Pl. 5, figs. 8, 10), conical or cylindro-conical in form. On the outer surface are fine longitudinal ribs, and on some specimens, rugae. Strong attachment scars occur on the convex side of the corallite, as in the present specimens. The calice has a broad, flat bottom, and gently sloping walls.

The major septa reach almost to the axis. They are thin, radial, and relatively straight, as in *C. typus*. The minor septa are over half a radius in length and are not contratingent. The somewhat shortened cardinal septum lies in a fossula on the convex side of the corallite. The axial end of the fossula is a round pit at the border between the tabularium and dissepimentarium. Two lateral fossulae are less distinct. In a diameter of 25 mm there are 50 septa of each order, roughly the same as found in the sectioned specimen of the present species.

The tabulae are incomplete, and broadly convex. The dissepimentarium consists of 30 to 40 rows of dissepiments, apparently zoned by size, and distinctly more numerous and smaller than those in my form and in *C. typus*. This is reflected in the abundant infilling of the septocoels by the dissepimental plates seen in transverse section. The dissepimental rows are all inclined axially. The dissepiments in the septocoels do not all extend completely across the septocoels, but rather lap on one another, so that in transverse section they produce a pattern similar to that seen in Smith's (1945) figure of *Ptychophyllum? whittakeri* (pl. 10, fig. 1).

In both *C. sedentarius* and *C. typus,* the dissepiments are concave to the axis, while the incomplete tabulae are concave toward

the periphery when seen in transverse section. Throughout ontogeny, the septa are equally thin.

C. tenuiseptatus differs from *C. sedentarius* in having more
abundant dissepiments and a round pit at the axial end of the
cardinal fossula. The similarities between these forms are several,
but for now, they are regarded as distinct species.

Distribution. — Kentucky, mid-Llandovery (Brassfield Formation).

Brassfield occurrence. — Vicinity of localities 53 and 53a (north
of Irvine, Kentucky); also reported from locality 1 (1/2 mile east
of Panola, Kentucky), and from 1/2 mile east of Waco, Kentucky;
all in the Waco bed, Noland Member, Brassfield Formation.

Genus **PROTOCYATHACTIS** Ivanovskiy, 1961

1961. *Protocyathactis* Ivanovskiy, p. 205.

Type species. — *Protocyathactis cybaeus* Ivanovskiy (1961, p.
206).

Type locality and horizon. — Upper Ordovician of the Siberian
Platform.

Diagnosis. — Identical to *Cyathactis* Soshkina (*q.v.*), except
that in *Protocyathactis*, the septa are thickened by stereome in all
stages of ontogeny.

Original diagnosis. —

Coral solitary, sub-cylindrical or ceratoid, with striate epitheca. Septa
thickened stereoplasmally through their entire extent; in central space axial
complex absent. Tabulae convex. In periphery are developed numerous small,
swollen dissepiments.

Discussion. — See discussion under *Cyathactis.*

Protocyathactis cf. P. cybaeus Ivanovskiy   Pl. 5, figs. 2-5; Pl. 18, figs. 4-8

Cf. 1928. *Cyathophyllum cormorantense* Twenhofel, pp. 118-119, pl. 3, figs. 2-4.
? 1961. *Protocyathactis cybaeus* Ivanovskiy, p. 206, pl. 3, figs. 3a, b.

Type material. — Ivanovskiy's holotype of *P. cybaeus* is
SNIIGGIMS 517/4.

Type locality. — (of *P. cybaeus*): Late Ordovician (Upper
Caradocian) Dolborskiy Stage, 4 km upstream from the mouth of
the Nizhnyaya Chunku River, in the basin of the Podkamennaya
Tunguska River, Siberian Platform.

Number of specimens examined. — Two.

Diagnosis. — *Protocyathactis* with talons restricted to the base, and exhibiting rejuvenescence.

Description. — The specimens studied are solitary, compressed laterally, and have little distinct curvature. Both specimens have small basal talons and display rejuvenescence. Because of the rejuvenescence, the maximum diameter is not at the top of the corallite, but at the point just below the rejuvenescent contraction.

Measurements (in cm):

	Maximum length	Elevation of maximum diameter	Maximum diameter
larger specimen (UCGM 41956)	2.5	1.8	1.45 × 1.25
smaller specimen (UCGM 41985)	1.5	1.1	1.2 × 1.0

The calice of the larger specimen is well preserved in a rejuvenated outgrowth of the corallite (Pl. 5, figs. 2-4). Here the diameter at the lip is 1.2 × 0.9 cm, and the calice has a depth of 5 mm. Two septal orders are present, the major septa having greater relief (about 0.25 mm) than the minor. Twenty-eight septa of each order were counted. No fossula is discernible in the calice. The walls of the calice slope downward from the rim at about 45°. At a point about two-thirds the way to the calice floor the slope increases to near vertical. The floor of the calice is concave and about 4.5 × 3 mm in diameter, at the level at which the slope of the wall changes.

The larger specimen (UCGM 41956) was studied in three transverse sections with diameters of about 6 to 10 mm (Pl. 18, figs. 4-6). No significant change in internal structure was noted through this interval. The major septa reach the axis, where some of their ends fuse with some contortion in the very center of the corallite. There is no noticeable whorl, nor is any axial structure produced. The minor septa are about one-fifth the length of the major septa in the earliest stage studied and reach about half the length of the major septa in the latest stage. They are fairly straight, and do not appear to contact the sides of the adjacent major septa. The septa of both cycles are slightly thickened where they meet the corallite wall. The major septa of the lowest section are peculiarly kinky in a certain region (fig. 6); the significance of this is not known. In this section, taken 8 mm above the base of the corallite, there are 23 major septa. In the middle section (fig. 5), taken 11 mm above the base, there are 26 major septa. In the highest transverse section (fig. 4), taken 15 mm above the base, there are 28 major septa.

The position of the cardinal septum is uncertain. In the highest section, however, at a point 270° from the angular location of the basal talon, where there should be a minor septum there is, instead, a long septum nearly reaching the axis. It is cut off from the axis by the distal juncture of the major septa to either side of it. Possibly this is the cardinal septum.

The nature of the theca is unclear, except that it does not seem to be a septal stereozone. It is thin, about the same thickness as the septa, and does not appear to be affected by stereomal thickening. On the outer surface of the corallite are distinct longitudinal ribs with narrow grooves separating them. The latter correspond to the septa on the inside, and the former to the septocoels.

The inner wall of the corallite is covered by a dissepimentarium consisting of rather small, abundant dissepiments, which appear as curved crossbars when the septocoels are viewed in transverse section (Pl. 18, fig. 8). These are apparently restricted, at least in later stages, to the portion of the interior from the distal ends of the minor septa out to the theca.

The tabularium has not been clearly seen in longitudinal section.

Discussion. — Ivanovskiy (1961) described *Protocyathactis cybaeus* as representing a new genus, as well as species. The following is his characterization of the species (paraphrased):

> The three specimens studied are ceratoid or subcylindrical in form, with weak ribbing (presumably longitudinal) on the outer surface. The length does not exceed 3.5 cm, with a maximum diameter of 19 to 20 mm. The septa are of two orders. The major septa go straight to the axis, some joining distally with other septa, but do not produce an axial structure. The septa of the second order are one-third to one-half the length of the major septa, and their distal ends often contact the side of a major septum adjacent. All the septa have stereomal deposits on their sides, especially at the proximal *end*, and taper axially. In a section diameter of 11 mm are found 28 septa of each order. The tabulae are thin, complete, and may be horizontal (?) or weakly convex. In the axis they are spaced no more than 1 mm from each other vertically. The dissepimentarium consists of abundant, small vesicular dissepiments.
>
> The early ontogeny of the species is unknown, due to incomplete preservation of the specimens. By midstage, all the mature characters are present.

The differences between the Brassfield form and Ivanovskiy's specimens are few, but they leave sufficient question as to their relationship to make it advisable to stress their strong similarities, without declaring them conspecific. The minor septa in the Siberian form are contratingent, a relationship rare in my specimen. In addition, Ivanovskiy made a point of the stereomal thickening of the septa in his specimens, whereas this does not clearly take place in mine. His transverse section (fig. 3a) does not, however, indicate septa appreciably thicker than mine. The imperfect preservation of the internal features in the Brassfield specimen may have obscured some similarities in these features.

Particularly significant is the fact that Ivanovskiy did not mention rejuvenescence, or the presence of a basal talon. He mentioned that the imperfect preservation of his material made it impossible to study the earliest ontogenetic stages, and this might include destruction of the talons. It is more difficult to rationalize his ignoring rejuvenescence, if it were present. Whatever the relationships between these two forms, it seems likely that they represent at least closely related taxa on either side of the Ordovician-Silurian boundary.

It is worthwhile comparing the Brassfield form with *Cyathophyllum cormorantense* Twenhofel (1928) from the Jupiter Formation (Late Llandovery) of Anticosti Island. These are large corallites (the longest collected, a broken specimen, measures 150 mm long and 52 mm in diameter, while a second specimen has a diameter of 60 mm). Twenhofel reported that the length and diameter do not appear to be strongly correlated. The calice is quite deep, 14 mm in a specimen 46 mm in diameter. This calice is 8 mm across the bottom and 25 mm across the top. The calice floor is flat or slightly convex.

The septa are of two orders, though the minor septa sometimes appear nearly as long as the major septa. They are straight, and do not produce an axial structure (Twenhofel's transverse section shows them reaching close to, but not quite attaining, the axis). It is not clear if the minor septa are contratingent. The tabularium is as wide as the calice floor, and contains tabulae that appear to be mostly complete and slightly concave to slightly convex. These are spaced from 1 to 2 mm apart vertically. The vesicles of the dissepimentarium are small and abundant.

In general form, this species is quite similar to mine, and is possibly congeneric with it. The major points of difference lie in the

great size of Twenhofel's corals (an unimportant difference taken
alone), the great length of the minor septa, and the lack of rejuvenescence. What may be a talon is present in the specimen
shown by Twenhofel in his figure 2, a feature this form might, then,
share with the Brassfield material.

It is unfortunate that the septa do not reach the axis in Twenhofel's transverse section. Perhaps if the section were taken a bit
lower, the nature of the axis would be more clear.

Distribution. — Kentucky, mid-Llandovery (Brassfield Formation).

An apparently closely related form occurs in the Late Ordovician (Upper Caradocian) of the Siberian Platform.

Brasssfield distribution. — Locality 46 (near College Hill, Kentucky) in the top of the Noland Member at this site.

Family **TRYPLASMATIDAE** Etheridge, 1907

(*Nom. correct.* Hill, 1956, p. 310, *pro* Tryplasmidae Etheridge)

Rugosa lacking dissepimentarium; corallites solitary or fasciculate; septa acanthine, of two orders, or sometimes absent; trabeculae
sometimes joined in longitudinal rows by stereomal deposits; theca
a septal stereozone; tabulae generally complete, predominantly horizontal.

Genus **TRYPLASMA** Lonsdale, 1845

1845. *Cyathophyllum (Tryplasma)* Lonsdale, p. 613.
1871. *Pholidophyllum* Lindström, p. 125.
1873. *Acanthodes* Dybowski, p. 334 (*non* Agassiz, 1833; *nec* de Haan, 1833;
      nec Baly, 1864).
? 1882a. *Polyorophe* Lindström, pp. 16, 20.
Cf. 1883. *Coelophyllum* Roemer, p. 409 (*non* Scudder, 1875).
Cf. 1889. *Cyathopaedium* Schlüter, p. 263 (*pro Coelophyllum* Roemer, *non*
      Scudder, 1875).
1894. *Spiniferina* Penecke, p. 592 (*pro Acanthodes* Dybowski, 1873; *non*
      Agassiz, 1833; *nec* de Haan, 1833; *nec* Baly, 1864).
1904. *Aphyllostylus* Whiteaves, p. 113.
1927. *Stortophyllum* Wedekind, pp. 30, 31.
? 1937. *Aphyllum* Soshkina, pp. 45, 94 (*non Aphylum* Bergroth, 1906, p. 604).
1940. *Pholadophyllum* Lang, Smith, & Thomas, p. 99 (*nom. van. pro Pholi-
      dophyllum* Lindström, 1871).

Type species. — (SD Etheridge, 1907, p. 42), *Tryplasma
aequabile* Lonsdale (1845, p. 613).

Type locality and horizon. — Silurian, from the basin of the
Kakva River, east side of the northern Ural Mountains, Russia.

Diagnosis. — Solitary and fasciculate corals lacking dissepimentarium; theca a septal stereozone; septa acanthine, not reaching axis; axial structure absent; tabulae both complete and incomplete, generally of low relief; talons usually small if present.

Description. — These are rugose corals, usually solitary and cylindrical, though some specimens are fasciculate. *Tryplasma* and its relatives are remarkable for their septal structure, which consists of longitudinal rows of trabeculae, generally projecting obliquely upward toward the axis, rather than solid sheets as are usual in the rugosans. These are referred to as "acanthine" septa.

The theca is a septal stereozone, and the septa are represented by grooves on the outer surface of the theca.

The interior of the corallite contains tabulae, nearly all complete, and generally horizontal. In some species, the upper surfaces of these tabulae bear trabeculae.

The septa are of two orders, differentiated by the length and stoutness of the trabeculae. There is no dissepimentarium.

Rejuvenescence is a common feature of this genus. Generally small basal talons in some specimens provide support in standing.

Discussion. — The name *Tryplasma* was first used (as a subgeneric connotation) by Lonsdale (1845, p. 613) in describing corals found in Russia by Murchison, de Verneuil, and von Keyserling. He regarded *Tryplasma* as a subgenus of *Cyathophyllum* Goldfuss, and applied it to two species (one new, the other previously described by Hisinger). Lonsdale's figures and descriptions coincide with current concepts of this genus.

In 1907, Etheridge (p. 42) chose the new species introduced by Lonsdale, *T. aequabile*, as the type of the genus. This is a solitary form, with a broad tabularium, and short acanthine septa of two orders. In their critical revision of Lonsdale's corals, Lang and Smith (1927, p. 461), apparently unaware of Etheridge's selection, duplicated his choice of *T. aequabile* Lonsdale as type of the genus. Unfortunately, they only listed the two species of *Tryplasma* found by Murchison, *et al.* in Russia, and gave a brief diagnosis of the genus. They added no descriptive details on the type species.

Lindström (1871) erected the genus *Pholidophyllum* for *Cyathophyllum loveni* Milne-Edwards and Haime (1851, p. 364) from the Silurian of Gotland. This species reportedly occurs in other Northern

Hemisphere localities as well. Apparently, Lindström was unaware of Lonsdale's Russian specimens, as *C. loveni* has all the earmarks of *Tryplasma*. Lang, Smith and Thomas (1940, p. 99) invalidly emended the generic name to *Pholadophyllum*.

Roemer (1883) introduced the generic name *Coelophyllum* into the literature. This German Devonian rugosan is a colonial form characterized by "wholly rudimentary" septa, complete, widely spaced tabulae, and the absence of a dissepimentarium. Roemer remarked that these characters were first considered distinctive by Schlüter (1880, *fide* Roemer, 1883) who described a species based upon them, attributing that species to *Calophyllum* Dana (1846). Roemer considered it doubtful that Schlüter's species (*Calophyllum paucitabulatum*) truly belonged to Dana's genus, and also pointed out that *Calophyllum* was a source of some confusion, and had been differently interpreted by various authors. Together with a recommendation that *Calophyllum* be dropped from use, due to the insufficient data afforded by Dana (no illustrations and no species assigned to the genus), Roemer erected a new generic name, *Coelophyllum*, for Schlüter's species.

Schlüter (1889) then noted that the term *Coelophyllum* was preoccupied, and proposed the name *Cyathopaedium* in its stead. Roemer's description (there are no figures) gives the strong impression that this genus is identical to *Tryplasma* Lonsdale, an opinion shared by Sokolov (1962, p. 308). On the other hand, Hill (1956, p. F298) listed *Cyathopaedium* Schlüter as a genus separate from *Tryplasma*. Hill's figures of *Cyathopaedium* (which she attributed to Schlüter) appear quite different from *Tryplasma*, especially with regard to septal structure, and the coloniality of the former, so that it seems likely they are not congeneric.

From the Silurian of Gotland, Dybowski (1873) described as new the genus *Acanthodes*, a group of solitary and phaceloid species. His figures and descriptions, however, suggest probable synonymy with *Tryplasma*. Penecke (1894) noted that *Acanthodes* was already in use as the generic name of a fish (by Agassiz), and so, changed the name to *Spiniferina*. Penecke's figures are clearer than Dybowski's, and plainly show the acanthine septa (with trabeculae inclined upward axially), the largely complete tabulae in a broad tabularium, and the absence of a dissepimentarium, all typical of

Tryplasma. Though Penecke and Dybowski did not figure the same species, Dybowski's rather stylized drawing agrees sufficiently well with Penecke's illustrations to make it clear that this genus is synonymous with *Tryplasma.*

Aphyllostylus Whiteaves (1904) is based upon specimens from the Silurian of Manitoba, Canada, and while he included no figures in his paper, the description of this fasciculate coral agrees well with *Tryplasma.*

From the Silurian of Gotland, Wedekind (1927) introduced the genus *Stortophyllum,* based upon several solitary species. These corals seem to fit readily into *Tryplasma,* having acanthine septa, no dissepimentarium, and tabulae which, though often incomplete, still retain the general aspect of a broad tabularium lacking the axial structure typical of the genus described by Lonsdale. *Stortophyllum* can also bear a small supportive basal talon, another feature found in *Tryplasma.*

Soshkina (1937) described the new genus *Aphyllum* from Wenlock beds on the eastern slope of the Ural Mountains. As the generic name implies, she regarded the weak septal development as an important feature of this genus. Septa are usually absent from her specimens, and when they do occur they are acanthine septa, with the trabeculae inclined upward axially. The tabulae are distinct, more or less horizontal, and (as suggested by Soshkina's figures) usually complete. There is no dissepimentarium. It seems quite possible that this form is a *Tryplasma.* The common absence of septa is unusual, and this form might fit in with *Zelophyllum* Wedekind (1927, pp. 34, 35) from the Silurian of Gotland, in which the trabeculae are also restricted to the wall and do not protrude into the tabularium. On the other hand, future studies may show that both *Aphyllum* and *Zelophyllum* belong to *Tryplasma,* as junior synonyms.

Polyorophe Lindström (1882a) from the Silurian of Gotland, is similar to *Tryplasma,* but is generally considered as a separate genus, by virtue of the highly developed talons it characteristically bears, often reaching far up along its side. As noted above, *Tryplasma* can also bear talons (at least, corals with all the earmarks of *Tryplasma* are found both without talons, and with small ones restricted to the very base (e.g., *T. radicula,* Pl. 6, fig. 4)). The degree to which talon development reflects a generic difference can only be judged by further study.

Tryplasma radicula (Rominger)   Pl. 6, figs. 1-4; Pl. 21, figs. 6, 7

  1876. *Cyathophyllum radicula* Rominger, pp. 109-110, pl. 39, fig. 3 (*partim*, see discussion).
? 1927. *Pholidophyllum hedströmi* Wedekind, pp. 27-28, pl. 3, figs. 1-4; pl. 29, fig. 1.
? 1927. *Pholidophyllum hedströmi* var. *attenuata* Wedekind, p. 28, pl. 3, figs. 6, 7; pl. 29, fig. 2.
? 1927. *Pholidophyllum intermedium* Wedekind, p. 28, pl. 3, figs. 8, 9.
? 1927. *Pholidophyllum intermedium* var. *articulata* Wedekind, p. 28, pl. 3, figs. 10, 11.
? 1927. *Pholidophyllum tenue* Wedekind, p. 28, pl. 3, fig. 12; pl. 29, fig. 3.
? 1927. *Pholidophyllum coniforme* Wedekind, p. 28, pl. 3, fig. 14; pl. 29, fig. 5.
? 1949. *Amplexus brownsportensis* Amsden, pp. 106-107, pl. 27, figs. 7-13 (*partim*).
  1952. *Tryplasma radicula* (Rominger), Stumm, p. 842 (*partim*), pl. 125, figs. 1, 2, 7, 8, 9 (*non* figs. 3-6).
  1965. *Tryplasma* cf. *T. radiculum* (Rominger), Sutherland, pp. 31-32 (*partim*), pl. 23, figs. 1, 3 (*non* fig. 2); pl. 24, figs. 1-6; pl. 25, figs. 1-6.

Type material. — Lectotype (Stumm, 1952, p. 842, pl. 125, figs. 1, 2): UMMP 8582a; Paralectotypes: UMMP 8582b-8582e, 8582g(?).

Type locality and horizon. — Point Detour, Michigan; from the Manistique Group (Late Llandovery to Middle Wenlock).

Number of specimens examined. — Four from the Brassfield Formation.

Diagnosis. — *Tryplasma* species of solitary, trochoid corallites exhibiting frequent weak rejuvenescence at fairly regular intervals; septa in two orders, their trabeculae not diverging noticeably from the plane of the septum (as occurs in *T. cylindrica*, q.v.); tabulae without spines on upper surfaces; talon present in some specimens, restricted to base.

Description. — The best preserved specimen (Pl. 6, figs. 2-4) attained a length of 20 mm and a final maximum diameter of 9 mm. The corallite increases in width through the first 1 cm of its length, beyond which it grows no wider than 9 mm. In this upper region, there is little if any net increase in width, although there are frequent variations. Weak rejuvenescence occurs at the following elevations (in mm) above the base: 6, 8, 9.5, 11, 13, 15, 16.5. The maximum diametric contraction involved is 1.5 mm.

At the base of this specimen is a skeletal deposit with a flattened bottom, 8 mm long, which is apparently a talon. It extends only a few millimeters up the convex side of the corallite.

The calice wall has apparently been broken away at its lip but what remains is vertical. The calice is 3.5 mm deep, with an essentially flat floor, as wide as the opening of the calice (about 5.5 mm across). The theca is just over 1 mm thick at the top, slightly thinner near the base. The calice is circular in cross-section (Pl. 6, fig. 2; Pl. 21, fig. 6) and contains acanthine septa of two orders, 28 septa of each order. The major septa consist of spines (trabeculae) up to 0.4 mm in length, spaced about 4 per mm. The spines of the minor septa are about half as long, with about the same spacing. These spines do not diverge noticeably from the plane of the septum distally, as they do in the Brassfield form of *T. cylindrica* (Wedekind).

Both complete and incomplete tabulae are present, with no appreciable variation in their thickness (Pl. 21, fig. 7). The complete tabulae are spaced about 1.5 mm apart and are horizontal, or slightly convex or concave. Their surfaces are free of spines, in contrast with the radiating rows of spines on the tabulae of *T. cylindrica* (Pl. 5, fig. 15).

Only UCGM 41951 has a basal talon, though at least one other specimen is sufficiently complete at the base to show a talon, if one had existed. With the exception of the talon, all specimens studied agree well with the one just described.

Discussion. — Rominger (1876) described the species *Cyathophyllum radicula* from the Manistique Dolomite (Late Llandovery to Middle Wenlock) at Point Detour, Drummond's Island in Lake Huron. He noted the frequency of rejuvenescence, and the presence of a talon. He also mentioned the two septal orders, and that in some specimens the flat calice floor is spinose. The specimens with spinose tabulae may be *T. cylindrica*. Other specimens have a more rounded, concave calice floor.

The description was accompanied by figures of ten specimens, concerning which Rominger said (pp. 109-110):

The left-hand, outer vertical row of specimens is from Point Detour; the central row represents specimens from Masonville, Iowa; the right-hand row may be a different species. It occurs in the Niagara group of Louisville, and of Charleston, Indiana; the stems are of longer cylindrical growth, often curved and geniculated; their calyces rarely exhibit a naked diaphragm in the bottom, and the crenulated lamellae generally reach to the center. The surface of the stems is in both kinds longitudinally ribbed by septal striae.

The specimen selected as lectotype by Stumm (1952) is UMMP 8582a, probably that one figured in the upper left-hand portion of Rominger's plate 39, figure 3. This specimen is from Point Detour. Two more specimens from this locality are UMMP 8582e and 8582g. These and three other specimens (UMMP 8582b, c, and d) from the Hopkinton Dolomite (Late Llandovery to Late Wenlock) at Masonville, Iowa are unfigured specimens of Rominger.

The lectotype of *Cyathophyllum radicula* Rominger (which is figured by Budge, 1972) is 1.6 cm long (as preserved), a trochoid corallite displaying three episodes of rejuvenescence, about equal in interval. Its diameter at the top is 1.25 cm (with a circular cross section), but its maximum diameter is attained in an earlier rejuvenescent stage, and measures 1.45 cm. Each rejuvenescent segment is conical. The earliest stage is broken, but an apparent attachment scar remains at the base.

The calice is 8 mm deep, with a flat floor 6.5 mm across. Its walls are vertical near their base, and slightly less vertical higher up. The calice floor is free of spines.

The septa are of two orders, thirty to each order. The spines of the major septa are 0.5 mm long, spaced 2.5 per mm. The spines of the minor septa are 0.25 mm long, and spaced 2.5 to 3.0 per mm. The spines of the major septa are somewhat thicker in the base than are those of the minor septa. None of the spines diverge noticeably at their distal ends from the plane of the septum.

No information is available at this time about the interior of the lectotype.

The main differences between the Brassfield form and the lectotype of *T. radicula* lie in the greater width-to-length ratio of the latter, and the somewhat greater thickness, and resultant wider spacing of its septal spines. While these differences clearly do carry some weight, there are many more important similarities between Rominger's specimen and mine: frequent, rhythmic rejuvenescence, basal talon, biordinal septa, nondivergent septal spines, flat, spine-free calice floor, and steep calice sides. The two unfigured specimens from Rominger's assemblage, collected from the same locality as the lectotype, are basically similar to that specimen: The calices have the same configuration as in the lectotype. The septa (visible only in 8582g) are of two orders, with major spines up to 0.7 mm long, and minor spines 0.25 to 0.3 mm long, both nondivergent distally. The calice floor and tabulae, where visible, are free of spines. They differ from the lectotype, however, in having rootlets, rather than a solid talon for attachment. Also, 8582g is not clearly rejuvenescent. Some tabulae are visible on the inside of this specimen. They do not lie parallel, and the edge of one nearly touches that of the other. In the center of the corallite, they are separated by about 2 mm, a distance similar to that found in my specimens. The width-to-length ratio for these two specimens of Rominger's is about 0.7 and 0.9, as compared with 0.9 for the lectotype, and 0.45 for the best-preserved of my specimens.

The differences between Rominger's specimens and mine seem subspecific rather than specific in degree and the two lots are here considered conspecific.

In his work on the rugose corals of Gotland, Wedekind (1927) described and figured many specimens showing similarities to mine. Unfortunately, his descriptions often provide little information, and the figures do not always make up for this deficiency. Nonetheless, in several cases, enough information may be obtained to make some interesting comparisons.

Pholidophyllum hedstroemi (*P. hedströmi* Wedekind) resembles my form in possessing two orders of septal spines, and lacking spines on the tabulae. The corallite is relatively long and thin, as are mine, and displays frequent rejuvenescence. There is no indication of a talon, but as noted above, a talon is not always present in my specimens, and its state of development probably reflects the attachment needs of the individual. The distinct upward-pointing spines of the septa (as compared with mine, which are normal to the wall) raise some question as to the identity of Wedekind's specimens with the Brassfield form. His specimens are from the *Dinophyllum-Chonophyllum* horizon (Högklint beds, Lower Wenlock, according to Manten, 1971, p. 43) at Kneippbyn and Stenkyrke, Gotland.

A variety of this form, *P. hedstroemi attenuata* Wedekind, is from the higher levels of the upper *Dinophyllum-Chonophyllum* horizon at Kneippbyn. It is supposed to be distinctive by its more attenuated form, but the figures show no impressive difference between this and the species *P. hedstroemi* proper. It is not possible to determine from the information given if there are two orders of septa.

Pholidophyllum intermedium Wedekind (1927) from Snäckgardet and Kneippbyn, Gotland (Lower Wenlock Högklint beds, according to Manten's geological map) is similar to the above species, except that Wedekind was here impressed by occasional strong concavity of the tabulae, sometimes sufficient to bring a tabula into contact with its lower neighbor. The spines are directed upward, but the lack of a transverse section makes it impossible to detect two septal orders. Here, as above, the tabulae are free of spines.

Pholidophyllum tenue Wedekind (1927) comes from the highest level of the *Dinophyllum-Chonophyllum* horizon (Högklint beds, Lower Wenlock) near Kneippbyn, Gotland. It is relatively long and thin (averaging 11 mm in length and 6 mm in width). It appears to exhibit rejuvenescence (Wedekind, pl. 29, fig. 3), though not as regularly as my form, and may possibly have a small talon on its convex side. Unfortunately, there is insufficient information from which to determine the presence of two septal orders. This fact, and the clear upward-pointing of the septal spines leaves the relationship between this form and *T. radicula* uncertain.

Pholidophyllum coniforme Wedekind (1927) is from the same horizon and locality as the previous species. Outwardly it resembles *Tryplasma radicula* from the Brassfield. It is relatively long and thin, reaching a length of 30 mm, and a width of 10.5 mm, displaying frequent and evenly spaced rejuvenescence (at least seven times in 3 cm). Further, there is a talon at its base, similar to that in my specimen. Also present are rootlets, which were not noted in my form (though they are present in the paralectotypes UMMP 8582e and g, from the Hopkinton Dolomite). The corallite is trochoid and convex on one side. In Wedekind's plate 29, figure 5, the talon is on the convex side, but the rootlets seem to originate on the concave side. The tabulae are free of spines. Here again, though the resemblances are strong, Wedekind's specimens cannot be confidently identified with my form, due to the presence of upward-pointing septal spines, and the absence of any indication of the number of septal orders.

It seems quite possible that all of these Gotland forms belong to the same species, and may be a subspecies of *Tryplasma radicula* (Rominger). Where Wedekind showed transverse sections, they always indicate two septal orders, with apparently nondiverging spines. Final judgment will have to await detailed examination of the types.

Amplexus brownsportensis was described by Amsden (1949) from the Brownsport Formation (Early Ludlow) of western Tennessee. The holotype (YPM 17666) is scolecoid and cylindrical, apparently with frequent weak rejuvenescence. Originally it measured 3.6 cm long and about 1 cm in width, but most of the specimen was used to make longitudinal and transverse thin sections. The specimen has been silicified, resulting in difficulty in discerning the finer structure of the calice and interior. It could not be confidently determined if secondary septa lie between those visible in the calice. If these are present (two possible minor septa are visible in an unfigured transverse section of the holotype), they are quite small, and have been all but obliterated. Adding to the difficulty of understanding the specimen is the fact that, at cessation of growth, the corallite was apparently rotating its axis through a 90° turn, so that the topography of the calice floor is distorted. All that can be determined is that the tabulae apparently are without spines, and are incomplete more frequently than in my specimens. The calice walls appear less steep than in the Brassfield form, and the septal spines show no sign of axial divergence.

Two figured paratypes (YPM 17668 and 17670) were also studied. The former is scolecoid, quite cylindrical, but with no clear rejuvenescence. Through a preserved length of 2.5 cm, it maintains a diameter of approximately 1 cm. Its calice floor is narrow, with steeply sloping, but nonvertical walls. The septa are of two orders, and there is no indication of distal spine divergence. The calice is too poorly preserved to be certain, but it appears that some of the septal spines may continue for a short distance toward the axis of the calice floor. This specimen is certainly not conspecific with the Brassfield form and may not be conspecific with the holotype of *Amplexus brownsportensis* Amsden.

The second paratype (YPM 17670) is too poorly preserved to give much information on the calice and septa, but in general form, it more closely resembles the holotype than does the other paratype. Its base is tapered, and there is a flat area at its lower end, suggesting an attachment scar.

While Amsden's holotype specimen has several points in common with my material (spine-free tabulae, possibly two orders of septa without distal divergence of the spines), its state of preservation leaves too many questions unanswered to allow confident identity with the Brassfield specimens.

Sutherland's (1965) *Tryplasma* cf. *T. radiculum* [*sic*] from the Henryhouse Formation (Ludlow in age) of Oklahoma, bears strong resemblance to the Brassfield specimens, to the extent that they are probably conspecific. This is most clearly seen in his plate 23, figure 1. This specimen shows frequent, fairly regular rejuvenescence (fig. 1a), two orders of septa (fig. 1e), and spine-free tabulae (fig. 1f). The septal spines apparently do not diverge distally from the plane of the septum (figs. 1b, d). The specimen apparently is without a basal talon (fig. 1a), but this is not critical.

Distribution. — Northern Michigan, Late Llandovery or Early to Middle Wenlock; Kentucky, mid-Llandovery (Brassfield Formation); Oklahoma, Ludlow.

Budge (unpublished Ph.D. dissertation, 1972) described and figured specimens of *T. radicula* from the Portage Canyon Member of the Laketown Dolostone in Nevada and Utah. He estimated the age of this member as Early and Middle Silurian.

Possible occurrence. — Forms that may be conspecific with mine are found in the *Dinophyllum-Chonophyllum* horizon of the Högklint beds (early Wenlock) of Gotland.

Brassfield occurrence. — Localities 1 (near Panola) and 53 (near Irvine) in central Kentucky.

Tryplasma cylindrica (Wedekind)   Pl. 5, figs. 12-15; Pl. 21, figs. 4, 5

? 1876. *Cyathophyllum radicula* Rominger, pp. 109-110, pl. 39, fig. 3 (*partim*).
Cf. 1882a. *Polyorophe glabra* Lindström, pp. 16, 20.
Cf. 1896. *Polyorophe glabra* Lindström, Lindström, pp. 43-47.
    1906. *Polyorophe radicula* Foerste, p. 313, pl. 5, figs. 3A-E (*non C. radicula* Rominger, 1876).
    1927. *Pholidophyllum cylindricum* Wedekind, p. 28, pl. 3, figs. 5, 13.
? 1965. *Tryplasma* cf. *T. radiculum* (Rominger), Sutherland (*partim*), pp. 31-32, pl. 23, figs. 2a-f (only).
Cf. 1971b. *Polyorophe glabra* Lindström, Lavrusevich, pp. 80-81.

Type material. — The location of the specimen or specimens figured by Wedekind is uncertain. Dr. Stefan Bengtson of the Department of Palaeobiology, Royal University of Uppsala, Sweden informed me that Wedekind's types are in two places: the Senckenberg Museum in Frankfurt-am-Main in Germany, and the Swedish Museum of Natural History at Stockholm.

Type locality and horizon. — Specimen(s) figured by Wedekind are from the *Dinophyllum-Chonophyllum* horizon of the Högklint beds (early Wenlock) at Snäckgärdet, Gotland.

Number of specimens examined. — Five from the Brassfield Formation.

Diagnosis. — Solitary *Tryplasma,* scolecoid in form, with occasional weak rejuvenescence (less frequent and even-spaced than in *T. radicula*); acanthine septa of two orders, in both of which the spines diverge distally from the septal planes; tabulae both complete and incomplete, of varying thickness, with spines on upper surfaces.

Description. — The corallites of this species are generally scolecoid, though there is gradual upward expansion through the first centimeter of growth. Corallites up to 2 cm in length show little net expansion beyond this point, though locally there is frequent increase and decrease in width. In one specimen, within a space from 5 to 10 mm from its base, the diameter goes from 6 to 5.5 to 8 mm in an upward direction. The corallites are curved, but it was not possible to relate concavity or convexity to any internal features, such as a cardinal septum.

The theca is thick, tapering upward in one individual from 1.0 to 0.6 mm through a length of 1.5 cm. The structure of the theca is unclear, but transverse sections suggest a septal stereozone. The outer surface of the corallite has longitudinal grooves and rounded ridges, the latter corresponding in position to the internal septa.

The calice has been observed only in a partially preserved condition. Its walls appear to have been steep (about 60° to 90° to the horizontal), and its floor was probably slightly convex, judging from the topography of the tabulae. The septa are acanthine, and two distinct orders can be differentiated by the length and thickness of their spines (Pl. 5, figs. 13, 15). Those of the major septa are about 0.3 to 0.4 mm long, and spaced about four per millimeter. Those of the minor septa are about 1/3 as long, and somewhat more slender, but have about the same spacing as the major spines. The spines of both septal orders have a tendency to diverge distally from the plane of the septum in which they originate.

The tabulae are for the most part gently convex, with some horizontal. In a longitudinal section (Pl. 21, fig. 4), about half are incomplete. The complete tabulae are about 1.5 to 3.0 mm apart vertically, with the incomplete tabulae between them. Characteristically, these structures are quite variable in thickness, ranging from about 0.4 mm to less than 0.1 mm thick. Their upper surfaces are studded with spines in a radial pattern, continuous with the acanthine septa of the walls (Pl. 5, fig. 15).

If talons for attachment were originally present, they must have been restricted to the basal area of the corallite, as they are not observed in the preserved portions of the corallites studied.

Discussion. — The lectotype of Rominger's *Cyathophyllum radicula* differs from *T. cylindrica* in several ways: *C. radicula* is more strongly and rhythmically rejuvenescent. *C. radicula* has talons, restricted to the basal portion, which seem to be lacking in *T. cylindrica*. Its septal spines do not diverge distally from the plane of the septum, as they do in *T. cylindrica*. The upper surfaces of the tabulae are free of spines in *C. radicula*, but not in *T. cylindrica*. But Rominger, describing the calice floor, said "bottom of cells flat, formed of a smooth or faintly carinated or granulose transverse diaphragm". This suggests that his assemblage may include both *Tryplasma cylindrica* and *T. radicula* though the former appears not to be from the type locality of *T. radicula* (*q.v.*). *T. cylindrica* is not included among the type suite of *Cyathophyllum radicula* Rominger at the University of Michigan Museum of Paleontology.

Polyorophe radicula Foerste, from the Waco bed, Noland Member of the Brassfield Formation (mid-Llandovery) in central Kentucky, appears to be identical to *T. cylindrica*. This is not unexpected, as Foerste's specimens come from the same horizon and general area as mine. They have the same scolecoid form and occasional weak rejuvenescence. The upper surfaces of the tabulae are invested with spines. Foerste made no mention of two orders of septa (his count of 60 to 70 septa agrees with mine if both orders are counted), nor did he describe distal divergence of the septal spines from the plane of the septum. It is more likely that Foerste did not notice these features if they were present, than that there was any major discrepancy between his specimens and mine. Unfortunately, the location of his types is unknown. Foerste's (1906) is the first description of *T. cylindrica* as it is currently understood. He considered his specimens as distinct from those of Rominger, and apparently chose to reflect this by using the same trivial name as Rominger had, but assigning his species to a different genus. Foerste's specimens are here considered referable to the genus *Tryplasma*, which makes it a secondary junior homonym of Rominger's species. Hence, the name next assigned to it (*Tryplasma cylindrica* [originally called *Pholidophyllum cylindricum* Wedekind, 1927]) must replace Foerste's designation. This is in keeping with the International Code of Zoological Nomenclature Article 57 (Stoll *et al.*, 1964).

Wedekind's specimens are from the upper *Dinophyllum-Chonophyllum* horizon (Högklint beds, Lower Wenlock) at Snäckgärdet, Gotland. They are cylindrical in form, and lack (well-developed?) talons. The corallite dimensions are in agreement with mine. Further, the acanthine septa occur in two distinct orders, and the upper surfaces of the tabulae are spinose. There is a suggestion in Wedekind's figure 13 that the tabulae (both complete and incomplete) are of different thicknesses, though not to the extent seen in my material. The thickness of the theca compares well with mine. The only major difference between Wedekind's form and mine is that the septal spines in his material are longer than in mine. On balance, it seems that this is more of subspecific than specific importance. It is not possible to determine from the information given if in Wedekind's specimens the septal spines diverge axially from the septal plane.

Polyorophe glabra, a species described first by Lindström, from the Silurian of Gotland (1882a, 1896), and later by Lavrusevich (1971b) from the Late Llandovery of Uzbekstan, is similar to the present species in its generally cylindrical form, and has spinose tabulae. On the other hand, it has a strongly developed talon, characteristically extending far up on one side of the corallite, and is not particularly scolecoid. *P. glabra's* acanthine septa do not appear to be distinctly differentiated into two orders, as is the case with the

Brassfield form. Finally, the spines on the tabulae in Lindström's material align in a labyrinthine pattern, while those in the Brassfield form radiate distinctly outward from the center of the tabula. Lavrusevich chose as lectotype of *P. glabra* the specimen in Lindström's (1896) pl. 8, fig. 106, and gave its age as Early Wenlock.

In his description and figures of *Tryplasma* cf. *T. radicula* from the Ludlovian Henryhouse Formation of Oklahoma, Sutherland (1965) may have included a specimen of *T. cylindrica* (Sutherland's pl. 23, figs. 2a-f). Figure 2e clearly shows the spines on the tabulae, and figure 2c suggests two orders of septa. The specimen (figure 2a) appears to have undergone repeated, rhythmical rejuvenescence, however, and it cannot be determined if there is distal spinal divergence. Thus, the identity of this specimen is open to question.

Distribution. — Gotland, Early Wenlock; Kentucky, mid-Llandovery (Brassfield Formation); Possibly also in Oklahoma (Ludlow) and northern Michigan (Late Llandovery or Early or Middle Wenlock).

Brassfield occurrence. — During this study, specimens were collected from locality 1 (Noland Member of the Brassfield, near Panola, Kentucky). Foerste's specimens are from this general area. Those specimens which he figured are "from along the road north of Estill Springs, north of Irvine, Kentucky" (my localities 53-53a), and "from half a mile east of Waco, where the road to Cobb Ferry turns off from the pike to Irvine" (my locality 45). He also reported specimens occurring "a mile southwest of Indian Fields, where the road from Kiddville joins the road from Indian Fields to Clay City" in central Kentucky.

Tryplasma sp.   Pl. 6, fig. 5

Description. — A specimen (UCGM 41987) of *Tryplasma* is here recorded, which cannot be neatly fit into either of the two species described from the Brassfield Formation. Insufficient infomation was obtained to assign it to another species.

The specimen consists of two corallites, one budded from the other (the only instance of such budding found in Brassfield *Tryplasma*). The parent has a diameter of 6 mm and is curved in a plane roughly perpendicular to the axis of the bud. Its length, along the curved axis, is estimated to be at least 1.5 cm, with more hidden at either end in the matrix. The tabulae are both complete and incomplete in the parent (but are not seen in the bud), and appear to be free of spines. There seem to be two orders of acanthine septa, but it is not clear if the trabeculae diverge axially from the plane of the septum (as they do in *T. cylindrica*).

The spacing of the tabulae cannot be seen in enough instances to be meaningful. The surfaces of the parent and bud are rugose, but there is no evidence of rejuvenescence.

Brassfield occurrence. — Locality 46, at the top of the Noland Member, near College Hill, Kentucky.

Family **CALOSTYLIDAE** C. F. Roemer, 1883

Rugosa lacking dissepimentarium; including solitary and colonial (at least one) forms; septa of both orders perforate, often mutually connected across the septocoels in marginarium by synapticulae; major septa commonly joining axially to form spongy axial structure; tabulae thin, domed; theca a thin coating, often worn away from all but lowest portions of corallite.

Genus **CALOSTYLIS** Lindström, 1868

1868. *Calostylis* Lindström, p. 421.
1887. *Hemiphyllum* Tomes, p. 98.
1930a. *Calostylis* Smith, pp. 261-262.

Type species. — (By monotypy), *Calostylis cribraria* Lindström (1868, p. 421) (= *Clisiophyllum denticulatum* Kjerulf, 1865, pp. 22, 25).

Type locality and horizon. — (of Lindström's material), Silurian (Salopian) of Gotland.

Diagnosis. — Solitary corals lacking dissepimentarium; theca generally quite thin, often missing from all but the base; septa perforate, but still retaining their integrity; septa of two orders, joined laterally by synapticulae; cardinal fossula sometimes developed, but only weakly; axial structure spongy; tabulae thin, usually convex.

Description. — This genus consists of solitary corals whose septa are perforate. In all other respects, the corallite bears the characters of the Rugosa. In some species a basal talon is present. The theca is usually missing (worn away?) from all but the lowest portions of the corallite, leaving exposed the inner spongy texture of the proximal ends of the septa. The calice is generally broadly convex. There are two orders of septa, and these behave as in most rugose corals. The major septa extend to the axial region, and the minor septa stop some distance from the axis, their distal ends marking the limits of the calice wall. In some specimens, a cardinal fossula and septum can be discerned.

The perforate nature of the septa of both orders, however, produces a distinctive internal aspect in this coral. In vertical section, a septum looks like a piece of swiss cheese, with large circular and elliptical holes through it. In the peripheral region of the calice, the septa are joined laterally by synapticulae, which together with the perforate nature of the septa, produces the distinctive spongy appearance. The middle portions of the septa tend to be more complete, and appear in transverse section as solid lines without interruptions. Once the major septa reach the axis, they again break up and intertwine, producing the spongy axial structure.

The tabulae are horizontal or broadly convex. They have an uneven appearance in longitudinal section, due to skeletal projections on both surfaces, but are complete, with no evidence of holes or other interruptions in their surfaces.

Discussion. — Lindström (1868) introduced the name *Calostylis* for corals with perforate septa typified by *Calostylis cribraria* from Silurian beds in the vicinity of Visby, Gotland. In 1870, Lindström (*fide* Smith, 1930a) reported that he had received some specimens from Theodor Kjerulf that the latter (1865) had figured as *Clisiophyllum denticulatum,* and that these specimens were conspecific with *C. cribraria.* Hence, *Calostylis cribraria* Lindström, 1868, the type species of *Calostylis,* should be called *Calostylis denticulata* (Kjerulf). This was accepted by Smith (1930a) in his work on the Family Calostylidae, but I must reserve judgement, never having seen the original material. Certainly, Kjerulf's figures (textfig. 32 on p. 25) are insufficient for a sound comparison with Lindström's.

Tomes (1887) proposed the generic name *Hemiphyllum* for coral material from the Silurian at Colwall and Wenlock, England. Nicholson (1887) and Smith (1930a) considered this a junior synonym of *Calostylis,* with apparent justification.

Smith (1930a, pp. 258-260) gave an interesting account of the controversy surrounding the affinities of *Calostylis* and its relatives.

The major disagreement concerned whether this genus belonged to the Rugosa or to the Scleractinia. The former position was championed by Roemer (1883), while the latter was supported by Lindström (1870). Through the years, many of the most prominent paleontologists found themselves in one camp or the other. As noted above, except for the perforate septa, *Calostylis* corallites have all the prerequisites for belonging to the Rugosa, including two orders of septa, a cardinal fossula, and most important, a septal development pattern and arrangement typical of the Rugosa, reported by Frech (1890) and confirmed by Smith (1930a, p. 260).

Calostylis spongiosa Foerste   Pl. 6, figs. 6-12, 14, 15; Pl. 22, figs. 8-10

Cf. 1865. *Clisiophyllum denticulatum* Kjerulf, pp. 22, 25, text-fig. 32 on p. 25.
Cf. 1868. *Calostylis cribraria* Lindström, p. 421, pl. 6, figs. 1-3.
        1906. *Calostylis spongiosa* Foerste (*partim*), pp. 322-323, pl. 7, figs. 3B-G (*non* 3A); pl. 7, figs. 1A, B.
Cf. 1917. *Calostylis parvula* Foerste, p. 200, pl. 8, figs. 2a-f; Pl. 9, fig. 5.
        1930a. *Calostylis spongiosa* Foerste, Smith, pp. 270-271, pl. 12, figs. 1-7.
Cf. 1930a. *Calostylis tomesi* Smith, pp. 269-270, pl. 11, figs. 12-17.
        1931. *Ptychophyllum riboltense* Foerste, p. 185, pl. 19, fig. 11.
        1931. *Calostylis spongiosa* Foerste, Foerste (*partim*), p. 187, pl. 20, fig. 1b (*non* 1a).
Cf. 1958. *Calostylis concavifundatus* Reiman (*in* Kal'o and Reiman), pp. 28-29, pl. 1, figs. 1, 2.
? 1958. *Calostylis luhai* Kal'o (*in* Kal'o and Reiman), pp. 29-30, pl. 1, figs. 3-6.

Type material. — Smith (1930a) said that Foerste's figured syntypes were lost in the Dayton, Ohio flood of 1913. Foerste sent him other specimens, which he placed in the British Museum under the numbers R26528 to R26533. Foerste's (1906) figures give a good picture of the external features of this species, but provide no information on the internal structure. As the material collected during this study is not from the type locality (from the locality of Foerste's figured specimens), a neotype should be selected from the British Museum material.

Type locality and horizon. — Foerste's figured specimens are all from the Waco bed, Noland Member of the Brassfield Formation, along the "road north of Estill Springs, north of Irvine, Kentucky." This is equivalent to my localities 53 and 53a. Foerste reported specimens from other localities in this area of central Kentucky, but as his concept of the species apparently included specimens here identified as *C. lindstroemi* Nicholson and Etheridge (*q.v.*), it cannot be certain that *C. spongiosa* does occur at these other localities.

Number of specimens examined. — Four from the Brassfield Formation.

Diagnosis. — *Calostylis* with solitary corallites of trochoid or sub-cylindrical (but not scolecoid) form, theca usually well-preserved, but absent from the most distal region, leaving an upper zone of exposed peripheral septal edges (cf. *Montlivaltia* Lamouroux and *Thecosmilia* Milne-Edwards & Haime, two scleractinian genera); cardinal fossula weakly-developed, on convex side of corallite; septa not contratingent.

Description. — Of the two Brassfield species of *Calostylis,* this is the more regular in form. The corallites are typically trochoid or subcylindrical, though the rate of expansion varies between specimens. Following are diameters of three corallites at specific elevations above the base, measured along the convex side (all measurements in mm):

	Elevation above base						Number of Major septa
	5 mm	10 mm	15 mm	20 mm	25 mm	30 mm	
Specimen 1	5	10	8	12	19	19	51
Specimen 2	8	9	11	13	14	—	—
Specimen 3	7	11	—	—	—	—	39

In the three largest specimens, the theca is well preserved, and reaches within a few millimeters of the calice rim. Above this level, the proximal edges of the septa are exposed. In the smallest specimen, the theca is preserved only at the base. Two of the specimens appear to have basal attachment scars (Pl. 6, fig. 11).

The calice rim is generally broadly rounded, though not so broad as the calicular depression. The depth of the calice seems to decrease relative to its diameter as the corallite grows. In the smallest corallite, the calice depth is 2.5 mm, while the diameter across the calice rim is 10 mm. In the largest specimen, the depth is 4 mm and the diameter 17 mm. The two intermediate specimens both have calice depths of 2 mm and diameters of 10 and 13 mm. The smallest specimen has a relatively narrow convex axial structure, with a distinct moat separating it from the steep calice wall (Pl. 6, fig. 15). In the larger specimens (Pl. 6, figs. 6, 7, 11) the convex axial structure is less peaked, and wider relative to its height, and the calice walls are more rounded at the crest.

The septa, as viewed in the calice of a whole specimen, are more distinct than in *C. lindstroemi*. They occur in two orders, of which the major septa are somewhat broader, and of greater relief than the minors. Most of the minor septa appear to be restricted to the spongy peripheral zone of the calice walls. The major septa continue alone across the corallite's calice floor, tapering toward the axial structure. In the peripheral zone, the synapticulae cross the septocoels, joining the sides of adjacent septa.

Perhaps the most distinctive feature of this species is the presence of a cardinal fossula. This appears as a rounded notch in the calice rim on the convex side of the corallite (Pl. 6, figs. 6, 11), and sometimes continues as a groove across the calice floor. In transverse section, the area of the fossula shows a centrally placed cardinal septum, flanked by two minor septa (Pl. 22, fig. 8, upper part). On either side of these latter is a major septum whose distal end curves away from the cardinal septum, and nearly contacts the side of the major septum next to it. This feature was not reported by Foerste, nor is it mentioned in any other *Calostylis* species description. Some of Foerste's figures, however, show a notch in the calice rim that may represent the fossula.

The septa bear circular or elliptical perforations up to 0.3 mm across (Pl. 22, fig. 10). Sometimes the perforations are connected, producing elongate patterns. Distally, the ends of the major septa appear to break up into paliform lobes that form the spongy axial structure. These lobes fuse and flex, giving rise to a reticulum reminiscent of sponge spicules.

The tabulae are convex, and appear as distinct plates, spaced about 1 mm apart, between the peripheral spongy zone and the axial structure (Pl. 22, fig. 10). They are obscure in the axial structure, where they appear as bridges between the spicule-like elements of this region.

Discussion. — Smith (1930a) discussed and figured specimens which Foerste sent him as examples of *C. spongiosa*. Smith probably did not differentiate between the two Brassfield species of *Calostylis*, continuing Foerste's polythetic concept of *C. spongiosa*, but his figures appear to agree with my understanding of this species. Foerste first described *C. spongiosa* in 1906. In 1931, he refigured it, both as *C. spongiosa* (illustrated by figures 3A and 3G of his 1906 description), and as *Ptychophyllum riboltense*. This latter specimen, from the Late Llandoverian Dayton Formation of Lewis County, Kentucky, is strikingly similar in appearance to my specimen (Pl. 6, fig. 6) and clearly belongs to *C. spongiosa.*

In 1865, Kjerulf described *Clisiophyllum denticulatum* from Silurian strata on the Norwegian island of Malmo. His figure shows a rather contorted, trochoid corallite with steep, clearly septate calice walls and a broad, convex axial structure that occupies the entire calice floor. Lindström (1868) later described the species *Calostylis cribraria* from the vicinity of Visby, Gotland (probably from the Lower Wenlockian Högklint beds), and upon it, based the new genus, *Calostylis*. This species was characterized by a cylindro-conical form, basal attachment and an incomplete, patchy theca. The calice region consists of a rounded rim, and a floor occupied by a broad, convex axial structure. The major septa, shown in Lindström's figure 1, are fused in pairs distally prior to continuing to the axial structure.

As described in the "discussion" section of the genus *Calostylis*, Lindström later examined specimens of Kjerulf's species, and came to regard his own as a junior synonym.

Smith (1930a) described *Calostylis denticulata* from material at the Stockholm Riksmuseet that had been identified by Lindström, and from material in his own collection. His figures show that the septa of both cycles in these specimens are far more distinct in the spongy peripheral region than they are in *C. spongiosa* from the Brassfield, and his specimens do not appear to possess a cardinal fossula. These observations and the distal fusion of septa into pairs shown in Lindström's figure 1, but not obvious in Smith's figures, indicate that this European form and the Brassfield species are distinct, but their many similarities in other respects do suggest a rather close relationship.

Smith (1930a) erected the name *Calostylis tomesi* for a coral described by Tomes (1887) from the Wenlock Limestone (Late Wenlockian, according to Gignoux, 1950) of Wenlock, England. The septa of this species apparently fuse distally into pairs, as in the specimen figured by Lindström. *C. tomesi* differs in this respect from *C. spongiosa*, and in apparently lacking a cardinal fossula as well, but the presence of a theca over most of the corallite (except the zone just below the calice rim) is reminiscent of *C. spongiosa*. The corallite of *C. tomesi* is trochoid or sub-cylindrical, as in *C. spongiosa*. *C. tomesi*, then, is another form similar, but apparently unrelated to the species described by Foerste.

From the upper Laurel Formation (Middle Wenlockian) of Ohio, Foerste (1917) described *Calostylis parvula*. It differs from *C. spongiosa* in possessing what Foerste referred to as porous lamellose structures. These are apparently different from the synapticulae of *C. spongiosa*, and connect the sides of the septa in the manner of vesicular dissepiments. Also, this species frequently displays contratingency of minor on major septa. Foerste spoke of a "cardinal side" of the corallite, but it is unclear if a cardinal septum or fossula are distinguishable.

In a joint paper, Kal'o and Reiman (1958) described two new species of Estonian *Calostylis*, which warrant comparison with *C. spongiosa*. The first, *C. concavifundatus*, was described by Reiman from the Porkuny Stage (Late Ordovician or Early Llandovery) of Estonia. These are trochoid and subcylindrical corallites, with the theca absent from the uppermost portion of the sides. The perforate septa are of two orders, with the minor septa about half the length of the major septa. Only the latter reach the axis, and they do not form an axial structure (unusual for this genus), but their distal ends may join. The concave tabulae are one distinctive feature of this species, by which it differs from *C. spongiosa*. Other differences include the absence of a well-developed axial structure and the absence of a cardinal fossula in *C. concavifundatus*.

The second species is *C. luhai*, described by Kal'o from the Adavere Stage (Late Llandovery, according to Manten, 1971) of Estonia. This species is also dealt with in the "Discussion" section of *C. lindstroemi* (*q.v.*). It bears a strong resemblance to *C. spongiosa*, a fact that was also recognized by Kal'o. Kal'o felt that the only important difference between the two species is that the septa of *C. spongiosa* are more clearly divided into two orders. Kal'o's figures, however, suggest that the difference is not nearly so great as he implied. Synonymizing of these two species is prevented only by the absence of clear evidence of a cardinal fossula in *C. luhai*. Kal'o's type suite should be examined closely to see if this structure can be detected.

Distribution. — Kentucky: mid-Llandovery (Brassfield Formation); Late Llandovery (Dayton Formation).

A similar form, *C. luhai* Kal'o, which may prove to be conspecific with *C. spongiosa,* has been reported from the Late Llandovery (Adavere Stage) of Estonia.

Brassfield occurrence. — During this study, · specimens were found at localities 1 (near Panola) and 46 (near College Hill), both in central Kentucky. Foerste's figured specimens are from the Irvine region, in the vicinity of localities 53-53a.

Calostylis lindstroemi Nicholson and Etheridge   Pl. 6, figs. 13, 16, 17; Pl. 22, figs. 1-7

    1878. *Calostylis lindströmi* Nicholson & Etheridge, pp. 65-67, pl. 5, figs. 2-2c.
    1906. *Calostylis spongiosa* Foerste (*partim*), pp. 322-323, pl. 7, fig. 3A (*non* pl. 7, figs. 3B-G; pl. 8, figs. 1A, B); *non* Smith, 1930a.
    1930a. *Calostylis lindströmi* Nicholson & Etheridge, Smith, pp. 265-266, pl. 10, figs. 18-22.
    1931. *Calostylis spongiosa* Foerste, Foerste (*partim*), p. 187, pl. 20, fig. 1a (*non* 1b).
? 1958. *Calostylis luhai* Kal'o (*in* Kal'o & Reiman), pp. 29-30, pl. 1, figs. 3-6.

Type material. — Lectotype (Smith, 1930a, p. 266): BM(NH) R26210 (figured by Nicholson & Etheridge, pl. 5, fig. 2, and by Smith, (1930a) pl. 10, fig. 18). Paralectotypes: BM(NH) R26210-26253, 26296-26271.

Type locality. — The greenish mudstones of Penkill, near Girvan, Scotland, of Late Llandoverian age, according to Smith.

Number of specimens examined. — Seven from the Brassfield Formation.

Diagnosis. — Scolecoid, solitary *Calostylis,* often displaying marked increases and decreases in diameter; no fossula present; tabulae represented by extremely thin lamellae connecting elements of the spongy axial structure; theca seldom preserved, except at base.

Description. — The scolecoid form of the corallite and the almost total absence of theca, are the most obvious features of *C. lindstroemi.* The crests of the calice walls are relatively narrow in contrast to their roundness and usual broadness in *Calostylis spongiosa* (*q.v.*). In well preserved specimens of *C. lindstroemi,* the ratio of calice depth to calice diameter is greater than in *C. spongiosa.* The absence of the theca exposes the spongy interior of the corallite. At the base of the corallite, where some theca may remain, one specimen (Pl. 6, figs. 16, 17) has retained a small talon.

The smallest free corallite has a straight axial length of 17 mm, but is 22 mm long when measured along its convex side. Its diameter at the top is 7 mm. The largest specimen has a length of 45 mm. This corallite (Pl. 6, figs. 16, 17) has a talon about 6.5 mm long at its basal extremity. Through its lowest centimeter of length it gradually expands to a diameter of 7 mm. Above this is a swollen region, 1.5 cm long, reaching a diameter of 13 mm at its center. Beyond this, the diameter abruptly decreases to 7 mm. Half a centimeter above this, the diameter increases to 10 mm, beyond which it remains constant.

The calice floor is usually poorly preserved, but appears to be convex. The spongy texture tends to camouflage the radial structure of the septa in the walls, more so than in other Brassfield species of *Calostylis.* Internally, the ontogenetic pattern is *Streptelasma-*like: as the corallite grew, the area occupied by the axial structure increased proportional to the zone of radiating septa.

The septa are perforated by holes up to 0.2 mm in diameter. Sometimes these holes interconnect, forming elongate gaps (Pl. 22, fig. 1). There are two septal orders, the minor septa remaining in the zone of the calice wall, and the major septa extending across the calice floor to mingle their distal ends in the axial structure of complexly fusing paliform lobes.

The peripheral zone, containing both major and minor septa, is the most dense, and synapticulae crossing the septocoels to connect adjacent septa accent the sponginess of the internal structure. The septa are broadest at the radial position where the minor septa come to an end, and from this point the major septa continue toward the axis.

In longitudinal section (Pl. 22, fig. 1) the tabulae are horizontal or convex upward. They appear extremely thin, but solid, and continue through the axial structure, connecting the interfused paliform lobes.

Both Nicholson and Etheridge, and Smith spoke of specimens displaying buds, but none were found in Brassfield Formation specimens of *C. lindstroemi.*

An interesting specimen was collected (Pl. 6, fig. 13) consisting of a young *C. lindstroemi* growing on the side of a worn fragment of *Schizophaulactis densiseptatus* (Foerste), a solitary rugosan.

Discussion. — Both Nicholson and Etheridge, and Smith stressed the taxonomic importance of the scolecoid form with varying diameter, and the near-absence of a theca. The spongy axial structure accounted for about one-third of the total diameter (as in my specimens). The spongy aspect of the interior, due to discontinuity of the septa, and presence of synapticulae is also noted. In addition, these authors observed that such features are visible along the sides of the corallite because of the absence of the theca. Nicholson and Etheridge stated that the corallite length varied from 0.5 to 1 inch (1.3 to 2.5 cm), and that the mean diameter was 2 lines (4.2 mm). Smith gave the length of the lectotype as 25 mm, and its width as 5 mm, dimensions comparable with most of my specimens. In both references, the septa are described as being of two orders, with only the major septa extending into the axial structure. Nicholson and Etheridge noted the presence of plates representing tabulae, that occur between the wall and the axial structure. These do not occur in my form, and Smith was unable to find these structures in his examinations. There is a possibility that the axial structure of Nicholson and Etheridge's specimens is more separated from the peripheral spongy zone than in my specimens, which might produce the isolated tabulae visible between these two regions. This is suggested in plate 5, figure 2c of the original reference, and in plate 10, figure 22 of Smith's work. Should this be so, the validity of identifying my specimens as *C. lindstroemi* might be questionable. The generally good agreement however, in all other respects suggests this difference may be on the subspecific level. Unfortunately, the lectotype has apparently not been sectioned.

Foerste (1906) erected the species *Calostylis spongiosa* for specimens from the Brassfield Formation north of Irvine, Kentucky. Judging from his description and figures he based it upon specimens that represent both species of *Calostylis* thus far identified from this unit. Only one illustration (fig. 3A) appears to be referable to *C. lindstroemi*, and this is a scolecoid corallite with rather rounded calice rim, and nearly devoid of theca. Assuming that Foerste's figure is natural size, the specimen is about 48 mm long, with an upper diameter of 14 mm. The calicular depression is about 7 mm across. The maximum diameter, 17.5 mm, is at mid-length. In a later publication (1931), Foerste again figured this specimen (and the one shown in figure 3G of his 1906 publication) under the name *Calostylis spongiosa*. (There is insufficient information available from Foerste's figure to ascertain the relationships of this latter specimen.) The scolecoid form shown in Foerste's (1906) fig. 3A, and absence of a theca over nearly the entire corallite strongly suggest *C. lindstroemi*, but the broad, rounded calice rim is more like that in *C. spongiosa*. It seems likely that this specimen does belong to the species described by Nicholson and Etheridge, especially in view of the fact that this species has been found, during this study, at the same locality as Foerste's specimens.

Kal'o (*in* Kal'o & Reiman, 1958) described the species *Calostylis luhai* from the Adavere Stage (Late Llandovery according to Manten, 1971) of western Estonia, based upon specimens of ceratoid or subcylindrical form with much reduced epitheca. The septa are not clearly of two orders, and are well defined only on the inner surface of the spongy peripheral zone. The axial structure appears similar to that in our specimens, as do the convex tabulae which are wavy in cross-section. Synapticulae occur between the septa. Kal'o considered the weak development of the septa, and the supposed absence of two easily distinguished septal orders to be important features of this species. Some of our specimens of *C. lindstroemi* show these features, while others have more distinct septa. Kal'o's figures indicate no appreciable difference between the septa of his specimens and mine. Important differences may, however, lie in the apparently greater separation of the axial structure from the spongy peripheral zone in *Calostylis luhai* (also noted by Nicholson and Etheridge in their material), and especially in the absence of a clear reference by Kal'o to scolecoid form in his specimens. The combination of these two factors suggests this species is best regarded as distinct from the Brassfield specimens of *Calostylis spongiosa*, pending further study of the Estonian material.

Distribution. — Scotland, Late Llandovery; Kentucky, and possibly Ohio, mid-Llandovery (Brassfield Formation).

Brassfield occurrence. — Localities 1 (near Panola), 46 (near College Hill) and 53 (near Irvine), all in the Waco bed, Noland Member in central Kentucky. In addition, this species may be present on a slab from the Upper Brassfield Formation (UCGM 41859) from locality 13 (Todd Fork, near Wilmington, Ohio).

Family CHONOPHYLLIDAE Holmes, 1887

Solitary and colonial Rugosa, with dissepimentarium; rejuvenescence common; marginarium lonsdaleoid; theca a thin septal stereozone; tabulae broadly domed, sometimes with supplementary plates.

Genus STROMBODES Schweigger, 1819

1819. *Strombodes* Schweigger, Table VI.
1873. *Donacophyllum* Dybowski, p. 80.
1874. *Donacophyllum* Dybowski, p. 46.

Type species. — (SD M'Coy, 1849), *Madrepora stellaris* Linnaeus (1758, p. 795).

Type locality and horizon. — According to Lang, Smith, and Thomas (1940, p. 126): "(Silurian) on the shore: Kyllei and Slite, Isle of Götland, Sweden".

Diagnosis. — Colonial rugose corals with dissepimentarium; corallum alternately phaceloid and cerioid, due to repeated expansions of the dissepimentarium touching similar expansions in adjacent corallites, the two separated by a thecal deposit; marginarium lonsdaleoid; tabulae complete and incomplete, often axially depressed and accompanied by supplementary plates.

Description. — This genus consists of colonial corals with a growth-form alternating between phaceloid and cerioid. This results from expansions of the dissepimentarium of one corallite, each joining a similar expansion from a neighbor. In this respect, the corallum superficially resembles a rather stout, irregular colony of *Tubipora,* a modern alcyonarian popularly known as the "pipeorgan" coral. At the point of juncture between these expansions, the two corallites are separated by a thecal deposit (Pl. 23, figs. 3, 4).

The dissepimentarium, with lonsdaleoid septa, has many large dissepiments. The septa are of two orders, and reach to or near the axis, without forming an axial structure. The tabularium contains both complete and incomplete tabulae, commonly depressed axially and near their juncture with the dissepimentarium, with supplementary plates in the peripheral moat.

Discussion. — Schweigger (1819, table 6) described the genus *Strombodes* through the excellent expedient of a flow-sheet, thus showing its morphologic relationships to other genera. He included two species in this genus, *Madrepora stellaris* and *M. truncata,* both of Linnaeus.

M'Coy (1849, p. 10) considered the second of these species to be a true *Cyathophyllum* (*C. dianthus* Goldfuss, 1826), and consequently chose *M. stellaris* as the type species of *Strombodes.* Linnaeus (1758[?]), prior to his introduction of this species, reportedly illustrated it in figures 11 and 14 of his plate on p. 312, but I have not personally seen these figures. Therefore, the concept of *Strombodes* Schweigger used here is based upon the illustration in Hill (1956, p. F301, fig. 2) that purports to show *Strombodes stellaris* (Linnaeus).

Dybowski (1873, 1874) described the genus *Donacophyllum,* from the Ordovician and Silurian of the Baltic. Lang, Smith, and Thomas (1940, p. 54) selected as type species *D. middendorffi* from the Ordovician of Estonia, the first species described by Dybowski (1874, pp. 46-50) under this genus. Dybowski's description and figures strongly suggest that *Donacophyllum* is a junior synonym of *Strombodes.*

The long history of confusion of the species of *Arachnophyllum* Dana for members of the genus *Strombodes* Schweigger, is briefly discussed under *Arachnophyllum.*

Strombodes socialis (Soshkina)   Pl. 7, fig. 1; Pl. 23, figs. 1-4;
Pl. 24, figs. 6, 7

Cf. 1876. *Diphyphyllum huronicum* Rominger, p. 121, pl. 45, fig. 1.
Cf. 1876. *Diphyphyllum rugosum* Rominger, p. 122, pl. 45, fig. 2.
1955. *Cyathactis socialis* Soshkina, pp. 124-125, pl. 12, figs. 2a, b.
1970. *Strombodes (Strombodes) socialis* (Soshkina), Flügel & Saleh, pp. 280-281, pl. 2, figs. 1, 2.
1972. *Strombodes* sp. Bolton & Copeland, p. 30, pl. 4, figs. 5, 6.

Type material. — Holotype: PIN 587/2465.

Type locality and horizon. — Kochumdekskoy Suite (Middle and Upper Llandovery), layer B5, upper level of the Suite, on the right bank of the Podkamennaya Tunguska River, opposite the mouth of the Listvenichnoy River, Siberian Platform.

Number of specimens examined. — Approximately six from the Brassfield Formation.

Diagnosis. — *Strombodes* with a wide lonsdaleoid dissepimentarium occupying one-half to two-thirds of the corallite radius and consisting of rather large, elongate dissepiments; outermost dissepiments distinctly larger (in cross-section) than the more central ones; septa not reaching axis; tabulae mostly complete, often axially concave and peripherally convex; narrow zone between tabularium and dissepimentarium occupied by supplementary plates; increase apparently non-parricidal.

Description. — This species consists of large coralla, generally about 12 cm across. The corallites are closely packed and mostly large. Individual corallites attain a length of 4 cm and more, and a diameter of about 1.5 cm (a few reach 2 cm in width). Increase appears to be nonparricidal. At a given level, the interstices between these broader corallites are filled with the calices of younger corallites, 0.3 cm or less in diameter. The calices are so closely packed that in surficial view they have the form of a massive, cerioid corallum, with scattered phaceloid regions (Pl. 7, fig. 1; Pl. 23, figs. 3, 4).

Sections show that thecal material, about 0.3 mm thick, is always present in the contact zone between corallites, with a thin midline dividing the theca of the contiguous corallites (Pl. 23, figs. 3, 4).

Close packing distorts a basically circular corallite cross-section into polygons. Calices observed were generally poorly preserved. One large example with a diameter of 15 mm had a depth of 3 mm, and was more concave in the axis than at the margins.

Externally, the thecae are longitudinally striated (Pl. 7, fig. 1), with irregularly spaced rugae about as distinct as the striations. In one slightly worn corallite, the striations seem to be alternately deep and shallow, but this is not evident on other well-preserved corallites in the same corallum.

The septa are lonsdaleoid peripherally, with minor septa only slightly shorter than major septa. No fossula has been seen distinctly, but in Plate 23, figure 2, two septa in the lower portion of the figure appear to curve together in the axis and encircle a third septum, reminiscent of the cardinal fossula in *Dinophyllum semilunum*, sp. nov. (*q.v.*). In a diameter of 1.5 cm (not counting the zone of coarse dissepiments into which the septa do not reach peripherally), there are 67 septa (approximate total for both orders). The corallite interior consists of three zones (Pl. 23, figs. 1, 3): an outer dissepimentarium, an inner tabularium, and an intermediate (though not ubiquitous) narrow zone of supplementary plates. Roughly, the tabularium accounts for one-third the diameter, the dissepimentarium for one-half the diameter, and the intermediate region of supplementary plates (where present) for one-sixth the diameter.

The dissepimentarium consists of as many as 12 layers of vesicles, and makes an angle of about 40° with the corallite axis. In longitudinal section, the vesicles of one corallite are from 1 to 3 mm long, with a height:length ratio ranging from one-half to one-seventh. There is no clear vertical zonation of dissepiments by size.

The tabularium structure appears chaotic in most views, but clearly consists of "cupid-bow" tabulae . . . that is, tabulae concave in the axis, convex on all sides surrounding the central depression, and depressed peripherally to form a moat against the dissepimentarium. The tabulae are often incomplete, occurring as plates tilted against the sides of the axial depression or the moat. In some specimens, the central depression is absent, leaving flat tabulae surrounded by a moat.

The relief of the moat is often moderated by supplementary plates, which may have produced a flat calice floor at times.

The complexity of the tabularium makes it difficult to determine the tabular spacing, but the tabulae appear to be about 0.3 to 1 mm apart vertically. The supplementary plates of the intermediate zone are spaced 0.3 to 0.4 mm apart vertically.

Discussion. — Rominger (1876) reported a coral, *Diphyphyllum huronicum* from the Manistique Group (Late Llandovery to Middle Wenlock) of Point Detour, Chippewa County, Michigan, which has much in common with my specimens. The holotype (UMMP 8592) could not be located, so that it is not possible to make a detailed analysis of the similarities and differences.

Rominger reported that his form consisted of aggregate, flexuose corallites, one to two cm in diameter, bearing delicate longitudinal striae and deeper rugae. The corallites connect laterally by stout rugose prominences which, "in all the stems of a colony are uniformly directed to one side." The calices are moderately deep,

"dish-shaped with explanate margins", with about 60 septa merging
into an axial "fascicle". There is a dissepimentarium, and the axial
region of the corallite contains large vesicles, but apparently nothing
that Rominger considered a tabula, so that the outer and inner
regions are ill-defined. Increase is by gemmation from the margins
and the center of the calicular area. Offsets occurring at the margin
remain slender for a longer time than do the others.

Superficially, my form agrees with Rominger's in most important respects. Only one specimen (Pl. 7, fig. 1) shows a good side
view of the corallites. This does not clearly show the rugose processes of the corallites directed uniformly to one side (though not
enough of the corallum is exposed to permit proper analysis). The
number of septa generally agree, but in the Brassfield specimens they
do not form an axial structure as they appear to do in Rominger's
form. The description of the internal structure of the Michigan
specimen cannot be readily reconciled with features seen in the
Brassfield material. In the latter, the tabularium and dissepimentarium may be differentiated without too much difficulty, though the
complexity of the tabular surfaces, the abundance of incomplete
tabulae, and the marginal narrow zone of supplementary plates
bounding the tabularium may have presented the confused appearance that Rominger apparently encountered. Also, Rominger's
example had coarser cell spaces in the central portion, whereas in
my specimens, the dissepimental cells are coarser. Finally, the Brassfield specimens do not show parricidal budding, but Rominger's
remark about gemmation from the center of the calicular area implies that this does occur in *D. huronicum*. Pending examination of
the holotype, there seem to be no grounds for regarding these two
forms as conspecific.

A specimen in the collection of the University of Michigan's
Paleontology Museum (Pl. 24, figs. 6, 7) indicates that *S. socialis* did
extend into the same regions where Rominger's material was found.
This specimen is better preserved internally than mine, and appears
to be conspecific with it. It comes from the top of the Fiborn Limestone Member of the Hendricks Dolomite (early Late Llandovery,
coeval with the Noland Mbr.) about 3 miles north of Hunts Spur,
Mackinac County, Michigan. It agrees in all respects with the structure of my specimens. The "cupid-bow" tabulae, and zone of sup-

plementary plates separating the tabularium from the dissepimentarium are clearly visible. Because of the more regular structure of its tabularium, the different internal zones are more distinct than in my material, possibly an indication of the variability of this species. The absence of an axial structure in this specimen adds to the likelihood that it is conspecific with the Brassfield form.

In the same paper, Rominger described a second species under the generic name *Diphyphyllum*, *D. rugosum*, giving the locality as Louisville, Kentucky in the Niagara Group. Possibly, his reference is to the Louisville Limestone of Late Wenlock to Early Ludlow age. Rominger recognized a strong resemblance between this species and *D. huronicum*, and saw them differing in the more slender corallites (1 cm) of the former. Of perhaps greater importance is the failure of the septa in *D. rugosum* to join distally to form an axial structure, in which respect it resembles my form. Rominger also noted the prolific gemmation from the calice (parricidal budding) shown by *D. rugosum*, which, along with the slender form of the corallites, is probably the greatest difference between this species and the Brassfield material, making their conspecificity unlikely.

Soshkina (1955) described *Cyathactis socialis* from the Middle or Upper Llandovery in the Siberian Platform. This is a shrublike corallum with no evidence of parricidal increase. The corallites are elongate, and expand distally with frequent and sharp lateral prominences. Daughter corallites arise at the rim of the calice, and diverge from the mother corallite as they grow upward. The major septa are thin, weakly bent and slightly twisted at the axis. They are discontinuous in the dissepimentarium (lonsdaleoid). The cardinal septum is shortened, lying in a parallel-sided fossula, and isolated from the axis by the distal juncture of neighboring septal ends. The minor septa are over half a radius in length, extending to the axial limit of the dissepimentarium. In a corallite diameter of 14 to 30 mm occur 33 to 36 septa of each order. The plates of the tabularium are cupid-bow in form, like those in my specimens, and are usually quite concave axially, with a deep moat against the dissepimentarium. The vesicles of the dissepimentarium are large and flattened, slope axially, and make a dissepimentarium about 10 to 15 layers thick.

From an early ontogenetic stage, the septa are all equally thin. They do not appear to form an axial structure. In all important respects, *C. socialis* Soshkina appears to be conspecific with the Brassfield specimens.

Flügel and Saleh (1970) reported *Cyathactis socialis* Soshkina [under the name *Strombodes* (*Strombodes*) *socialis*], from the Silurian Niur Formation at Ozbak-Kuh and E Gatch-Kuh in east Iran. The corallum is phaceloid, with fairly cylindrical corallites, 45 to 60 mm long and 20 to 30 mm in diameter. Their outer surfaces are finely striate longitudinally and distinctively rugose. In a diameter of 20 mm there are 70 septa (35 of each order). The major septa reach close to the axis, and a longer septum is considered the cardinal septum. The minor septa are barely half as long as the major septa.

The dissepimentarium is about as broad as in my specimens, and the vesicles are of similar size. Reportedly there is no lonsdaleoid septal development in the Iranian form, but the figures suggest it is in fact present, certainly to the degree found in my specimens. The tabularium is as in my form, with cupid-bow tabulae fringed by a moat against the dissepimentarium. There are numerous incomplete tabulae.

The Iranian form differs from mine in no important respects. The nature of the cardinal(?) septum in my form is not very clear, as the projections of the complex tabulae in transverse section cloud the central features. Probably this is true in both Soshkina's and Flügel and Saleh's figures. In all cases it is difficult to ascertain whether isolation of the axial end of the cardinal septum from the axis is caused by distal linking of neighboring septa, or by a depression in the tabula, skirted at the rim by the neighboring septa. The shorter minor septa in the Iranian form are more likely to be of subspecific than specific importance in differentiating this form from the Siberian and Brassfield material.

Bolton and Copeland (1972) figured transverse and longitudinal sections of what they termed *Strombodes* sp., from the Thornloe Formation (Late Llandovery and Early Wenlock) of Mann Island, near the northeast shore of Lake Timiskaming, Ont. In the figures, the corallite diameter is about 1.25 cm, and the length at least 3.5 cm. Its dimensions thus agree generally with those of the Brassfield specimens, and with the UMMP specimen from the Fiborn Limestone of Michigan, which it most resembles.

The outermost portions of the dissepimentarium contain distinctly larger dissepiments than those closer to the axis, a common pattern in the present species. The peripheral ends of the septa are lonsdaleoid and the septa do not quite reach the corallite axis, so that there is no axial structure.

The Timiskaming form appears to have fewer septa in a given, diameter than either my form or that from the Fiborn Member, having perhaps 44 septa of both orders in a diameter of 1 cm, as compared with approximately the same number in a diameter of about 0.6 cm in the Fiborn specimen. The importance of this is difficult to judge, however, as it is possible to compare the corallites as to total diameter, or to exclude the nearly septum-free peripheral region of the dissepimentarium (as was done in this case).

The tabulae are mostly flattish in the Timiskaming specimen, but some show evidence of peripheral convexity and axial depression. There is no clear intermediate zone of supplementary plates in a moat, but as noted above, that condition is not ubiquitous in this species. The similarities suggest that this specimen is conspecific with *Strombodes socialis* (Soshkina).

Distribution.— Southwest Ohio, mid-Llandovery (Brassfield Formation); northern Michigan, Middle Llandovery; Lake Timiskaming, Ontario, Late Llandovery or Early Wenlock; Siberian Platform, Middle or Upper Llandovery; Iran, Silurian.

Brassfield occurrence.— Localities 7 and 7a, near Fairborn, Ohio.

Family ARACHNOPHYLLIDAE Dybowski, 1873

Solitary and colonial Rugosa with dissepimentarium; major septa meeting axially to join with steeply domed tabulae in an axial boss, or otherwise somewhat withdrawn from axis where tabular domes sag axially.

Genus ARACHNOPHYLLUM Dana, 1846

1846. *Arachnophyllum* Dana, p. 186.
1873. *Darwinia* Dybowski, p. 148 (*non* Bate, 1857).
? 1885b. *Nicholsonia* Schlüter, p. 53 (*non* Kiär, 1899, p. 37; *nec* Pocta, 1902, p. 184).
1940. *Arachniophyllum* Lang, Smith, & Thomas, p. 19 (*nom. van. pro Arachnophyllum* Dana, 1846); *non Arachniophyllum* Smyth, 1915, p. 558).

Type species. — (by monotypy, see discussion): *Acervularia baltica* Schweigger of Lonsdale (1839, *partim*: plate 16, figs. 8b-e; *non* 8, 8a), *non Acervularia baltica* Schweigger (1819, table 6). = *Strombodes murchisoni* Milne-Edwards & Haime (1851, p. 428).

Type locality and horizon. — According to Lang, Smith, and Thomas (1940): "Silurian, Wenlock Limestone: Wenlock, Dudley, etc., England." Ziegler, Rickards, and McKerrow (1974) gave the age of the Wenlock Limestone as Late Wenlock.

Diagnosis. — Colonial Rugosa with dissepimentarium; corallum massive, with thamnasterioid septal arrangement; tabularia narrow, with incomplete, convex tabulae; dissepimentarium quite broad, continuous between adjacent tabularia; septa not dilated or carinate in tabularium, but dilated in the marginal areas between tabularia, resulting in reduction of septocoels in this region to sutures between septa.

Description. — *Arachnophyllum* consists of massive, thamnasterioid rugosan coralla. The surface of the corallum bears calicular pits, each generally containing an axial boss. From these pits radiate dilated septa that are separated laterally by sutures. Most of these septa are continuous from one calicular pit to that of a neighbor, producing the thamnasterioid pattern. No theca separates neighboring corallites. Though there may be a bounding ridge marking the limits of each corallite [as in *A. pentagonum* (Goldfuss)], yet the septa are continuous across it.

The platform within which the calicular pits lie is underlain by vesicular dissepiments and is, in effect, a coenosteum. The pits may or may not be elevated above the level of this platform. In *A. mamillare* (Owen) [*q.v.*], they are each set at the peak of a mound. The septa are of two orders, with the minor septa going no further than the lip of the calicular pit (thereby defining it) and the major septa extending to the axis, where they intertwine to form an axial structure. This structure commonly projects above the floor of the calice as an axial boss.

The septa are in some instances represented by trabeculae projecting upward from the surfaces of the dissepiments of the coenosteum, but piercing no more than one or two successive layers of vesicles.

Discussion. — Dana (1846) introduced the genus *Arachnophyllum* with specific reference to *Acervularia baltica* Schweigger of

Lonsdale (1839). This specimen was subsequently selected as the basis of *Strombodes murchisoni* by Milne-Edwards and Haime when they erected this species in 1851. As Lonsdale's was the only species described by Dana under his genus *Arachnophyllum*, this species (and thus, *S. murchisoni*) is the type by monotypy.

Hill's (1956, p. F274) reference to a subsequent designation of this species as the type, by Lang and Smith (1927) is thus erroneous. Lang and Smith's contribution was to note that Lonsdale's figures included two species of *Arachnophyllum*, and to restrict the type to those specimens shown in Lonsdale's figures 8b-e, which they considered to be *A. baltica, sensu* Lonsdale.

In the past, numerous authors have confused *Arachnophyllum* Dana with *Strombodes* Schweigger. This is apparently because, as explained by Lang, Smith, and Thomas (1940, pp. 126-127), Milne-Edwards and Haime (1850, p. lxx) erroneously cited *Strombodes pentagonus* Goldfuss (1826, p. 62, pl. 21, figs. 2a, b), which morphologically belongs to *Arachnophyllum*, as the type of *Strombodes*. They were apparently unaware of the fact that M'Coy (1849, p. 10) had previously designated *Madrepora stellaris* Linnaeus as the type of *Strombodes*. The misuse of the name *Strombodes* persisted for quite some time.

From the Silurian of the Baltic Area (Estonia), Dybowski (1873, p. 148) described the genus *Darwinia*. Though a bit sketchy, the description, together with the figures, strongly suggests that this genus is a junior synonym of *Arachnophyllum* Dana. Dybowski stressed the colonial nature of the corallum, with calicular pits separated from one another by coenosteum, and that the septa are "perfect" only in the calicular pits (where the major septa form an axial structure), and broaden outside this area where they extend into the coenosteum.

From the Middle Devonian of Germany, Schlüter (1885b, p. 53) introduced the genus *Nicholsonia*, based upon two species he had previously described as *Darwinia*, and upon *Strombodes diffluens* Milne-Edwards and Haime (1851), which is clearly an *Arachnophyllum*. Lang, Smith, and Thomas (1940, p. 89) designated one of these species, *Darwinia perampla*, as the type species of *Nicholsonia*, and stated that they considered this species congeneric with the type species of *Arachnophyllum*, thus making the former genus a subjective junior synonym of the latter. I have not seen the original description of *D. perampla,* and so cannot be confident that the decision of Lang, Smith, and Thomas is valid. Based upon the nature of *Strombodes diffluens,* and on the synonymy of *Darwinia* with *Arachnophyllum,* however, it seems likely that they were correct.

Arachnophyllum mamillare (Owen)   Pl. 7, figs. 2, 3; Pl. 25, figs. 1, 3

1844. *Astrea mamillaris* Owen, p. 70, pl. 14, fig. 3.
1876. *Strombodes pygmaeus* Rominger, pp. 132-133, pl. 48, fig. 3.
1876. *Strombodes mamillatus* [*sic.*] (Owen) Rominger, p. 133, pl. 48, fig. 4.
? 1878. *Lyellia striata* James, p. 10.
1887. *Strombodes mammillaris* (Owen), Davis, pl. 123, fig. 4.
1888. *Strombodes pygmaeus* Rominger, Foerste, p. 120, pl. 13, fig. 18.
1901. *Arachnophyllum mamillare* (Owen), Lambe, p. 182, pl. 15, fig. 4.
1906. *Arachnophyllum (Strombodes) mamillare-distans* Foerste, pp. 319-320, pl. 3, figs. 2A-C.
1906. *Arachnophyllum mamillare-wilmingtonensis* Foerste, p. 320.
1964. *Arachnophyllum mammillare* (Owen), Stumm, p. 31, pl. 20, figs. 7-9.
1966. *Arachnophyllum mammillare* (Owen), Bolton, pl. 6, fig. 15.

Type material. — Repository of Owen's material is unknown.

Type locality and horizon. — Owen reported this species from the coralline beds in the top of the Magnesian Cliff Limestone of Iowa and Wisconsin. Judging from his map (pl. 3), this area consists of the drainage basins of the Makoqueta and Wapsipinicon Rivers, southwest of the Mississippi River. As this species is reported by other authors from Silurian beds, it is likely, though not certain, that the horizon of the original material is the Kankakee or the Hopkinton Dolomite.

Number of specimens examined. — Three from the Brassfield Formation, including the holotype of *Arachnophyllum (Strombodes) mamillare-distans* Foerste.

Diagnosis. — *Arachnophyllum* with calices located on mammiform protuberances rising from a vesiculose surface; septal striae on corallum surface connect the calices in a thamnasterioid pattern, with no clear boundary between adjacent calice regions.

Description. — *A. mamillare* consists of flat, expansive, holothecate coralla, built up by and consisting of generally broad, flattened vesicular dissepiments, with maximum thicknesses of 0.6 mm and breadths of 0.3 mm or more. The dissepiment walls are about 0.5 mm thick. Possible spines were observed on the upper surfaces of the dissepiments in thin-section, and these may correspond to the "granules" reported by Foerste (1906, p. 320) on the septal striae of the corallum surfaces. These "granules" have not been observed on the surfaces of my specimens, but my material appears less well-preserved than Foerste's.

The surface of the corallum is strewn with mammiform processes, each bearing a calicular pit of variable width at its apex. From the rim of this pit, the septa extend out to the flat surface of the common corallum in thamnasterioid pattern, where they flatten and broaden to about 0.7 mm, compared with less than 0.1 mm at the lip of the calice. Characteristically there is no border of any sort to set off the region of one mammiform process from that of its neighbors.

The septa appear to be present in two, or possibly three orders, with about 18 major septa. Outside of the calicular pit, the septa of the different orders are indistinguishable, in this sense resembling *Schlotheimophyllum* (*q.v.*). These relationships are most clearly seen in the holotype of *A. mamillare-distans* Foerste (USNM 113639; Pl. 7, fig. 2), where a ropy axial structure is apparently formed of the distal ends of the major septa and the interstitial vesicular or tabular dissepiments (Pl. 25, fig. 1). The rims of the calicular pits are formed by a concentration of high-domed vesicles, and the calicular space between the wall and the axial structure appears to contain concave tabulae, which rise sharply toward both the axis and the wall. In the intercorallite regions of the uppermost zone in this specimen, the vesicular dissepiments are much smaller than at earlier stages. This may reflect a decrease in growth-rate, culminating in the death of the colony.

In this specimen, the mounds have a relief of about 7 mm and are spaced about 1.0 to 1.5 cm apart. Their bases are about 1.0 cm wide, and the calice openings are about 3.0 to 3.5 mm across.

Two specimens were found in the course of this study. One, from locality 1, agrees with Foerste's specimen, except that the calices are more widely separated, and no axial boss was observed. The surface is poorly preserved.

The second specimen is from locality 53. Here the calices are more densely distributed, about 1.5 cm apart, as in Foerste's specimen. The calicular pits are about 5.0 to 6.5 mm across their rims, and while there is no clear axial boss, weathered mammiform processes show a resistant axial structure of some sort. The surface of this specimen is quite uneven, and the axes of the calices are not all parallel. In this sense, they resemble the specimen figured by Owen. The local relief of the mammiform processes varies from about 0.5 to 1.2 cm.

Discussion. — Owen's original description (1840, p. 70) of this species is: "Stars very much elevated, so as to form mamillary processes, irregular both as to size and distance apart; rays confluent." His figure resembles the more irregular of the two specimens found in the field (from locality 53).

Strombodes mamillatus Rominger (1876) agrees well with my concept of *A. mamillare*. Mammiform processes rise from a common vesiculose surface, with no boundaries between them. Their centers are 1 to 2 cm apart. At the peak of each process is a calicular pit, 4 to 5 mm across, with a "small styliform projection". From this pit the septa radiate outward to form a thamnasterioid pattern on the upper surface of the corallum, between the calices. In this region, the septa reportedly have granulose surfaces.

Rominger's description indicates that his figured specimen (UMMP 8605) is from Point Detour, Drummond Island, in Lake Huron. If this is so, the age would probably be Llandovery or Wenlock. The specimen label, however, indicates the specimen is from Niagaran beds at Masonville, Iowa. Rominger did supplementary collecting in this area for such species as *Tryplasma radicula*, which came from the Hopkinton Formation (Late Llandovery to Late Wenlock), so that it is possible that the museum label is correct.

James (1878) described *Lyellia striata* from undetermined strata in the region of Cincinnati, Ohio. Probably, "region" in this context refers to quite a wide area, as the Ordovician beds that underlie and surround the city of Cincinnati bear few corals, and almost certainly none of the type reported by James. It is likely, then, that James obtained his material from Silurian beds (possibly Brassfield) some distance from Cincinnati. *L. striata* is horizontally expansive with a vesiculose coenosteum. Mammiform processes arise from the upper surfaces with their centers from about 4.2 mm to over 12.7 mm apart. At their crests are calicular pits, about 2.1 mm in diameter, with an axial prominence in each. Twelve to twenty septa are present at the crests, but reportedly do not extend into the body of the tube. The corallum surface is invested with coarse sinuous striae, probably like the thamnasterioid pattern of my specimens. From James' description, it seems likely that he was dealing with *A. mamillare*.

Davis (1887) figured a specimen identified as *Strombodes mamillaris* from the "Upper Niagara red clay near Louisville", Kentucky (probably strata of Late Wenlock or Early Ludlow age). This specimen closely resembles my material, having mammiform processes with rather broad apertures. An axial prominence appears to be present in the calicular pits. One point of difference between this specimen and mine, however, is that the septal ridges do not attain the thickness that mine do in the coenosteum. In spite of this, it seems reasonable to identify this specimen as belonging to the species described by Owen.

From the Llandovery or Wenlock beds of Grand Manitoulin Island, Lake Huron, Lambe (1901) reported a specimen which he identified as *A. mamillare*. The corallum is holothecate, with mammiform processes, each bearing a broad calicular pit. These prominences are spaced 1 to 2 cm apart. The septa are of two orders, totalling 36, with the major septa attaining the axis, while the minor septa are restricted to the calice walls. The thamnasterioid septal pattern of the coenosteum includes septa with breadths of nearly 1 mm. It is probable that this specimen does indeed belong to *A. mamillare*.

Foerste's *A. mamillare-distans* is from the southeast region of the Brassfield outcrop, in central Kentucky. The holotype, described above, was collected from near locality 1 (near Panola, Kentucky) in the Noland Member of the Brassfield Formation. Foerste felt his specimens were distinct from other forms of the species in having narrower apertures and more widely spaced mammiform processes. Considering the variability seen in other members of the species, Foerste's form does not seem to be sufficiently different to warrant a separate subspecies.

Stumm (1964) reported this species from the Louisville Limestone (Late Wenlock and Early Ludlow) near Louisville, Kentucky. It has all the important features of the Brassfield form. The mammiform processes are about 3 mm high, and contain broad calicular pits, 4 mm across and 1 mm deep, some of which have weak axial bosses. The coenosteum is vesiculose, and its surface bears the usual thamnasterioid septal pattern.

Bolton (1966) reported this species from the Fossil Hill Formation (Middle Llandovery to Middle Wenlock) of Manitoulin Island. The mammiform processess are 3 to 4 mm in relief, and contain calicular pits about 4 mm across. About 38 septa (of all orders) were seen on the lip of one calice. The specimen agrees well with our concept of *A. mamillare*.

There appears to be a smaller form of this species. It was first described as *Strombodes pygmaeus* by Rominger (1876) from the Llandovery or Wenlock beds of Point Detour in Lake Huron. Rominger's specimen is obviously worn. Its flattish surface is dotted with calicular pits, about 2 mm wide and spaced 6 to 8 mm apart. Within these pits, the septa form a "small styliform projection". The coenosteal surface appears to have a thamnasterioid pattern of very fine septal ridges, each ridge bordered on both sides by a row of circular pores. The nature of these pores is unclear, but they may reflect erosion of the internal vesicular deposits.

A. pygmaeus (Rominger) is characterized by its narrow apertures and the close spacing of the mammiform processes. The coenosteal septal striations tend to be quite fine. *Lyellia striata* James may belong to this group.

Foerste (1888) reported this smaller form from the Dayton Formation (Late Llandovery), apparently in the vicinity of Wilmington, Ohio. He referred it to *Strombodes pygmaeus* Rominger, but later (1906) proposed the name *Arachnophyllum mamillare-wilmingtonensis*. As it appears that this form is identical to *S. pygmaeus*, it seems appropriate that Rominger's subspecies name should be used rather than Foerste's. Foerste described the mammiform processes as more prominent than they appear to be in Rominger's figure, with calice diameters of about 2 mm, spaced from 4 to 10, but usually 6 mm apart. The calices are elevated about 2 mm above the coenosteum, and the coenosteum has a thamnasterioid pattern of septal ridges, each about 0.3 mm wide.

At locality 12 (Ohio Brush Creek), a sample of this subspecies was found loose near the top of the Brassfield Formation (Pl. 7, fig. 3). It may have come from the Dayton Formation, as that unit overlies the Brassfield in southwest Ohio, and the condition of the specimen suggests transport. The Dayton Formation, however, was not found in this vicinity, so that the possibility that it comes from the Brassfield itself cannot be eliminated.

The calicular pits are 2 mm in diameter, and are spaced with their centers about 5 to 7 mm from the nearest neighbor. The mammiform processes are quite worn, but appear to have had reliefs of about 2 to 2.5 mm. In the few calices that are reasonably-well preserved, there are about 16 septa and an axial protuberance about one-third the diameter of the calice. The septal striae of the coenosteal platform are fine, with breadths of about 0.3 mm (compared with 0.7 mm in the two specimens found in the Kentucky Brassfield during this study, and in Foerste's holotype of *A. mamillare-distans*).

Distribution. — Northern Michigan region, Llandovery or Wenlock; Kentucky, mid-Llandovery (Brassfield Formation), and Late Wenlock or Early Ludlow; Ohio, Late Llandovery; Upper Mississippi Valley: Silurian (Owen's material).

Brassfield occurrence. — Localities 1 (near Panola) and 53 (near Irvine) in central Kentucky. Foerste also reported this species from the vicinity of Indian Fields, in the same general area as the above localities, in Kentucky. All these specimens are from the Waco bed, Noland Member of the Brassfield Formation.

A subspecies of this form, *A. mamillare pygmaeus*, was found at locality 12, Ohio Brush Creek in southwest Ohio. Its precise stratigraphic position is uncertain.

Arachnophyllum granulosum Foerste

Cf. 1826. *Strombodes pentagonus* Goldfuss, p. 62, pl. 21, figs. 2a, b.
Cf. 1876. *Strombodes alpenensis* Rominger, p. 133-134, pl. 38, fig. 1.
    1906. *Arachnophyllum (Strombodes) granulosum* Foerste, pp. 318-319, pl. 3, fig. 1.

Type material. — The repository of Foerste's specimen is not known.

Type locality. — Waco bed, Noland Member of the Brassfield Formation, "half a mile east of Waco, Kentucky, where the road to Cobb Ferry turns off from the Irvine Pike".

Number of specimens studied. — None found in the field. Analysis of species is based on Foerste's description, and illustration of one specimen.

Description. — Foerste's description follows:

Corallum compound, explanate, thin; only a part of the corallum is preserved in the specimen figured, and this shows a maximum width of 70 millimeters and a thickness of scarcely 3 millimeters. The lower surface is not exposed. Polyparia opening above in shallow calices meeting along polygonal out-

lines rising but slightly above the middle parts of the calyx [sic]. A very narrow, filiform line forms a border between the calices; this line is interrupted
at intervals by minute pits or pores, situated most frequently at the ends of the
grooves between the septal ridges. Calices about 7 to 10 millimeters wide along
their longer diameters, and from 6 to 7 millimeters wide transverse to this
direction. Very shallow depressions, from which the septal ridges radiate, occur
in an excentric position, so that the calices have the appearance of opening
obliquely upon the surface. It is not known to what extent this oblique position
of the slight depression is characteristic of the species. Septal ridges average
about 14 or 15 in number. These begin at the margins of the shallow depressions and radiate toward the margins of the calices; those toward the nearest
or rear border widen rapidly and are more triangular in outline; those toward
the farther or anterior border are more nearly linear; those on the sides curve
toward the anterior border. These septal ridges are quite flat or only gently
convex in a transverse direction, and are separated by narrow and rather shallow, but very distinct grooves. The surfaces of these septal ridges are marked
by very minute granules. Somewhat coarser granules, very irregularly arranged, occupy the excentric depressed area. In some calices the margins of the
depressed areas are slightly, or even distinctly raised, producing the effect of
a sharp circular border. In some cases the weathering away of one of the
calicular layers leaves the depressed area of this layer distinctly above the
general level of the next lower calicular surface. In vertical sections the corallites are seen to consist of successive calicinal floors supported by a rather
coarse vesicular tissue, the cavities of which are rather wide but very shallow.
Horizontal tabulae, closely arranged, appear to cross the region corresponding
to the central, depressed part of less excentric calices, but this part is not well
exposed in the specimen at hand. Among the specimens so far described, this
coral appears most nearly related to *Strombodes alpenensis*, Rominger.

Discussion. — Of the described species of *Arachnophyllum*,
Foerste's specimen most closely resembles *A. pentagonum* (Goldfuss)
1826, from the Silurian rocks of Drummond Island, Lake Huron.
In both forms, the calicular pit is within a shallowly depressed
region, isolated from neighboring calicular areas by a fine ridge that
forms a polygonal border. Septal striae radiate from the pit out to
the bounding wall. The major differences are that the septal ridges
of Foerste's specimen are generally broader than in *A. pentagonum*
(though those of Goldfuss' figure are also rather broad); and the
calicular pits of *A. pentagonum* do not seem to have the characteristic eccentricity of *A. granulosum*. Further, Goldfuss' material
does not appear to have the granules on its septal ridges and in its
calicular pits reported by Foerste in his specimen.

Forste's reference to *Strombodes alpenensis* was to a form reported from the Middle Devonian beds at Thunder Bay, Michigan
by Rominger (1876). This is similar to *A. granulosum*, the only
obvious differences being that its calicular pits are not markedly
eccentric, and the axial ends of the septa are more attenuated.

It is possible that *A. granulosum* is a subspecies of *A. pentagonum*, but in the absence of a specimen for study, it seems best
to accept Foerte's contention that it represents a new species.

Distribution. — Kentucky, mid-Llandovery (Brassfield Formation).

Brassfield distribution. — From the Waco bed, Noland Member of the Brassfield, near Waco, Kentucky.

Genus **PETROZIUM** Smith, 1930

1930b. *Petrozium* Smith, p. 307.

Type species. — (Original designation): *Petrozium dewari* Smith (1930b, pp. 307-308). This species is discussed under *Petrozium pelagicum* (Billings).

Type locality and horizon. — Llandovery (Lower Silurian) of England.

Diagnosis. — Colonial Rugosa with dissepimentarium; corallum fasciculate; corallites slender; septa of two orders; axial structure formed by stereomal deposits among distal ends of major septa; dissepimentarium narrow, generally no more than three to four dissepiments thick; tabulae sometimes depressed axially, and may have supplementary plates.

Description. — This genus consists of species with fasciculate growth-habit. The theca is a septal stereozone. A dissepimentarium is present, rarely more than three or four layers thick. This generally has the form of large vesicles with limbs at nearly right angles to one another, perched upon each other like steps in a stairway.

The septa are thin, and sometimes bear carinae. The major septa extend to the axis where they form an axial structure consisting of their distal ends and deposits of stereome.

The tabulae are mostly complete and are often depressed axially and peripherally to produce a lateral profile similar to that of a volcano with a moat around its base. This peripheral moat contains supplementary plates.

Discussion. — The presence of an axial septal structure and invariable coloniality of *Petrozium* serve to distinguish it from the Silurian genus *Entelophyllum* Wedekind (1927). In addition, the dissepimentarium of the latter genus typically consists of much smaller vesicles in more layers, than is common in *Petrozium*.

From *Palaeophyllum* Billings (1858), *Petrozium* differs in having a dissepimentarium, and in the presence of an axial septal structure.

Petrozium pelagicum (Billings)   Pl. 7, fig. 8; Pl. 22, figs. 11-14;
                                           Pl. 25, figs. 2, 4

1862. *Cyathophyllum pelagicum* Billings, p. 108.
1865a. *Cyathophyllum pelagicum* Billings, Billings, p. 108.
1874. *Donacophyllum lossenii* Dybowski, pp. 50-51, pl. 4, figs. 6, 6a, 6b.
1901. *Diphyphyllum caespitosum* (Hall) Lambe, p. 158, pl. 13, figs. 3, 3a,
        3b (*non Diplophyllum caespitosum* Hall, 1852).
Cf. 1930b. *Petrozium dewari* Smith, pp. 307-308, pl. 26, figs. 20-28.
1958b. *Petrozium losseni* (Dybowski), Kal'o, pp. 114-115, pl. 4, figs. 11-17.

Type material. — Lambe (1901, pl. 13, figs. 3, 3a, 3b) figured portions of what he described as the type of *Cyathophyllum pelagicum* Billings. Bolton (1960) in the catalogue of types in the collection of the Canadian Geological Survey listed the holotype of this species as missing, but a specimen in that collection (GSC 2351a) is labelled as the "holotype-lectotype". This specimen is figured in Pl. 25, figs. 2, 4.

Type locality and horizon. — Becscie Formation (Middle Llandovery) at Becscie River Bay, Anticosti Island, Canada.

Number of specimens examined. — Eight, including the type specimen.

Diagnosis. — *Petrozium* undergoing parricidal budding; tabulae complete, depressed axially, but convex peripherally, with a moat separating tabularium from a single layer of strongly convex vesicular dissepiments lining the theca.

Description. — The corallum is dendritic and apparently stood erect on the sea floor like a shrub. The largest specimen studied (Pl. 7, fig. 8) was originally at least 16 cm in maximum breadth and 13 cm in height, expanding markedly from its relatively narrow proximal region. The colony increased asexually by parricidal budding, with usually two, and occasionally three buds forming at intervals of approximately 1.5 to 2 cm. The corallites are longitudinally striate and somewhat rugose. Their diameters change but little throughout their length, and range from 4 to 8 mm, most falling about midway in this range.

The theca is a septal stereozone, about 0.25 mm thick.

The septa taper distally. They are of two orders, the major septa reaching almost to the axis, and the rather long minor septa often one-half to two-thirds the length of the major septa. The minor septa sometimes become contratingent on either the cardinal or the counter side of a neighboring major septum. The cardinal septum is not obvious, so that it is difficult to determine the sense of contratingency. In a corallite 5 mm in diameter, a total of 43 septa (both major and minor) were counted. In a section from another corallum, 50 septa were counted in a diameter of 8 mm.

The tabulae are distinctive, being complete, strongly convex, but clearly depressed in the axis. Laterally they bend downward sharply, leaving a moat between the convex axis and the periphery of the corallite. Lining the walls of the corallite is a single layer of strongly convex vesicular dissepiments, standing one upon the other. This layer may not be ubiquitous, however. In vertical sections it sometimes appears on one side of the corallite, but not on the other (Pl. 22, fig. 14).

Between the tabularium and this vesicular lining of the theca, concave supplementary plates lie in the moat. These may be continuous with the peripheral edges of the central tabulae, though this is uncertain. Their concavity and the depression of the peripheral edges of the tabulae form the moat. In one specimen the vertical spacing of the convex portions of the tabulae is about seven in 5 mm. In the depressed central areas, the vertical spacing is about two per millimeter.

Where parricidally-budded portions of the corallum are observed in longitudinal section, the wall formed between buds has a distinct dark midline.

Discussion. — This species was first described by Billings (1862) in a preliminary report, and again in 1865 (Billings, 1865a). Unfortunately, I have been able to obtain neither of these references.

Lambe (1901) described this species under the name *Diphyphyllum caespitosum* (Hall), and figured portions of what he called the type specimen of *C. pelagicum* Billings in his 1865a paper. He described the species as an aggregate of upright, cylindrical corallites, increasing by lateral gemmation. The corallites were 5 to 8 mm in diameter, frequently contacting each other laterally (note the two corallites in Pl. 7, fig. 8, top center), and were marked by longitudinal striae and annular growth-lines. The septa were of two orders, the major septa nearly reaching the axis, the minor septa about one-half as long. Depending upon the corallite diameter, the two orders totalled 40 to 50 septa. A single layer of vesicular dissepiments usually lined the inside wall of the corallite, appearing in transverse section as a second wall, less than 1 mm inside the theca. The tabulae were spaced about 2 per millimeter apart vertically, flat or concave in the axis, and downflexed at the periphery.

GSC 2351a (Pl. 25, figs. 2, 4), labelled as the type of the species, agrees well with Lambe's description. The corallites are cylindrical, about 6 to 7 mm in diameter, with a theca about 0.3 mm thick. The septa number about 47 of both orders in a corallite 7 mm in diameter, and have a distinct midline. The minor septa are about half the length of the major septa, and are occasionally contratingent upon them. The major septa reach nearly to the axis. The tabulae are complete, convex, with a flat or depressed axis, and sides that are downflexed. They touch a single layer of vesicular dissepiments lining the corallite wall. The tabular axes are spaced slightly less than 1 mm apart vertically, and the lateral dissepiments stand about 1 mm apart at their upper surfaces. This specimen agrees remarkably well with my material, the only difference being Lambe's statement that increase is by lateral gemmation. In my material, it is clearly parricidal, and the type material figured by Lambe does not clarify the mode of asexual reproduction in the Canadian form. Agreement is so close in all other respects that my inclination is to consider the two forms conspecific.

Dybowski (1874) described the species *Donacophyllum losenii* from the Tamsalu Stage (Middle Llandovery) of Dago Island (though Kal'o, 1958b, gave the locality as Hiiumaa Island, Estonia) in the Baltic. The specimens described are fragments 7 to 8 cm long, and 2 to 7 mm in diameter, in coralla resembling bundles of sticks. The corallites are commonly cylindrical, but through crowding often display a tetragonal cross section (not seen in my form). Asexual increase is reportedly by lateral budding. The theca is very thin(?) with distinct longitudinal striae and growth-lines.

In longitudinal section, the theca is lined with a single layer of vesicular dissepiments, each about 1 to 2 mm high, and 0.5 mm broad. The central 5 mm of a corallite with a total diameter of 7 mm is occupied by convex tabulae, spaced about 0.5 mm apart. They are axially flat or concave, and peripherally downflexed to meet the dissepimentarium. The septa total 50 in a diameter of 7 mm. The major septa do not quite reach the axis, and the minor septa are only a little shorter. As Lambe noted, the transverse view of a corallite suggests the presence of two concentric walls.

Dybowski's description and figures match my concept of *P. pelagicum*, except for the description of the mode of increase as lateral budding. Like Lambe, he showed the moat as floored with direct extensions of the downflexed tabular peripheries, whereas in my form the continuity is not so clear. The agreement in all other details is so good that it seems safe to consider Dybowski's form as conspecific with that in the Brassfield.

Kal'o (1958b) redescribed Dybowski's species from the type locality. The corallites are united in a faggot-like colony, and are cylindrical, 6 to 9 mm in diameter. On their outer surfaces, they show a regular pattern of swelling. The epitheca is thin, longitudinally-striate and wrinkled. The calice is shallow. The septa are thin, slightly wavy, and most of the major septa nearly reach the axis. The minor septa are half to three-quarters the length of the major septa. In the diameter range of 6 to 9 mm, the number of septa varies from 20 to 23 of each order. The septa are weakly carinate. The dissepimentarium is not present in all portions of the corallite, as was noted in my form. The dissepiments are oblong in vertical profile, about the same size as mine, or slightly smaller, and are usually present in one or two layers. The tabulae appear (as in the Brassfield specimens) convex and usually depressed axially, with their peripheral edges downflexed.

It is well to compare this species with the type species of *Petrozium*, *P. dewari* Smith, 1930b, from the Late Llandovery of Buildwas, Shropshire, England. The corallum of the latter species is dendroid, with cylindrical corallites that typically attain a diameter of 8 to 10 mm when fully-developed, but in one specimen reach 20 mm. The epitheca is thin, and lightly striate longitudinally. Asexual increase is reportedly marginal, and non-parricidal. The septa are thin, taper distally, and are moderately carinate. (Kal'o noted that *P. lossenii* is less carinate than *P. dewari*). The major septa sometimes reach the axis, and an axial structure of stereome may form among the ends of the major septa. The minor septa are about half as long as the major septa. In a diameter of 8 to 10 mm, there are 28 to 30 septa of each order, while in the corallite with a diameter of 20 mm, there are 40 of each order. The tabulae appear to be as in my form, strongly convex, but with occasional depressed axial areas. They are downflexed peripherally, and either they, or a separate series of concave supplementary plates, connect with a broad dissepimentarium that consists of three or four layers of dissepiments. *P. dewari* differs from my form in the much greater thickness of the dissepimentarium, the carination of the septa, and the formation of an axial structure of septal ends and stereome.

Distribution. — Anticosti Island, Middle Llandovery; Estonia, Middle Llandovery; Ohio, mid-Llandovery (Brassfield Formation).

Brassfield occurrence. — Localities 7 (near Fairborn), 14 (near Piqua), 15 (near West Milton), and 55 (near Ludlow Falls), all in southwestern Ohio.

Genus CRATEROPHYLLUM Foerste, 1909

1909a. *Chonophyllum (Craterophyllum)* Foerste, pp. 101, 102 (*non* Barbour, 1911; *nec* Tolmachev, 1931).
1926. *Naos* Lang, pp. 428-429.

Type species. — (SD Lang, Smith, and Thomas, 1940, p. 42): *Chonophyllum (Craterophyllum) vulcanius* Foerste (1909a, p. 101).

Type locality and horizon. — Brownsport Formation (Lower Ludlow) of Tennessee.

Diagnosis. — Large, solitary Rugosa with dissepimentarium; calice explanate, forming an outward-sloping apron around the tabularium; in this apron, major and minor septa indistinguishable, all dilated, turning septocoels into sutures; septa not extending into tabularium, where incomplete tabulae predominate, complete ones rare.

Description. — These are solitary corals of patelloid form. The calice wall is everted, except for a central calicular pit. The septa of the outer portions of the calicular platform are dilated, such that each is bounded laterally by sutures rather than septocoels. The septa occur in vertical layers, with vesicular dissepiments between them forming a dissepimentarium. The calicular pit is aseptate, as both orders reach only as far as the walls of the pit. Thus, the axial region of the coral consists of complete and incomplete tabulae, with no septal ends. This character is the most obvious difference between *Craterophyllum* and *Chonophyllum* Milne-Edwards and Haime, in which the major septa reach the axis.

Discussion. — While Foerste (1909a, p. 101) first used the name *Craterophyllum* parenthetically as *Chonophyllum (Craterophyllum) vulcanius*, suggesting a subgeneric status, at the end of his description of this species (p. 102) he stated

For these chonophylloid corals which have a flat base, upon which the calycular side arises in a manner resembling a low volcanic crater, the designation *Craterophyllum* is proposed. This term includes *Craterophyllum vulcanius* from the Brownsport bed, *Craterophyllum canadense* from the Anticosti Group, and an undescribed species from the Devonian limestone at Louisville, Kentucky.

Foerste's taxon is today used as a genus, whatever his original intention.

Lang (1926) described a new genus, *Naos,* based upon *Ptychophyllum pagoda* Salter (1873, *fide* Lang, 1926, p. 428) from the Silurian of Arctic Canada. This species has all the earmarks of *Craterophyllum* Foerste, including a vesiculose dissepimentarium containing successive layered platforms of dilated septa, with the septa reaching only to the margin of the calicular pit. The tabularium consists solely of complete and incomplete tabulae with septal ends entirely absent. As Hill (1956, p. F275) concluded, *Naos* appears to be synonymous with *Craterophyllum* Foerste.

Craterophyllum is compared morphologically with other corals of similar form in the discussion of *Schlotheimophyllum* Smith.

Craterophyllum (?) solitarium (Foerste)

1906. *Chonophyllum solitarium* Foerste, pp. 317-318, pl. 4, fig. 2.

Type material. — Location of the material figured and described by Foerste is not known.

Type locality and horizon. — Waco bed, Noland Member of the Brassfield Formation, from the road running north of Estill Springs, north of Irvine, Kentucky. This is approximately equivalent to localities 53-53a.

Number of specimens examined. — None found in the field. Apparently, Foerste's description and figure are the only occurrence of this species in the literature.

Description. — It is difficult to get a clear understanding of this species from the little information available. At first it was believed to be *Schlotheimophyllum patellatum* (Schlotheim), a species that occurs in the Brassfield Formation in Ohio. Judging by Foerste's description, these two species share a *Fungia*-like patellate form, and two orders of septa dilating and flattening toward the periphery of the horizontal calice wall, with a central calice depression. Foerste, however, reported dissepiments between the septa, whereas the septocoels of *S. patellatum* (and indeed, of nearly all members of the genus) are constricted to suture thinness by the septal dilation, and the internal structure is trabecular. It seems implicit in Foerste's description that there are no septa in the central calicular pit. This, together with the vesicular internal structure, suggests a species of *Craterophyllum* (*q.v.*). *Craterophyllum* was erected as a subgenus by Foerste in 1909, and had no formal status at the time Foerste described this species.

As described by Foerste, his specimen is at least 60 mm wide and about 25 mm high. A raised area with a relief of about 10 mm above the calice margin surrounds the central calicular pit, which is about 9 or 10 mm in diameter. Peripherally from this raised rim, the septa dilate and flatten, while axially they plunge down the wall of the pit, where they are thinner, and stand vertically. Between the septa are vesicular dissepiments. A total of about 60 septa, alternating in size, may be recognized. At the margins of the corallite, however, Foerste reported that the shallow grooves separating these septa number about 100, so that this is more likely the true septal number. The calice area outside of the central pit and rim is structurally " . . . a succession of calicinal plates resting upon intermediate vesicular tissue." The septa are represented by ridges on the calicinal plates. Foerste presented no information on the presence or absence of tabulae in the central portion of the corallite.

Distribution.— Kentucky: mid-Llandovery (Brassfield Formation).

Brassfield occurrence. — Waco bed, Noland Member, in the vicinity of localities 53 and 53a near Irvine, in central Kentucky.

Family HALLIIDAE Chapman, 1893

Solitary Rugosa with dissepimentarium of relatively small, globose dissepiments; septa thin in marginarium, at least in later growth-stages, but thickened and in axial contact in tabularium, thinning with upward growth first along axial edges and in counter quadrants; cardinal septum typically elongate.

Discussion. — In certain forms, such as *Pycnactis* Ryder, the septa are dilated throughout most ontogenetic stages, and appear to inhibit tabular and dissepimental development. Hill (1956) probably included *Pycnactis* in this family because Ryder (1926) regarded it as part of a continuum with *Phaulactis*, which fits more readily into the family Halliidae. Sokolov (1962), on the other hand, placed *Pycnactis* in the Streptelasmatidae, apparently due to its lack of a dissepimentarium.

As I have decided to follow Hill (1956) in family assignments for the Rugosa, *Pycnactis* (*q.v.*) is here placed with the Halliidae. At the same time, the problems that caused Sokolov to place this genus in a different family are noted.

Genus PYCNACTIS Ryder, 1926

1926. *Pycnactis* Ryder, p. 386.
? 1936. *Cymatelasma* Hill & Butler, pp. 516-517.

Type species. — (original designation of Ryder, 1926) *Hippurites mitratus* Schlotheim (1820, partim).

Type locality and horizon. — Silurian of Gotland.

Diagnosis. — Solitary Rugosa with dilated septa commonly occupying entire corallite cross section; cardinal septum longer than others; septa pinnately arranged; minor septa very short; tabulae present only at highest levels of corallite.

Description. — Members of this genus are solitary, trochoid rugose corals, with a septal stereozone.

From a very early stage, the septa are dilated, reducing the septocoels to sutures. This condition persists, as well as the occupation of the entire radius of the corallite by septa, until very near the base of the calice where the septa each come to an edge on their upper surfaces, allowing very shallow septocoels to form between them.

The cardinal septum is prominent from the brephic stage, and is the longest of the septa throughout most of ontogeny. It is located on the convex side of the corallite. The minor septa are quite short, less than half the length of the major septa. They occur as wedges between the proximal ends of adjacent major septa.

Tabulae are visible only in the highest levels of the corallite, but there is no evidence of a dissepimentarium.

Discussion. — Hill and Butler (1936) described *Cymatelasma* from the Silurian (Upper Llandovery to Lower Ludlow) of England. Noteworthy features of this genus are dilation of the septa from

the early stages of development, with gradual width reduction and
moderate withdrawal of the septa from the axis. The septa are
commonly wavy, and often carinate. In the later stages of development, as septocoels result from narrowing of the major septa (the
minor septa are quite short relative to the major septa), the septal
stereozone remains quite broad. The tabulae are complete and inversely conical. There are no dissepiments.

The similarities between *Pycnactis* and *Cymatelasma* are
strong. Both show marked dilation of the septa during most of the
ontogeny of the corallite, but in later stages this dilation is less in
Cymatelasma than in *Pycnactis*. Hill and Butler do not make it
clear if the cardinal septum in their genus is as prominent as in
Pycnactis. The material from the Brassfield resembles *Cymatelasma*
in the pattern of dilation-reduction of the septa, and the retention
of a broad stereozone, and resembles *Pycnactis* in the prominence
of the cardinal septum throughout development of the coral. It is
tempting to regard the Brassfield specimen as a link between these
two genera, justifying the contention of Soshkina and Kabakovich
(1962, p. 318) that they are synonymous. In the absence of further
information, however, it seems best to reserve judgement on this
matter. Largely on the basis of the development of the cardinal
septum, my material is assigned to *Pycnactis* Ryder, though *Cymatelasma*-like features are present.

Pycnactis tenuiseptatus, sp. nov.   Pl. 7, figs. 4, 5; Pl. 26, figs. 1-8

Cf. 1926. *Pycnactis mitratus* (Schlotheim), Ryder, pp. 386-390, pl. 9, figs. 1-7;
text-fig. p. 389.
Cf. 1926. *Pycnactis mitratus,* var. *grandis* Ryder, p. 390.
Cf. 1926. *Pycnactis rhizophylloides* Ryder, p. 390, pl. 9, fig. 8.

Type material. — Holotype: UCGM 41963.

Type locality and horizon. — Locality 48, southwest of Owingsville, central Kentucky, in the Brassfield Formation.

Number of specimens studied. — One (the holotype).

Diagnosis. — *Pycnactis* with the dilation of the major septa de-
creasing in the ephebic stage, leaving a more open axial area than
is usual in this genus; elongate carinae present on sides of septa.

Description. — The corallite is canine tooth-shaped, compressed
laterally, trochoid, with a sharp, edged base. It is weakly rugose,
and longitudinally striate. The specimen studied measures 5 cm

along the convex side, which curves through about 90° of arc. The cross section is elliptical, of greatest diameter in the plane of curvature. Throughout the lower 3 cm of its length (along the convex side) the corallite increases gradually in diameter, until its cross section measures about 16 × 13 mm. Beyond that, there is little change in the diameter, except at the distal-most end, where the calice appears to have been crushed.

In transverse section, the theca is a septal stereozone, in which the sutures between adjacent septa are defined by zigzag lines. This reflects the presence of interlocking rib-like carinae that run from the peripheral edge of the septum to the axial edge in an upward direction. Their orientation is precisely what would be expected of the trabeculae making up the septa, and the carinae probably represent thickening of these trabeculae. Their spacing is about 8 to 9 in 3 mm. In one longitudinal section (Pl. 26, fig. 1), small circular features just distal to the ends of the carinae suggest spines projecting from the sides of the septa into the septocoels, similar to those in many of Sutherland's (1965) Henryhouse corals, and in *Ptychophyllum? cliftonense* Amsden (1949) from the Brownsport Formation. Such lateral processes, however, are not definitely homologous with the structures of my specimen.

The earliest stages revealed in thin-section show the septa reaching to the axis, with one septum on the convex side of the corallite (taken to be the cardinal septum) distinctly thicker and longer than the others. The septocoels are nonexistent, due to septal dilation. Septa appear to be intercalated along the sutures between extant septa. Just below the level where the corallite's rate of expansion decreases sharply, the septocoels first become apparent, but are closed distally by the axial juncture of the septa (Pl. 26, fig. 5). At this level there appears to be little or no twisting of the axial ends of the septa, and no involved axial structure is produced. The cardinal septum is no longer distinctive by its size.

Above this level, the distal ends of the septa draw away from the axis, leaving an open space. The major septa now are slender (maximum of 0.25 mm across, tapering distally) and curved, with the concave side toward the cardinal septum. The minor septa appear to be restricted to the stereozone, which has grown quite thick (up to 3 mm). The highest section reveals the axial space wider still, with the major septa retaining concave cardinal sides. Minor septa now appear to protrude more from the stereozone, sometimes one-third to one-half the length of the adjacent major septa. Both orders of septa retain a tapered form, narrowing axially. What appears to be stereome has fused many of the distal ends of septa on the cardinal side of the corallite. At no stage is there a prominent cardinal fossula.

In longitudinal section (Pl. 26, figs. 1, 2) it is seen that, beginning at the level where the septa withdraw from the axis and the septocoels become open, the interior of the corallite is occupied by abundant incomplete tabulae, rising toward the axis to produce what must have been a convex calice floor. These incomplete tabulae appear as transverse lines in the septoceols, that are usually concave toward the periphery of the corallite.

Discussion. — Ryder (1926) gave a fairly detailed description of *Pycnactis mitratus*, his type species for *Pycnactis*. This species was originally described as *Hippurites mitratus* by Schlotheim (1820), but as I was unable to obtain the original reference my comparison is with Ryder's specimens from the Wenlock of England. Most noteworthy in these is that septal dilation almost completely squeezes out the septocoels until the later stages of development, when the septa withdraw slightly from the axis. As a result, there apparently is no development of tabulae or vesicular dissepiments. These features have generally been considered characteristic of the genus. *P. mitratus*, then, does not have septocoels to the extent encountered in my specimen, nor a comparable development of tabulae. It also lacks the thin, curving septa seen in later stages of the Brassfield specimen.

Nonetheless, the similarities between my specimen and Ryder's genus are unmistakable. The early stages are identical, with dilated septa filling the whole interior, and a prominent cardinal septum attaining the axis. The dark midlines of the septa are identical in the Brassfield and English specimens. Even the more open stages of the interior of *P. mitratus* are similar to mine, with the septa tapering axially. The mode of septal insertion is also identical. Finally, but probably least conclusive, the exterior of the English form is slightly striate longitudinally, with feeble annulations occasionally developed, as in my specimen. It is concluded that the Brassfield form is indeed congeneric with *P. mitratus*, though not conspecific.

Also from the Wenlock of England, Ryder described a new variety of this species, *P. mitratus grandis*, differing from more typical specimens of the species in attaining a much larger size. From the Wenlockian of Gotland, Ryder described the species *Pycnactis rhizophylloides*, apparently distinctive only in the flattening of its convex (cardinal) side, somewhat in the manner of *Rhizophyllum*. The Brassfield specimen shows no such flattening, and in fact, is elliptical in all cross sections.

Distribution. — Kentucky; mid-Llandovery (Brassfield Formation).

Brassfield occurrence. — Locality 48, southwest of Owingsville, central Kentucky.

Genus SCHIZOPHAULACTIS, gen. nov.

Cf. 1926. *Phaulactis* Ryder, p. 392.

Type species. — *Cyathophyllum densiseptatum* Foerste (1906, pp. 314-315).

Type locality and horizon. — mid-Llandovery; Noland Member of the Brassfield Formation, central Kentucky.

Diagnosis. — As for species.

Description. — As for species.

Discussion. — Details of internal structure show that *Schizophaulactis* is closely allied with *Phaulactis* Ryder, the only significant difference between them being the mode of formation of new septa. In *Schizophaulactis* this appears to happen by the splitting of new septa from the proximal ends of extant septa (see Text-fig. 10), whereas in *Phaulactis* new septa arise through simple intercalation between the proximal ends of extant septa (e.g., *Phaulactis cyathophylloides* Ryder, 1926, the type species of *Phaulactis*, in Ryder's pl. 11, figs. 1, 2).

It is possible that what is taken as an unusual mode of septal origin in *Schizophaulactis* is really a form of septal contratingency. Contratingency can be of variable importance in the different species of a single genus (*e.g., Paliphyllum* Soshkina), so that *Schizophaulactis* may prove to be synonymous with *Phaulactis* Ryder.

Schizophaulactis densiseptatus (Foerste)   Pl. 7, figs. 6, 7; Pl. 8, figs. 1-3; Pl. 26, figs. 9-13; Pl. 36, fig. 6

1906. *Cyathophyllum densiseptatum* Foerste, pp. 314-315, pl. 6, figs. 2A-F.
Cf. 1969. *Streptelasma eccentricum* Neuman, pp. 25-28, figs. 20A-F, 21A-H.

Type material. — Foerste designated no type, but noted that the specimens he showed in his plate 6, figures 2A-C were "typical specimens". With these three specimens he showed three more which he described as "possibly young specimens of the same species". The latter three specimens most closely approximate my material. Foerste's figured specimens have apparently been lost. Several specimens in the collection of the U.S. National Museum, however, appear identical to mine, and as these museum specimens are labelled in what seems to be Foerste's handwriting, the Brassfield material is identified with this species. A neotype, UCGM 41944, was selected from the specimens found during this study (Pl. 8, fig. 3; Pl. 26, figs. 9-13).

Type locality. — Locality 1, approximately 1/2 mile east of Panola, Kentucky, in the Waco bed of the Noland Member, Brassfield Formation (same locality and horizon as Foerste's Pl. 6, fig. 2E).

Number of specimens studied. — Forty-one, all collected in the field during this study.

Diagnosis. — Solitary Rugosa in which septa form as protuberances on the proximal ends of pre-existing septa where they join the theca, and split off from those septa toward the axial end; dissepimentarium present; in ephebic stage, proximal ends of septa thin, so that septa are thickest near axis; ontogenetically, septa thin first on counter side of corallite.

Description. — This species includes ceratoid corals of varied form. Usually they strikingly resemble a canine tooth, curving through as much as 90° of arc in the cardinal-counter plane. This tooth-like aspect is emphasized in some specimens in which the cross section is elliptical, with greatest diameter in the plane of curvature, giving "edges" to the "tooth".

In a representative specimen, the diameter reaches 1.4 × 1.1 cm at a level 2.5 cm above the base of the corallite (measured along the convex side). Beyond this level, the diameter increases very little, measuring 1.5 × 1.2 cm at 6 cm elevation. Other specimens, however, are less cylindrical (Pl. 7, figs. 6, 7; Pl. 8, figs. 1, 2). One such corallite reaches a diameter of 2.2 $\times$ 2.0 cm in 4 cm of length.

The theca appears to be a septal stereozone, though it is difficult to discern its structure due to the generally worn condition of the specimens.

The corallites studied frequently showed proximal edges of the septa along their weathered sides, between which were series of much thinner vesicular dissepiments of the marginarium. These septal edges are often paired, due to the mode of septal formation: splitting of a new septum from the proximal edge of an extant one, with gradual distal separation until there is no longer contact between the two (Text-fig. 10). It is also possible to interpret this as contratingency of each septum in its early stage, with gradual elimination of its contratingency ontogenetically.

In well preserved specimens there is a slight axial boss formed of loosely intertwining axial ends of the septa, with what appear to be incomplete tabulae among them. The proximal edges of the septa, showing their characteristic bifurcating pattern, rise to the edge of the calice, leaving a low area of calice floor midway between the theca and the axial boss.

The cardinal septum is on the convex side of the corallite. The cardinal fossula is not especially prominent until the later stages, and is generally more evident in rapidly expanding individuals (Pl. 8, figs. 1, 2). In one specimen in which it is well-developed, the fossula contains three septa: the cardinal, which traverses the fossula below the level of the calice floor, and two other septa, one on either side of the cardinal, that fuse laterally to the cardinal side of their respective neighboring septa.

In the neanic stage of ontogeny (Pl. 26, fig. 13) the septa reach nearly to the axis, but usually leave a small open space in their midst, occasionally crossed by a single septum. As growth proceeds, the septa appear to withdraw from the axial area in a manner typical of streptelasmatids, leaving a greater proportion of the transverse aspect occupied by the open axis. This space is invested with a spongy structure formed by the thinned distal septal edges, occasionally uniting, with interspersed incomplete tabulae. Between the septa in the marginarium is a zone of vesicular dissepiments forming a dissepimentarium of rather narrow proportions.

During the neanic stage, as the corallite expanded regularly, the septa were broader than the septocoels. During the ephebic stage (Pl. 26, figs. 10-12; Text-fig. 10), however, as the septa begin withdrawing from the axial area, the septa on the counter side become thinner. The proximal portion of the septa on the cardinal side also thin from their proximal ends to about one-fourth the distance to the axis. This leaves a zone around the corallite of thin skeletal elements, less resistant to weathering than the inner area of thicker septa (Pl. 26, figs. 10, 11; Pl. 36, fig. 6), so that the distal end of the corallite is often rounded on its outer edges, and the calice area appears a bit narrower than the areas just below it. This pattern of peripheral thinning of the septa may be seen in other rugose genera (cf. *Phaulactis*, p. 507, fig. D, and *Aulophyllum*, p. 513, fig. G in Hill, 1935; *Bothrophyllum*, p. F293 in Hill, 1956).

The septa appear to begin as swellings on the proximal end of the cardinal side of an extant septum, and split away from that septum in a proximal-to-distal direction, until the new septum is completely free. The intermediate stages of septal development cause the extant septum to appear to bifurcate proximally.

The pattern of septal insertion was observed in one specimen from which ten transverse sections were cut (Text-fig. 10). It was possible here to trace many of the septa of the later stages back to their origins in earlier stages. The lowest level studied was about 0.7 cm above the base and contained 24 septa. The latest stage studied was about 2.5 to 3 cm higher and contained 34 septa. Septal insertion appears to occur in four positions: to either side of the cardinal septum, and at positions about 90°, 180° and 270° from this location.

Text-figure 10. — Series of ten transverse sections, more or less equidistantly spaced, from a specimen (UCGM 41998) of *Schizophaulactis densiseptatus* (Foerste) 1906 from locality 1, near Panola, Kentucky, in the top of the Waco bed, Noland Member, Brassfield Formation. All sections are mounted right-side up, with the convex side of the corallite toward the top of the figure. The sections are arranged in descending order, from *A* (highest in the corallite) down to *J* (lowest in the corallite). The septa have been numbered to show my interpretation of the pattern of insertion and development of the septa, and of other features of the internal structure. The numbers continue throughout the series of sections for the same septum. Septum number 1 appears to be the cardinal septum. Where a septum has first been fully inserted at a given level, its designating number is circled.

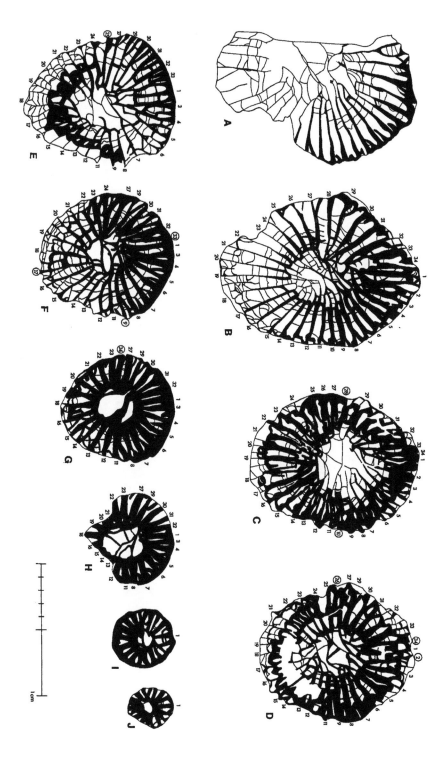

The presence of minor septa could not be clearly demonstrated. However, at a diameter of about 1.5 × 1.0 cm, probably well into the ephebic stage, the newly-formed septa seemed to remain short (about 1/5 to 1/6 the length of the extant septa), and occupied positions in the septocoels between the 34 pre-existing major septa. Unfortunately, the specimen was not long enough to show the further development of these possible minor septa.

Discussion. — Foerste (1906) described this species from specimens found near Panola (locality 1) and Irvine (localities 53-53a) in central Kentucky, in the top of the Noland Member, Brassfield Formation. His specimens appear to consist entirely of the more rapidly expanding variety of the species. He described this form as attaining a length of 75 mm measured along the convex side, and a diameter of 30 mm. His specimens, as mine, were worn on the exterior surface, exposing the proximal septal edges and vesicular dissepiments in the septocoels. The calice is rather broad and shallow, 8 to 9 mm deep in specimens 28 mm in diameter. What Foerste regarded as full-grown specimens had about 90 septa (presumably of all orders), with only the "primary septa reaching the center where the ends are more or less twisted and slightly raised above the immediately surrounding part of the bottom of the calyx." A somewhat inconspicuous cardinal fossula is located on the convex side of the corallite, containing one major and two "secondary" septa. The septocoels contain abundant dissepiments, curving upward and outward. The septa in cross section were described as varying "between slightly convex and V-shaped on the side facing the exterior of the corallum". Presumably, Foerste was here referring to the bifurcative appearance produced in the process of forming new septa. His mention of major and minor septa is difficult to understand, unless he felt that one of each pair of septa in a bifurcation was a minor septum. Foerste's specimens are not definitive in this respect, but do appear, in places, to show this "splitting" pattern. Specimens in the reference collections of the U.S. National Museum, identified by Foerste as *C. densiseptatum*, clearly are of the same form as my specimens. The number of septa in my form is comparable to that reported by Foerste. One of my specimens, the widest of these collected in this study, with a calice 2 cm in diameter at a level 4 cm above the base, has 76 septa. There is little question that the material collected is identical to the species described by Foerste.

Neuman (1969, pp. 25-28) reported *Streptelasma eccentricum,* a form bearing certain similarities to the present species, from the Late Ordovician 5a beds at Herøy, in the Langesund-Skien area of Norway. The minor septa are more distinct than in my specimens, and they are contratingent to the cardinal side of the adjacent major septum. There is no evidence, however, that the septa form as swellings on the sides of extant septa, as they appear to do in the Brassfield form. The axis seems to be vesiculose, consisting of highly-contorted incomplete tabulae, as in my material. But the Norwegian specimens lack the dissepimentarium, the differential thinning of the septa, the periphery of thin skeletal elements, and the mode of septal formation characteristic of *S. densiseptatus.* Also, Neuman's species has an eccentric axis, which mine lacks. Clearly these are distinct species, belonging to different genera, but sharing some superficial features.

Distribution. — Kentucky, mid-Llandovery (Brassfield Formation).

Brassfield occurrence. — Locality 1 (near Panola), and locality 53 (near Irvine), both in central Kentucky. Questionable occurrence at locality 34 near Raywick, Kentucky (UCGM 42000).

Family CYSTIPHYLLIDAE Milne-Edwards & Haime, 1850

Rugosa with dissepimentarium; septa acanthine, trabeculae typically holacanthine, often extending onto the surfaces of the tabularium; tabulae typically display axial sag, even attaining inversely-conical form; dissepiments usually smaller than tabellae of the tabularium.

Genus CYSTIPHYLLUM Lonsdale, 1839

1839. *Cystiphyllum* Lonsdale, p. 691.
1851. *Conophyllum* Hall, p. 399.

Type species. — (SD Milne-Edwards & Haime, 1850, p. lxxii): *Cystiphyllum siluriense* Lonsdale (1839, *partim, non* pl. 16 *bis,* fig. 2).

Type locality and horizon. — Middle Silurian of England; according to Lang, Smith, and Thomas (1940), Wenlock Limestone of Dudley, Worcestershire, England.

Diagnosis. — Solitary and compound rugose corals, sometimes bearing basal talons; corallites cylindrical or ceratoid, not tetragonal;

septa consist of separate trabeculae, restricted to the thickness of one layer of dissepiments; tabellae of tabularium not readily distinguishable from vesicles of dissepimentarium, so that these two zones appear continuous; calice inversely conical, without axial boss.

Description. — *Cystiphyllum* includes solitary and compound rugose corals of cylindrical or ceratoid form, in which the interior is filled with vesicular dissepiments and tabellae, with understated septa. Most specimens, in fact, appear to contain nothing but dissepiments internally under superficial examination. The calice is fairly deep, and lacks an axial boss or structure. The tabularium and dissepimentarium are not markedly differentiated.

The septa consist of rows of separate, trabecular elements radiating along the upper surfaces of the vesicular dissepiments and tabellae. These elements are short, and do not generally penetrate more than one layer of vesicles.

Basal attachment structures may be present.

Discussion. — Hall (1851), in a preliminary paper, described the genus *Conophyllum*, based upon his *C. niagarense* from the Silurian of New York State. In 1852 (pp. 114-115) he repeated his description of the genus, and gave his first description and figures for the type species. Apparently based on observations of weathered specimens, Hall interpreted the internal structure of the corals as a "cone-in-cone" of successive transverse "septa" (tabulae). His figures, however, suggest the presence of vesicular dissepiments in the peripheral zone, an interpretation more acceptable after the examination of weathered specimens of *Cystiphyllum*. While Hall recognized the similarity between *Cystiphyllum* and his genus, as indicated in both his 1851 and 1852 descriptions, he nonetheless considered them separate entities.

Today, *Cystiphyllum* and *Conophyllum* are generally regarded as synonymous, with priority belonging to the former. The evidence appears to support this view.

Cystiphyllum spinulosum Foerste

1906. *Cystiphyllum spinulosum* Foerste, pp. 321-322, pl. 5, figs. 1A-K.

Type material. — Location of Foerste's specimens is unknown.

Type locality and horizon. — Foerste's figured specimens are from two localities in central Kentucky: along the road north of Estill Springs, north of Irvine, Ky. (equivalent of my localities 53-53a), and half a mile east of Waco, where the road to Cobb Ferry turns off from the pike to Irvine; Waco bed, Noland Member, Brassfield Formation.

Number of specimens studied. — None found in field. Analysis based entirely on Foerste's description and figures.

Description. — Foerste's description follows:

> Corallum simple; moderately curved or nearly straight. Length of specimens found so far less than 50 millimeters. Exterior of corallum covered by an epitheca which is marked by longitudinal grooves, dividing the surface into low, narrow linear ridges as in species having conspicuous septa; septa, however, obsolete. Of these longitudinal ridges, about six occur in a width of 5 millimeters. Exterior also with fine transverse, radiating striae. Epitheca frequently absent, owing to weathering. In that case the vesicular structure is well exposed. Blister-like cavities, large and conspicuous, frequently 5 millimeters long and of equal or even greater width. The plates forming this vesiculose structure are convex above but concave as seen on the side of weathered surfaces of the coralla, since they curve upward and outward. The concave surface of these plates is marked by low longitudinal striations which correspond in frequency to the septal striae on the exterior of the corallum. Calyx only partially exposed in the larger specimens at hand, apparently of moderate depth; formed by the convex sides of the uppermost layer of plates. In most specimens the convex side of these plates as exposed in the calyx is covered with coarse granules interspersed with smaller granules; these may take the place of the obsolete septa; in a few specimens they are arranged in approximately radiate lines; in others this arrangement can not be readily detected. While the usual appearance of these markings on the upper side of the plates is that of coarse granules, on well-preserved specimens they frequently are detected as short spines fully a millimeter in length. These spines have been seen in so many specimens that it is evident either that the coarse granules of weathered specimens represent the bases of the weathered spines, or that both spines and coarse granules occur in this species.
>
> This species may be distinguished by the coarsely vesicular structure, the septal striations on the lower side of the plates, and the coarse granules and short spines on the upper side.

Discussion. — The information and figures provided by Foerste are sufficient only to determine that his specimens belong to the genus *Cystiphyllum* Lonsdale. The characters he pointed out as distinctive (coarsely vesicular structure, septal striations, coarse granules, and spines) are found in most species of *Cystiphyllum*. If additional material can be found at the Brassfield localities from which Foerste obtained his specimen, detailed study of the corallites should help determine their relationships on the specific level. For now, it is safe only to say that *Cystiphyllum* does occur in the Brassfield.

Distribution. — Kentucky, mid-Llandovery (Brassfield Formation).

Brassfield occurrence. — Foerste's figured specimens are from near Waco, Kentucky, and from near Irvine, Kentucky (the equivalent of localities 53-53a of this study). He also reported this species from near Panola, Kentucky (the equivalent of my locality 1). In all cases, the specimens are from the Waco bed, Noland Member of the Brassfield Formation.

TABULATA

Family **FAVOSITIDAE** Dana, 1846

(*Nom. transl.* Milne-Edwards & Haime, 1850 [*ex* Favositinae Dana, 1846]).

Tabulate corals with massive cerioid coralla, hemispherical to ramose, consisting of narrow, prismatic corallites; corallite walls perforate on wall faces, in corners, or in both positions; when present, longitudinal elements consist of spines or tubercles; squamulae may be present, not necessarily aligned longitudinally; tabulae mostly complete, with incomplete tabulae only occasionally present; intercorallite walls with distinct midline.

Genus **FAVOSITES** Lamarck, 1816

1816. *Favosites* Lamarck, p. 204 (*non* Nicholson, 1874, p. 253).
? 1851. *Astrocerium* Hall, p. 399 (*nom. nud.* — No described species included).
1852. *Astrocerium* Hall, p. 120.
Cf. 1906. *Boreaster* Lambe, p. 323.
? 1914. *Parallelopora* Holtedahl, p. 13 (*non* Bargatsky, 1881).
? 1918. *Actinopora* Vinassa de Regny, pp. 98-99.
? 1930. *Parafavosites* Orlov, p. 122.
? 1934. *Paralleloporella* Strand, p. 271 (*pro* *Parallelpora* Holtedahl, 1914, p. 13; *non* Bargatzky, 1881).
1934. *Sapporipora* Ozaki, p. 74 (*in* Shimizu, Ozaki, & Obata).
1936. *Eufavosites* Rukhin, p. 96.
? 1937. *Asteriophyllum* Porfiriev, pp. 30, 33.

Type species. — (SD Milne-Edwards & Haime, 1850, p. lx): *Favosites gothlandicus* Lamarck, 1816, p. 205.

Type locality and horizon. — Silurian of Gotland.

Diagnosis. — Corallum massive, with cerioid arrangement of corallites, producing a "honey-comb" appearance; dark midline in walls separates contiguous corallites; tabulae generally complete; septal spines vary in development, but never form lamellar septa; mural pores located only on faces of corallite walls.

Description. — This genus consists of coralla with simple tubes joined laterally, sharing common walls which are split visually by a midline. The tubes are traversed by tabulae which are mostly complete, and their wall faces are pierced by one to three longitudinal rows of mural pores.

The walls bear (to varying degrees) longitudinal rows of septal spines, which are never fused into continuous laminae. These spines can be short or nearly absent, or they can be quite long, rectilinear, or even curved (e.g., *F. hisingeri,* which see). Corallite diameter may be fairly uniform or show great variation even in a small area of the surface (e.g., *F. discoideus, q.v.*). Also, it may vary over "fields", with areas of narrow corallites grading peripherally into areas of broader ones (e.g., *F. hisingeri*).

Discussion. — *Favosites* is probably one of the most problematic coral genera, species descriptions suggesting that variation within this group is poorly understood. Additionally, the taxonomy of *Favosites* presents a difficulty shared by other tabulate groups: that the hard parts may inadequately reflect the specific variations of the soft parts.

Sutton (1966) tried to gauge the value of some traditional *Favosites* and *Paleofavosites* taxobases. In *Favosites gothlandicus* Lamarck he found that abundance of spines and thickness of corallite walls showed marked variation, even within individual coralla. Abundant spines and thick walls are distributed rhythmically, corresponding to concentric (seasonal?) bands of closely-spaced tabulae. These characters, then, are of inadequate diagnostic use, at least in *F. gothlandicus.* Of greater value for this purpose, Sutton determined, is the average diameter of the *adult* corallites (defined by Sutton (text-fig. 1) as the average between the shortest wall-to-wall diameter, and the longest corner-to-corner diameter within a corallite). Sutton (p. 257) felt that "the one hundred largest corallites, in transverse sections which contained upwards of one thousand corallites, could certainly be considered as adult". He found this average diameter to be fairly consistent within coralla that varied in other characters. This character also proved remarkably consistent between different coralla, whether from the same, or from different horizons of the Wenlock Limestone. Results of this study support Sutton's view of the value of corallite diameter as a taxobasis in *Favosites.*

The generic name *Favosites* was first used by Lamarck in 1816 (p. 204), with a description that is brief, but in accord with the modern concept of this genus. No mention, however, was made of mural pores, an important feature of the corallum. Lamarck described two species, *F. alveolata* and *F. gothlandica*. Milne-Edwards and Haime (1850) designated the latter as type species, and Jones (1936, p. 8) chose a neotype for this species.

Calamopora Goldfuss (1829) has traditionally been regarded as a junior synonym of *Favosites* Lamarck, but Oekentorp (1971; see discussion under *Paleofavosites*) has shown that Goldfuss' genus has a type species which belongs to *Paleofavosites* Twenhofel (1914). This automatically transfers *Calamopora* to synonymy with *Paleofavosites*. Oekentorp's (1971) applications to the International Commission on Zoological Nomenclature for suppression of *Calamopora* Goldfuss in favor of *Paleofavosites* Twenhofel (*via* the plenary powers) was successful (see discussion under *Paleofavosites*).

In 1852 (and 1851 as a *nomen nudum*), Hall introduced *Astrocerium*, a tabulate coral genus of the New York Silurian similar to *Favosites*, but containing long, thin spines, distally inclined upward. He included four species in this genus (1851, 1852), and examination of some of his type material at the American Museum of Natural History (see Pl. 29, figs. 1, 2, and discussion under *Favosites hisingeri*) suggests they may be conspecific. The relatively narrow corallite diameter, wide mural pores (on the wall faces) and elongate, upwardly pointing septal spines appear to be an inadequate taxobasis on the generic level. Hall's original species, and subsequently described "species" appear to intergrade in morphology to some degree, and more work is needed to determine if they really are separate species. If they are, they have more in common with one another than with any other species of *Favosites*, thus constituting a morphologic cluster of coral species that could reasonably be the basis of a separate genus. If, however, they intergrade with one another to the extent of likely constituting a single species, it is questionable if that species would differ sufficiently from the other *Favosites* species to justify a separate genus.

Lambe (1906) introduced the genus *Boreaster*, based upon material collected by A. P. Low in Arctic Canada in 1904 during a Canadian Dominion Expedition. Both Hill and Stumm (1956, p. F463), and Lang, Smith, and Thomas (1940, p. 28) attributed a

Silurian age to the material on which Lambe based his new species, *B. lowi*, the type species of this genus. Lambe noted that the corallum is generally similar to that of *Favosites*, except that it possesses lamellar septa of two orders. He showed mural pores on the walls on either side of the midpoint of each wall. Flat tabulae are also present. The nature of the septa seems to preclude the inclusion of this form in *Favosites* Lamarck, though the similarities are clear. In addition, the two orders of septa, if present, are unusual in tabulate corals.

Actinopora Vinassa de Regny (1918) is based upon Devonian coral species in which the point of juncture between several corallites takes the form of a "cross" or a five-rayed "star", which apparently is a slightly thickened area. The rays can extend for over one-third the width of a wall. The three species on which Vinassa de Regny based this genus (one from Central Asia, and two from the Carnic Alps on the northeastern border of Italy) agree with *Favosites* in all other respects. Another generic name was erected by Porfiriev (1937) (see below) for Russian favositids with this same peculiarity. The validity of a separate generic designation in the case of the Italian material will have to await further study.

From the Upper Silurian of Fergana (in Turkestan), Orlov (1930) described *Parafavosites*, a tabulate coral form which differed from *Favosites* by the presence of narrow tubes, which lack tabulae and spines, in the corallite walls at the junctures of 3 or more corallites. Opinions differ as to the status of this genus. Bassler (1944) and Hill and Stumm (1956) considered it valid, while Sokolov (1962) regarded it as a junior synonym of *Favosites*. It is possible that the small tubes represent parasites infesting the coral, but pending further study, the status of *Parafavosites* Orlov remains uncertain.

In 1914, Holtedahl had proposed the name *Parallelopora* for specimens with the same features as *Parafavosites*, from the Silurian of Ellesmereland, Arctic Canada. Strand (1934) noted that this name was preoccupied, and modified it to *Paralleloporella*. Bassler (1944) provided a discussion of this genus, which he considered a junior synonym of *Parafavosites*.

Ozaki (1934, *in* Shimizu, Ozaki, & Obata) introduced the name *Sapporipora* for a *Favosites*-like form from the Silurian of northwestern Korea, characterized by rather narrow corallites (about 0.5

to 0.7 mm), wide pores (about one-third the width of the wall) occurring mostly in a single row on the wall face, and the appearance of new corallites usually at the junction of four pre-existent corallites. None of these features would prevent inclusion of this form in *Favosites* Lamarck.

In his report on favositids from the Lower Devonian of Transbaykal (U.S.S.R.), Rukhin (1936) used the term *Eufavosites* as a subgeneric designation, apparently to denote those *Favosites* species which are close in form to Lamarck's original concept (as he applied the name to *F. gothlandicus*, the type species).

Porfiriev (1937) described *Asteriophyllum*, based on a new species from the Devonian rocks of the eastern slope of the Ural Mountains, U.S.S.R. It is characterized by the thickening of the juncture of three or four corallites, with a dark *Y*- or *K*-shaped figure appearing in the mass of calcareous deposits, as seen in transverse section (cf. *Actinopora*, discussed above in this section). As figured and described by Porfiriev, these dark "characters" do not necessarily have their lines coincident with the orientations of the midlines of the corallite walls, and so, are not necessarily thickened portions of the midline of the wall. As with *Parafavosites* cited above, the status of this genus with respect to *Favosites* remains uncertain. It possibly is a case of parasitically induced thickening of the walls, though it might also represent a meaningful generic taxobasis.

Favosites favosus (Goldfuss)   Pl. 8, figs. 4-6, 9; Pl. 27, figs. 1-4; Pl. 42, figs. 3, 4

1826. *Calamopora favosa* Goldfuss, pp. 77-78, pl. 26, figs. 2a-c.
Non 1852. *Favosites favosa?* (Goldfuss), Hall, p. 126, pl. 34A, figs. 5a-g.
? 1875. *Favosites favosa* (Goldfuss), Nicholson, p. 229.
1876. *Favosites favosus* (Goldfuss), Rominger, pp. 21-22 (*partim*), pl. 4, figs. 3, 4 (figs. 1, 2 are questionable. as is pl. 5, fig. 2).
1887. *Favosites favosus* (Goldfuss), Davis, pl. 8, fig. 1 (*non* fig. 2).
Non 1890. *Favosites favosus* (Goldfuss), Foerste, p. 333.
? 1919. *Favosites favosus* (Goldfuss), Williams, pl. 18, fig. 1.
? 1925. *Favosites favosus* (Goldfuss), Hume, pl. 9, fig. 1.
Non 1926. *Favosites* cf. *favosus* (Goldfuss), Wilson, p. 16, pl. 3, fig. 7.
? 1928. *Favosites favosus* (Goldfuss), Twenhofel, p. 127 (*partim*).
Cf. 1936. *Angopora hisingeri* Jones, pp. 18-19, pl. 2, figs. 4-7; pl. 3, figs. 1, 2.
Non 1937. *Favosites* cf. *F. favosus* (Goldfuss), Teichert, pp. 129-130, pl. 6, figs. 5, 6; pl. 7, fig. 2.
1937. *Favosites favosus* (Goldfuss), Teichert, p. 130, pl. 7, fig. 1 (redescription of the holotype).
? 1939. *Favosites favosus* (Goldfuss), Shrock & Twenhofel, pp. 254-255, pl. 28, figs. 3, 4.

1962a. *Favosites favosus* (Goldfuss), Norford, pl. 9, fig. 13.
1962b. *Favosites favosus* (Goldfuss), Norford, p. 28, pl. 5, figs. 6, 7, 11.
1964. *Favosites favosus* (Goldfuss), Stumm, pp. 62-63, pl. 68, fig. 4.
? 1965. *Favosites crassimuralis* Poltavtseva, p. 47, pl. 11, figs. 1, 2 (*non* Zhizhina, 1957, *in* Zhizhina & Smirnova).
? 1966. *Favosites favosus* (Goldfuss), Bolton, pl. 6, fig. 12.
1966. *Favosites favosus* (Goldfuss), Bolton, pl. 8, figs. 1, 5.
? 1968. *Favosites favosus* (Goldfuss), Bolton, pl. 13, figs. 19.

Type material. — The holotype is a silicified specimen in the collection of the Geological and Paleontological Institute at the University of Bonn, Germany, designated as the original of Goldfuss' plate 26, figure 2. It was refigured and redescribed by Teichert (1937, p. 130, pl. 7, fig. 1), and is also treated in the present work (Pl. 8, figs. 4-6, 9; Pl. 27, figs. 1, 2; Pl. 39, figs. 3, 4).

Type locality and horizon. — From the Silurian strata of Drummond Island, in Lake Huron.

Number of specimens examined. — Four collected in the field, plus the holotype.

Diagnosis. — *Favosites* with distinctly convex tabulae; spines abundant on corallite walls and on surfaces of tabulae; mural pores seem to occur in two rows on the wall faces; in many (but not all) cases, edges of tabulae are fluted by a series of marginal fossulae; corallite diameter apparently variable between coralla, ranging from 2 mm (and possibly as low as 1 mm) up to about 5 mm.

Description. — The coralla from the Brassfield appear to have had convex upper surfaces, and to have grown in a centrifugally radial manner. The base of one corallum shows periodic annulations, doubtless reflecting periodicity of tabular spacing. If a holotheca was originally present, it has not been preserved. The best preserved specimen (UCGM 41957) was originally about 2.5 cm thick and 11 cm in maximum width.

The corallites range in diameter from about 2.0 to 2.5 mm, measured from the midline of the wall. They are mostly six-sided, but other forms can be seen. The walls in some areas of each of the coralla appear thickened (up to 0.8 mm across), but this has proven to be due to lateral fusion (presumably secondary) of the rather abundant, thick septal spines. These spines (Pl. 27, fig. 3) are spaced about four per millimeter vertically, and lie both normal to the walls, and projecting upward axially. The walls themselves may be fairly thick, as much as 0.3 mm.

The most striking feature of this form is the strong upward convexity of the tabulae, in many cases with the vertical relief equaling the distance across the base of the tabula. They are fairly close-spaced, usually 1.6 to 2.0 per millimeter, though they may be as sparse as one per millimeter, or so close as to produce incomplete tabulae. The upper surfaces of the tabulae contain spines with axes normal to the tabulae surface (Pl. 27, fig. 3; represented as dots in the corallite lumens in fig. 4). It is unclear if there is an orderly arrangement to these tabular spines, but in at least one case, they seem to describe part of a circle concentric with the walls of the corallite.

Possible mural pores were observed in thin-section. They occur on the wall faces, rather than in the corners, but the number of rows is unclear. In some cases, the gaps in the walls are wide (0.6 mm) and centered on the wall face, suggesting uniseriality. In others, they are narrower and off-center, as though part of a biserial, or even triserial arrangement of pores. In some specimens the wall bordering the pore appears thickened, but the preservation of my specimens is too poor to allow confident interpretation of the fine skeletal structure.

It was not possible to establish the presence of peripheral fossulae in the tabulae of the Brassfield specimens, possibly due to recrystallization. Examination of a specimen identified as *Favosites favosus* (OSU No. 16199: from the Middle Silurian exposed at Monticello, Iowa; in nearly every way identical to the holotype) indicates that these fossulae vary in their development, and are not present in every corallite of a colony. A search through the literature indicates this form is quite distinctive, and the Brassfield specimens agree so well with the holotype that it seems reasonable to assign them to Goldfuss' species. Nonetheless, the variations in corallite diameter in different coralla may eventually justify splitting *F. favosus* into two or more species.

Discussion. — The holotype of *Calamopora favosa* Goldfuss is apparently the remnant of a broadly convex colony. What remains is about two-thirds of the specimen figured by Goldfuss. It measures slightly less than 3 cm in maximum thickness, with a maximum width of 7 cm. Probably, its original diameter was at least 10 cm. The corallites are 3 to 4 mm in diameter, most falling midway between these extremes. Their walls are not very thick, about 0.2 mm across, and they are generously furnished with spines, some of which project upward axially (though Teichert, 1937 reported an absence of septal spines). The tabulae are distinctly convex upward, their relief commonly up to one-third their diameter. In rare instances the periphery of the tabulae may be convex, while the center is depressed and sharply concave, resulting in a cone or bulb on the underside of the tabula (Pl. 8, figs. 4, 9). This appears to be the feature seen on a number of spots in Goldfuss' figure 2a, and it is thus likely that this figure shows the specimen upside-down. It is clearly also the feature seen on the tabulae in his figure 2b, and here the view is described as basal. Nonetheless, this axial depression of the tabulae appears exceptional and nearly all tabulae have smooth, convex upper surfaces. The upper surfaces of the tabulae, and the walls are invested with spines, though these are not visible in all cases. In a few corallites, seen from above, the periphery of the tabula are downflexed into about 12 marginal fossulae. The tabulae are generally spaced vertically 1.0 to 1.5 per millimeter. Their spacing seems to vary rhythmically, but not enough material is available to be certain. Mural pores are present, but it is not possible to add to Teichert's statement that they do not seem to be located in the corners. Goldfuss (fig. 2c) showed them arranged biserially on the wall faces, and other specimens apparently belonging to this species bear this out.

Hall's (1852) *Favosites favosa?* (*sic*), from the "Niagaran" (probably meaning fossiliferous Silurian) at Milwaukee, Wisconsin, clearly does not belong to *F. favosus* (Goldfuss). The tabulae are horizontal, with no evidence of spines either on their surfaces or on the walls. There is no indication of marginal fossulae on the tabulae, nor are mural pores mentioned. Hall stated that he chose to refer his specimens to Goldfuss' species because they came from a locality close to, and an age approximating that of the latter's fossils.

Nicholson (1875) reported a species from the "Niagara Formation" (probably Brassfield or Dayton Formation) at Dayton, Ohio whose characters generally match those of *Favosites favosus*: corallite diameter is 2.1 to 4.2 mm ("one to two lines"); tabulae spaced 1.5 per millimeter, "usually more or less conspicuously curved, with their convexities directed upwards"; mural pores are biserially arranged on the wall faces; septal spines are present. There is no mention of marginal fossulae, or of spines on the tabulae. Though his description parallels well the features of this species, necessary information is lacking, and so Nicholson's material can only be questionably considered conspecific with *Favosites favosus*.

Rominger (1876) described several specimens under the name *Favosites favosus*. Two of these (pl. 4, figs. 3, 4) are certainly conspecific with Goldfuss' species. They are from Drummond Island, Lake Huron (the type locality of the species), and probably are from the Manistique Group (Late Llandovery to Middle Wenlock). The corallite diameters are, in one case about 2 to 4 mm, and in the other, about 3 to 5. The tabulae are convex and spinose, and have distinct marginal fossulae. The walls bear septal spines.

Rominger figured three other specimens from the same locality (pl. 4, figs. 1, 2; pl. 5, fig. 2), all of which have tabulae with marginal fossulae. In these cases, however, the tabulae are not distinctly convex, and there is no mention of spines on the tabulae. These show a variation of corallite diameter (between colonies) ranging from 2 to 6 mm. If these are to be included in *Favosites favosus*, despite their more-or-less flat tabulae, it must be assumed that the species has an enormous variability in corallite diameter, tabular topography, and fossular development, so much that its limits are difficult to set. Rominger recognized the possibility of such variation, and made the purely functional observation that

... on careful examination of large collections of these forms, we find them all so much linked together by intermediate gradations that no line can be drawn between one and another.

Davis (1887) figured two specimens which he attributed to *Favosites favosus*. His plate 8, figure 2 is a lateral view, and shows horizontal tabulae. Stumm (1964) figured the specimen of Davis' figure 1 (USNM 52654), which has corallites 4 to 5 mm in diameter, convex, fossulated tabulae, and apparently biserial mural pores. Stumm noted that the tabulae are elevated above and depressed between the mural pores. The lack of mention of spines on the tabulae might be due to poor preservation, but it seems quite likely that this specimen belongs to *F. favosus*. Davis had recorded the locality as the Niagaran ferruginous clay near Louisville, Kentucky, but Stumm questioned this, noting that the lithology is very similar to that of the Manistique Dolomite of Michigan (Late Llandovery to Early Wenlock). This possibility seems greater, since Davis mentioned that the specimen had been collected by Nettelroth, rather than himself. Stumm also disagreed with the identification of the specimen in figure 2, and called it *Emmonsia tuberosa* Rominger. Foerste (1890) mentioned specimens that he attributed to *Favosites favosus*, found at Collinsville, Alabama, and at Fair Haven, Ohio. In the case of the Alabama specimens, not enough information is given to allow identification in the absence of an illustration. In the case of the Ohio specimens, the information is again too meager, but the mention of horizontal tabulae makes it unlikely that these specimens are *Favosites favosus*.

Williams (1919) figured a specimen identified as *F. favosus* with large corallites (about 4 to 6 mm across) and fossulate tabulae from Fossil Hill, Manitoulin Island (probably the Fossil Hill Formation, equivalent in age to the Manistique Group of northern Michigan). It is quite possible that the identification is correct, but this must remain in question until the specimen (GSC 5118) is examined.

The identification of Hume's (1925) specimen from the "Lockport formation" of the Lake Timiskaming region of Canada must also remain questionable. Most of the corallites are about 3.5 mm in diameter, and some appear to have fossulate tabulae, but without more information, the identification as *Favosites favosus* cannot confidently be accepted. Teichert (1937, p. 130) examined the holotype of *Calamopora favosa* Goldfuss, and saw good agreement between it and Hume's figure, but there is no indication that he had examined Hume's specimen.

Wilson's (1926) specimen from the Late Ordovician of British Columbia is almost certainly not *F. favosus*. Little information is given in the description and figure, and nothing is told of the tabulae. The great variation in corallite size (0.5 to 1.5 mm) in a small area (about 1 square centimeter) make conspecificity with the present species highly doubtful.

Twenhofel (1928) reported *F. favosus* from the Gun River, Jupiter and Chicotte Formations (Middle Llandovery to Early Wenlock) of Anticosti Island. He apparently gave a composite description, noting corallite diameters of 4 to 5 mm, with septal spines present. Marginal fossulae of the tabulae are variably developed, distinct in some, hardly visible in others. Mural pores are usually arranged biserially. The tabulae may be convex, flat, or concave. The information, though sparse, generally agrees with my concept of this species. The mention of concave and flat tabulae, however, leaves the identification open to question, especially in the absence of a figure. Further, Twenhofel did not mention in which of the above three formations the tabulae are flat or concave, so that while *F. favosus* probably does occur on Anticosti Island, it is not possible at this time to determine its horizon.

Jones (1936) described a new genus and species, *Angopora hisingeri*, from the "Övre Visby Margelsten at Högklint, south of Visby, Gotland" (probably the Högklint beds of early Wenlock age). The holotype has mostly hexagonal corallites that attain a fairly uniform diameter of about 1 mm and have moderately thick walls. Characteristic of this species are short lamellar septa which do not necessarily persist through the length of a corallite, and whose axial edges break up into long spines that sometimes reach more than halfway to the axis. The mural pores are biserially arranged, generally located near the corners of the corallite. Jones figured several specimens from England and Gotland, one of which (pl. 2, fig. 7) has convex tabulae, with a hint of the presence of spines on their upper surfaces. Except for the discontinuity of the septa, their partly lamellar nature, and the smaller size of the corallites, it is similar to the Brassfield form. But these differences seem important enough to warrant regarding Jones' specimens as similar to, but not identical to mine. Jones' European material is more similar to a Kazakhstan form which was described by Poltavtseva (1965), discussed below.

Teichert's (1937) *Favosites* cf. *F. favosus* from the Silurian of the Brodeur Peninsula on Baffin Island, Canadian Arctic, is clearly not referable to *F. favosus* (Goldfuss). The corallite diameter is about 2.8 mm (some smaller), but the tabulae are much more closely spaced (2.2 per millimeter) than has been reported in *F. favosus* (Goldfuss). Further, the tabulae are dominantly horizontal. There is no mention of marginal fossulae on their surfaces. Spines are apparently not well-developed on the walls (Teichert reports no spines present) and the tabulae seem to be free of spines. While this specimen does not seem referable to *F. favosus,* it is probably referable to another Brassfield form, *Favosites* sp. A (which see).

Shrock and Twenhofel's (1939) specimen from the Pike Arm Formation of northern Newfoundland (belonging to a group of sediments ranging in age from Llandovery to Middle Ludlow), superficially appears not to be conspecific with *F. favosus*. Its corallites are 4 to 6 mm in diameter and have generally horizontal tabulae spaced 1 to 2 per millimeter. Spines are well-developed on the wall, but there is no mention of them on the tabulae. The flatness of the tabulae and absence of tabular spines make it difficult to equate this specimen with the species described by Goldfuss, but mention is made of "peripheral pits" in the tabulae where they join the wall. It thus seems best to leave the identification open to question.

Norford (1962a, 1962b) figured and described apparent *Favosites favosus* from the Sandpile Group (considered by Norford on faunal evidence to be Late Early, or Early Middle Silurian) of northern British Columbia. The specimen figured in Norford's 1962a paper was shown only in surface view, but apparently has convex tabulae with marginal fossulae. In his other paper (1962b), Norford refigured this specimen (fig. 6) and showed a transverse and longitudinal section of a second specimen clearly displaying convex tabulae. The corallites were described as 4 to 5 mm in diameter, with circular mural pores arranged biserally for the most part, numerous septal spines spaced about 5 per millimeter longitudinally, and tabulae (some flat) about 1 mm apart. No mention was made of spines on the tabulae.

Perhaps the most interesting occurrence of a specimen possibly belonging to this species is that of *Favosites crassimuralis* Poltavtseva (a junior homonym of a species described by Zhizhina *in* Zhizhina & Smirnova, 1957), reported in 1965 from the Wenlock of Kazakhstan. The corallites are five- and six-sided, with diameters of 0.8 to 1.0 mm. Rare four-sided corallites occur as well. The walls are 0.2 mm thick, with a well-defined midline. The tabulae are convex in the figure, though Poltavtseva reported them to be horizontal or weakly sagging (perhaps axially, as seen in Goldfuss' specimen?), and are spaced from one to five, but usually three per millimeter. Though it is not mentioned in the text, spines appear to occur on the surfaces of the tabulae. The mural pores are 0.16 mm in diameter, arranged biserally. The acanthine septa consist of long (0.3 to 0.4

mm), thick spines pointing upward axially. There is no mention of marginal fossulae in the tabulae. For this reason and because of the unusually small diameter of the corallites, the identification of *F. crassimuralis* Poltavtseva with *F. favosus* Goldfuss' must remain tentative and questionable. In all other respects however, they are alike. There is even closer agreement with the Brassfield specimens, whose diameter of 2 mm is closer to the 1 mm of the Kazakhstan specimens, and in which the presence of fossulae has also not been demonstrated. It may be said that *Favosites crassimuralis* is more similar to *F. favosus* and to the Brassfield form, than to any other described species. If it does belong to Goldfuss' species, it is the only reported occurrence of *F. favosus* outside of North America.

Favosites crassimuralis is quite similar to *Angopora hisingeri* Jones (discussed above). The corallite diameter, mural pore distribution, and prominence of spines are all similar, and one of Jones' figures shows convex tabulae. No indication is given by Poltavtseva that the septa of her species are comparable with those of Jones' specimens, so that the relationship between them must await direct comparison of the specimens.

Bolton (1966, 1968) figured specimens of what appears to be *Favosites favosus* from the Fossil Hill Formation (Middle Llandovery to Middle Wenlock) of Manitoulin Island. In all cases the marginal fossulae can be seen, and in plate 8, fig. 1 (1966), where the view is lateral, the mural pores appear to be biserially distributed on the wall faces. The tabulae are convex. Corallite diameter ranges from about 4 to 7 mm.

Distribution. — Drummond Island (Lake Huron), Late Llandovery or Early or Middle Wenlock; Anticosti Island, within the range of Middle Llandovery to Early Wenlock; Manitoulin Island (Lake Huron), within the range of Middle Llandovery to Middle Wenlock; northern British Columbia, Late Llandovery or Early Wenlock; Ohio and Kentucky, mid-Llandovery (Brassfield Formation); Iowa, Middle Silurian (based upon OSU 16199 from Monticello, agreeing closely with the holotype).

Questionable occurrences include Lake Timiskaming (eastern Canada), Wenlock; northern Newfoundland, Silurian; Louisville, Kentucky, Late Wenlock or Early Ludlow (if Davis' locality data is correct); Kazakhstan, Wenlock.

Brassfield occurrence. — Localities 1 (near Panola), 46 (near College Hill) and 53a (near Irvine), all in central Kentucky, in the top of the Noland Member; and locality 13 (Todd Fork, near Wilmington, Ohio) in the top portion of the Brassfield Formation.

Favosites hisingeri Milne-Edwards & Haime Pl. 8, fig. 8; Pl. 28, figs. 1, 2; Pl. 29, figs. 1, 2; Pl. 37, figs. 5, 6; Pl. 39, fig. 4

1843. *Porites* ————? Hall, pp. 86, 91, pl. 22, figs. 3, 4.
1851. *Favosites hisingeri* Milne-Edwards & Haime, p. 240, pl. 17, figs. 2, 2a, 2b.
? 1851. *Astrocerium venustum* Hall, p. 400 (*nomen nudum*).
? 1851. *Astrocerium parasiticum* Hall, p. 400 (*nomen nudum*)
? 1851. *Astrocerium pyriforme* Hall, p. 400 (*nomen nudum*).
1852. *Astrocerium venustum* Hall, p. 120, pl. 34, figs. 1a-i.
? 1852. *Astrocerium parasiticum* Hall, p. 122, pl. 34, figs. 2a-i.
? 1852. *Astrocerium pyriforme* Hall, p. 123, pl. 34a, figs. 1a-e.
1854. *Favosites hisingeri* Milne-Edwards & Haime, Milne-Edwards & Haime, p. 259, pl. 61, figs. 1, 1a, 1b.
1875. *Favosites venusta* (Hall), Nicholson, p. 226.
1876. *Favosites venustus* (Hall), Rominger, p. 23, pl. 5, fig. 3.
1887. *Favosites niagarensis* Hall, Davis (*partim*), pl. 8, fig. 3 (*non* fig. 4; *non* Hall, 1852).
1887. *Favosites venustus* (Hall), Davis, pl. 9, figs. 7, 10.
? 1890. *Favosites venustus* (Hall), Foerste, p. 335.
1902. *Favosites fidelis* Počta (incorrectly ascribed *to* Barrande by Počta), (*partim*), pp. 227-228, pl. 105, figs. 5, 6; pl. 106, figs. 1, 2; (*non*? pls. 83, 88, 89, 94).
1902. *Favosites fidelis* var. *clavata* Počta (*partim*), pp. 228-229, pl. 83, figs. 11-15; pl. 105, figs. 7, 8; pl. 106, fig. 5 (*non* pl. 90).
1902. *Favosites intricatus* Počta (incorrectly ascribed *to* Barrande by Počta), pp. 233-235, pl. 88, figs. 11-18; pl. 91, figs. 10-12; pl. 95, figs. 1-12; pl. 102, figs. 2, 3.
1906. *Favosites hisingeri-aplata* Foerste, pp. 299-300, pl. 2, fig. 2; pl. 4, fig. 5.
? 1919. *Favosites hisingeri* Milne-Edwards & Haime, Williams, pl. 17, fig. 2.
1928. *Favosites hisingeri* Milne-Edwards & Haime, Twenhofel, p. 129.
1933. *Favosites hisingeri* Milne-Edwards & Haime, Tripp, pl. 13, figs. 1, 2a-c, 3, 4; text-figs. 35, 36.
1936. *Favosites hisingeri* Milne-Edwards & Haime, Jones, pp. 17-18.
1937. *Favosites hisingeri* Milne-Edwards & Haime, Teichert, pp. 128-129, pl. 5, fig. 3; pl. 6, fig. 2.
1939. *Favosites hisingeri* Milne-Edwards & Haime, Shrock & Twenhofel, p. 255, pl. 30, figs. 24, 25.
? 1939. *Favosites hisingeri* Milne-Edwards & Haime, Northrop, p. 150.
1964. *Astrocerium venustum* Hall, Stumm, p. 60, pl. 58, figs. 7-9, 14-16.

Type material. — Lectotype (by Jones, 1936, p. 17): the specimen figured by Milne-Edwards and Haime, 1854, pl. 61, figs. 1, 1a (*non* 1b). Repository unknown.

Type locality and horizon. — Benthall Edge, England. Age uncertain, possibly Wenlock.

Number of specimens examined. — Approximately 8, including the holotype of *Favosites hisingeri-aplata* Foerste.

Diagnosis. — Favosites with corallites of fairly uniform diameter and aspect, usually 1.0 mm or narrower, rarely 1.5 mm across; corallites of different diameters tend to occur in fields which grade at their borders into other fields of different diameters; spines well-developed, often reaching nearly to corallite axis, and upflexed distally; mural pores abundant, about 0.25 mm in diameter.

Description. — In the Brassfield Formation this species occurs as hemispherical coralla, the largest about 18 cm in maximum diameter, and 10 cm thick.

The corallites are characteristically quite narrow, more so than those of the other Brassfield favositids. The largest are slightly over 1 mm in diameter, but most are narrower, down to about 0.5 mm. The narrower corallites tend to occur in fields (Pl. 28, fig. 1), doubtless areas of rapid proliferation, though some can be seen scattered among the broader corallites. This occurrence of such fields seems to be an important diagnostic character of the species.

The corallites are irregular in cross-sectional form, some being six-, five-, or four-sided, while a few are round. In addition, some are compressed into an oblong shape, while others are equidimensional.

The tabulae are generally horizontal, and show periodicity of spacing (Pl. 28, fig. 2). In the denser areas they are at most 0.25 mm apart, while in the sparser regions they may be as much as 1.0 mm apart. The average spacing is about 0.5 to 0.6 mm, and there are generally an average of 10 to 14 tabulae in a vertical space of 5 mm.

The corallites contain spines which, when well-preserved, appear long and thin and sometimes reach the axis. The number and arrangement of the spines could not be determined, nor could it be ascertained whether they were horizontally directed, or projecting upward axially, as is characteristic in this species.

The mural pores are seldom seen in Brassfield specimens, but in transverse section, gaps that may represent these pores suggest they are located in the center of the walls.

Discussion. — The earliest valid use of this name is by Milne-Edwards and Haime (1851), who assigned the following characters to the new species: corallum with somewhat swollen surface; calices only slightly unequal, in general regularly polygonal, and separated by somewhat thickened walls; corallite diameters approximately 1.5 mm; transverse section of each tube shows 12 septal spines that are subequal, moderately thickened, gently curved upward at their distal ends, and nearly reaching the corallite axis; tabulae thin, rather close-spaced, horizontal or a bit undulose. The figures accompanying this description agree with the Brassfield specimens. The corallites shown are mostly less than 1.5 mm across, and are fairly uniform in diameter within a given area. There are about nine tabulae in a vertical distance of 5 mm, and while there is a suggestion that their spacing is rhythmically zoned, this is not especially prominent. The tabular spacing and arrangement are as in my specimens. The identification of the Brassfield material with this species seems secure.

Milne-Edwards and Haime, in this reference, reported this species from the "Silurien (inférieur)" of England (Tortworth), and the "Silurien (supérieur)" of Gotland, England (Wenlock Edge, Benthall Edge), and America (Niagara). They also reported it from what they questionably called the Devonian at Perry County, Tennessee. The first reference ("Lower Silurian") may actually be Ordovician, while the "Upper Silurian" may refer to the Wenlock. The Tennessee reference may possibly be to the Early Ludlow Brownsport Formation, though to my knowledge, this species has not been reported from the Brownsport.

Specimens from western New York referable to the present species were figured by Hall (1843) in his report on the geology of the Fourth District of New York State. His reference was to "*Porites* ————?" from the Niagara Group (possible Lockport Dolomite of Late Wenlock through Middle Ludlow age). Hall showed what appear to be two different specimens, one a probable natural vertical section. The corallites are a bit less than 1 mm in diameter, with close-spaced tabulae, apparently zoned. Most of the corallum of one specimen seems to have been dissolved, leaving the more densely tabulate zones as distinct layers with nothing between them.

In a preliminary exposition of the new genera of fossil corals formally presented in his 1852 work, Hall (1851) first used the name *Astrocerium venustum*. The name, however, was not elaborated upon or illustrated, but was simply listed with three other species names as examples of his new genus *Astrocerium*. As these were then undescribed species, *A. venustum* is a *nomen nudum,* and priority for the species name should go to *Favosites hisingeri* Milne-Edwards and Haime, 1851.

Hall's first use of the name *Astrocerium venustum* in conjunction with a description was in 1852. Along with other figures, he repeated his 1843 figure showing a corallum with the more sparsely tabulate areas dissolved away, and referred his 1843 material, including this specimen, to *A. venustum*. Hall mentioned the narrowness of the corallites, showing specimens with tube diameters slightly over 1 mm, and also remarked on the 12 "ascending spiniform rays" (axially rising acanthine septa), which are also illustrated. The tabulae are shown as close-spaced, and Hall noted the remarkable rhythmic zonation of their spacing.

The best-preserved of Hall's syntypes (AMNH 1470/2, shown here on Pl. 29, figs. 1, 2) is in agreement with Hall's figures as to corallite diameter. A vertical section (Pl. 29, fig. 1) shows the great abundance of the spines, about five in 1 mm vertically, and their upward distal curve is clearly seen. The tabulae are about half the thickness of the corallite walls. Mural pores are abundant, about 3 per millimeter vertically, and are about 0.25 mm in diameter. They appear evenly spaced, and in at least one instance, seem to occur biserially on the wall faces. Hall's material comes from the "Niagara Limestone" (probably the Lockport Dolomite, Late Wenlock through Middle Ludlow in age) at Lockport (site of the above specimen) and Rochester. This specimen (AMNH 1470/2) is here chosen as lectotype of *Astrocerium venustum* Hall, and is considered conspecific with my material, and with *Favosites hisingeri* Milne-Edwards and Haime.

Astrocerium parasiticum Hall (1852) is similar to *A. venustum*, but differs from it primarily in having markedly narrower corallites interspersed among the full-sized ones. Also, there appear to be more spines in a given cross-sectional area than in *A. venustum* (and in Brassfield *F. hisingeri*), and its corallites seem to be a bit broader. It is quite possible that *A. parasiticum* is identical to the present species, as variation in this form is poorly understood. Surely they are closely related, but it is best to consider the relationship open to question. *A. parasiticum* was reported by Hall from the Niagara Group, in the "limestone near Lockport" (Lockport Dolomite?).

A third species described by Hall (1852) is *Astrocerium pyriforme*. From my concept of *Favosites hisingeri*, it seems to differ only in its characteristic pear-shaped form. There is a suggestion that the tabulae are more wide-spaced than is characteristic in my form, as Hall's figure 1d shows them separated by a distance greater than the tube diameter. Only a very short length of three tubes is shown however, and similar spacing is seen in limited portions of my specimens. In light of the unusual shape of the corallum, it is best at this time to leave the relationship between *A. pyriforme* and *F. hisingeri* open to question. Hall's specimens come from his Niagara Group, in the Rochester Shale (Wenlock) of Rochester, Wolcott, and Lockport, New York, and more rarely in "the limestone" (Lockport Dolomite?, Late Wenlock through Middle Ludlow) at Rochester and Lockport, New York.

In 1854, Milne-Edwards and Haime repeated their 1851 description of *F. hisingeri* (this time in English, rather than the original French), and figured two specimens, one from Benthall Edge, and the other from Wenlock Edge, England, probably of Wenlock age. In this description, the authors expressed the opinion that *A. venustum* Hall was a synonym of their species.

Jones (1936) designated as lectotype of *Favosites hisingeri* the specimen figured by Milne-Edwards and Haime (1854, pl. 61, fig. 1). Figure 1a is another view of the same specimen, but figure 1b is a different specimen.

Nicholson (1875) reported *Favosites venusta* [*sic*] from the Clinton Group (probably the Brassfield Formation) of Yellow Springs, Ohio. The corallum was large, hemispherical or spheroidal, with layering that probably reflected tabular zonation. The corallites were slender, ranging from about 1 mm down to about half a millimeter in diameter. The smaller ones were intercalated among the larger. No mention was made of the smaller corallites occurring in fields. Nicholson noted that the septa were formed of "very short spiniform projections". The tabulae were complete, horizontal or flexuose, about two or three per millimeter. No information was given concerning mural pores. Doubtless this form is *F. hisingeri*, as it agrees with my specimens in all respects save the reported shortness of the spines and the lack of reference to "fields" of corallites of similar diameter. The latter is probably more important than the former, as the spines could easily have been affected by diagenetic alteration. The distribution of the corallites in fields according to diameter is not developed to the same degree in all specimens and might be missed. Consequently, it seems likely that Nicholson was dealing with *F. hisingeri.*

Rominger (1876) reported *Favosites venustus* from the Niagara Group (probably actually the Manistique Group, of Middle Llandovery through Middle Wenlock age) on Drummond Island, northern Michigan. The specimen he figured (UMMP 8442) appears identical to the Brassfield form, even showing fields of smaller corallites. Rominger described the corallites as narrow, not over 1 mm wide, and of fairly equal size. The septa consisted of 12 rows of long spinules. The tabulae were close-spaced, flat or gently convex. The mural pores occurred in one, or rarely two rows on each wall of the corallite. Rominger noted that the corallum occurred as a large, massive expansion, with an epitheca on the base. In all respects, the Drummond Island form appears to be identical to that of the Brassfield.

In his work on the fossil corals of Kentucky, Davis (1887) reported two species, *Favosites venustus* and *Favosites niagarensis,* of which the former, and some specimens of the latter are apparently identical to *F. hisingeri.* These specimens were from the Louisville Limestone (Latest Wenlock and Earliest Ludlow) of Louisville, Kentucky.

Stumm (1964) described the Louisville material and illustrated some of Davis' hypotypes. He reported that the coralla are discoid to hemispherical, with the corallites averaging 1 mm across. The septa are formed by 12 rows of distally rising spines, extending halfway to nearly the entire distance to the axis. Mural pores are usually uniserially arranged. The tabulae are both complete and incomplete, flat or convex, close-spaced (about 4 per millimeter). Stumm considered this form different from *Favosites hisingeri,* remarking that the corallites of the latter are wider, averaging 1.5 mm across, and its tabulae not as close-spaced. He referred his material to *Astrocerium venustum* (which I consider a junior synonym of *F. hisingeri*). Stumm overestimated the diameter of the corallites in *F. hisingeri* Milne-Edwards and Haime and the tabular spacing has not yet proven sufficient grounds for separating species. Consequently, the Louisville corals are considered conspecific with the present species.

Foerste (1890) reported *Favosites venustus* from the Brass-field Formation at Ludlow Falls and Fair Haven, southwestern Ohio. The corallites were 0.6 to 1.2 mm in diameter in specimens from Ludlow Falls, and reach a maximum of 1.3 mm at the second locality. In the Ludlow Falls material, the tabulae were five in 2.4 to 3.2 mm vertically, whereas in the second locality they are more closely spaced, 5 in 1.9 mm. The walls were "scarcely crenulated" in the Fair Haven specimens. Spines were reported only in a few of the specimens. The vagueness of Foerste's description makes it best to reserve judgement on the identity of these corals.

Počta (1902), in his continuation of Barrande's work on the Lower Paleozoic of Bohemia, described three coral forms that re-semble *F. hisingeri*:

Počta attributed *Favosites fidelis* to Barrande, but as there is no firm evidence of a description of this species in works by the latter, (except perhaps in manuscript), the authorship should be as-signed to Počta. This form is distinctive for its small corallites and the distribution of the pores. The corallum resembles a loaf of bread: it is massive, flattened, thick, and convex underneath. No trace of a holotheca remains. The corallites reach a diameter of 1.0 to 1.5 mm, forming fairly regular hexagons in cross-section. Their walls are thin, and finely wrinkled transversely, and contain a medial line running along their centers.

Spines are frequently encountered, and vary in number. They sometimes reach the axis of the corallite. Počta noted that these spines may be present in some tubes of a corallum, while absent from others in the same specimen.

The mural pores are numerous, very small, and rounded. They occur in three to five longitudinal rows, and are also aligned hori-zontally. The tabulae are complete and closely spaced, with 1.6 to 1.8 per millimeter. Their spacing is rhythmically zoned, as shown in some of the figures (*e.g.*, pl. 105, fig. 6; pl. 106, fig. 2). The figures of plates 105 and 106 can be confidently identified with *Favosites hisingeri* and the Brassfield material. These specimens are from horizon e2 of Barrande (which, because of errors in Barrande's stratigraphic interpretations, includes assorted rocks of the Wen-lock, Ludlow, Prídoli and basal Lochkov, fide Kriz & Pojeta, 1974, p. 491). Those of plates 83, 84, 89, and 94 are more questionable. It should be noted, however, that the mural pores are fairly large (especially on pl. 94), similar to those in *A. venustum* Hall (Pl. 29, fig. 1 of the present work).

The second of Počta's forms (this time not attributed to Barrande) is *Favosites fidelis* var. *clavata*, also from horizon e2 of Barrande. The corallum is massive, and convex distally. Počta referred to a "thick trunk" which gives the corallum the appearance of a club, but this is not shown in the figures. (This may be similar to the pear-shaped *Astrocerium pyriforme* Hall, 1852.) No basal holotheca was present on the specimens Počta studied.

The corallites are polygonal, of about the same size over the entire surface (0.8 to 1.3 mm across). The wall between tubes is 0.2 mm thick, and a dark midline runs through its middle, delimiting the portion of the wall belonging to each corallite. Some walls are undulose in transverse section.

Many of the corallite tubes are empty, while others have 5 to 12 septal spines (in transverse view), straight and fairly uniform in size. In broken corallites, the spines appear distributed without order, but they are absent in tubes with pleated walls, as if their function were served by the wall folds.

The tabulae are usually horizontal, but sometimes oblique. On average, there are 25 per centimeter. The mural pores are ordinarily in three rows, but where the tube is narrow, there are only two.

As this form shares the essential features of my own, they are considered conspecific. Only the specimens of plate 90, especially figure 4 (showing small knobs on the upper surfaces of the corallite walls) seem questionable.

The third form, *Favosites intricatus* (attributed by Počta to Barrande, apparently invalidly), is from the f2 horizon (Early Devonian, *fide* Kriz and Pojeta, 1974, pp. 490, 491, mostly Koneprusy and Slivenec Limestones of the Pragian). Characteristically, this form has an irregular surface, with corallite tubes oriented in many directions as shown by the remains of what must have been several protuberances on a corallum. This produced a complex pattern of corallite orientations on the resulting cobble. Počta noted that large specimens often consist of several coralla.

The corallites are of fairly uniform diameter, varying between 0.7 and 1.0 mm, with a mean of 0.8 mm. They are separated by walls about 0.5 mm or more thick. Often a dark median line is visible in the walls.

Počta noted that spine patterns fall into the following categories:

1) They are totally absent, or very short (0.12 mm), sparse and blunt.
2) They are well-developed, terminating in a blunt point. This category is further divided into two types:
   a) Spines stout, about 0.12 mm long, with broad bases, narrowing gradually distally toward a blunt point, their number about equalling the number of sides of the corallite.
   b) Spines more slender, longer (up to 0.18 mm), more numerous than the sides of the corallite (often 10 to 15 in transverse view).

Počta considered that these two categories of spinal pattern occur as variations in a single species, rather than marking separate taxa. This is because both patterns occur in a single corallum, and are not accompanied by correlated differences, other than the fact that those corallites with abundant spines have thicker walls than those devoid of spines. The spines are usually elevated distally, an important feature in *F. hisingeri*.

The tabulae are complete, often curved, and generally "bulging downward" (concave?). About 35 occur vertically in 1 centimeter. The mural pores are quite large, occurring in single and double vertical rows. Their round apertures are spaced vertically about 32 per centimeter.

This form appears to have all the important features of *Favosites hisingeri* and is considered conspecific with the Brassfield form.

Foerste (1906) described what he considered a "variety" of the present species, *Favosites hisingeri-aplata*, from the Waco bed of the Noland Member, Brassfield Formation, on the east side of the Cincinnati Arch in central Kentucky. The type specimen (USNM 87171, Pl. 8, fig. 8, and Pl. 37, figs. 5, 6) is from "along the road north of Estill Springs, north of Irvine", that is, from the vicinity of localities 53 and 53a. Other specimens, according to Foerste, are from the vicinities of Panola (locality 1), Indian Fields, and Clay City, all in the same area in Kentucky.

The holotype is about 6 cm in maximum diameter, with a thickness varying between 0.2 and 1.2 cm. The corallites of the basal side are positioned with their longitudinal axes horizontal. These upflex distally, so that the corallites of the upper surface have their axes vertical. The corallites of the base appear to radiate from a center which is not included in the surviving portion of the corallum, suggesting that the corallum was originally low and lens-shaped, and grew radially and centrifugally. The initial diameter was about 8 cm.

The corallites of the holotype are fairly uniform in size, nearly all under 1 mm in diameter, with only the largest attaining 1 mm. The most common diameter is about 0.7 mm. Most corallites are six-sided in transverse section, but some are four- and five-sided. The calices all contain abundant knobs on their walls which may represent spines, but which do not appear to be aligned in surface view.

When viewed in thin section, the specimen appears to have undergone considerable diagenetic disruption. The tabulae are more wide-spaced than in other Brassfield specimens, with about three or four per millimeter in the denser levels, and 1 mm apart in the sparser regions. In many instances, the space between tabulae is greater than the tube diameter, and the greatest portion of the vertical section reveals sparsely tabulated areas. Spines are almost absent from my sections, but this is likely due to alteration during the fossilization process. This is shown by a few small areas in the lower portion of the vertical section in which the plane of section transversely intersected several spines. There appear to be about six spines in a vertical distance of 1 mm. Mural pores have not been observed.

Foerste wished to distinguish this form from the main body of *F. hisingeri* on account of its flatness. Whether or not this is justified, the form is clearly conspecific with the Brassfield material.

Williams (1919) figured a coral he referred to *Favosites hisingeri* from Manitoulin Island in Lake Huron. This specimen, according to Bolton (1960) is from the Fossil Hill Formation (Middle Llandovery through Middle Wenlock). The catalogue records the number of this specimen as GSC 5076, while Williams gave it as GSC 5075. Only one view of the specimen is presented, a surficial view showing the corallite apertures. These are of uniform size, nearly all 1.5 mm across. Unfortunately, insufficient information is given to allow a valid judgement of the identity of the specimen without studying the specimen itself. The uncertainty of the identification is compounded by the absence of a descriptive text accompanying the figure, and by the fact that the corallites are in the upper size-range of *F. hisingeri*, while also fitting into the size-range of other forms (*e.g., Paleofavosites prolificus*, which see).

Twenhofel (1928) reported *F. hisingeri* from Anticosti Island, in the Gun River (Middle Llandovery), Jupiter (Late Llandovery), and Chicotte (Late Llandovery and Early Wenlock) Formations. The coralla are massive and generally laminate (reflecting rhythmic tabular zonation). The corallites are thick-walled, four- to seven-sided, and never exceed 1 mm in diameter. The tabulae are thinner than the walls, and are numerous (1 to 3 per mm), and flat, convex or concave. The mural pores are large, circular, and arranged in one or two longitudinal rows. In the latter case, the rows are close, and the pores alternate. The septa consist of upward-flexing spines, some extending nearly to the axis of the corallite. There are one or two rows of spines on each wall, and each spine is apparently situated just above a pore. No figures accompanied the description of this Anticosti form. Nonetheless, the description agrees sufficiently well with my specimens to consider them conspecific.

Tripp (1933) took a practical stand with regard to the favositids. He considered them best divided into form-groups, without commitment to their genetic relationships. His figures and descriptions of Gotland corals include some that are clearly *F. hisingeri*, with narrow tubes (1.5 mm and less in diameter), and very long, distally upflexed spines. The tabulae are close-spaced, about two per millimeter. The mural pores occur in one, two, and sometimes three rows on the walls of the corallites. Most clearly similar to my material are Tripp's text-figures 35 and 36 (representing his Groups V and VI, respectively), and the figures of his plate 13. The specimens of this plate are from Kneippbyn, with the exception of one from Lickershamn. Judging from Manten's (1971) geologic map of Gotland, the Kneippbyn material is probably from the Högklint beds (Early Wenlock according to Manten, p. 43), while the Lickershamn material could be from the Högklint, or from the Visby beds (Late Llandovery, according to the same source).

Teichert (1937) reported *F. hisingeri* from the Silurian of Southampton Island, in the northeastern Canadian Arctic. He studied a single specimen with corallites of fairly uniform size, none narrower than 1.2 mm, and a few larger than 1.5 mm. Most of the tubes were about 1.5 mm in diameter. The corallite walls were about 0.1 mm thick, and were heavily invested with spines. The maximum length of the spines was 0.5 mm, but most were shorter. It was not determined if the spines point upward distally. There were eight to ten tabulae in 5 mm, spaced from 0.3 to 0.8 mm apart. These tabulae were flat, convex, or concave. Teichert had no information about the mural pores, except that they were not located in the corallite corners. The corallites were larger than is common in this species, but judging from other material encountered in the field and in the literature, it seems best to regard this as a large variant of the present species.

Shrock and Twenhofel (1939) reported *F. hisingeri* from Newfoundland, in the Goldson Formation, the Pike Arm Formation, and the Natlin's Cove Formation. According to Berry and Boucot (1970), the stratigraphy of Newfoundland is still only broadly understood, so that these units might best be regarded simply as Silurian. The corallites ranged in diameter from less than 0.5 mm to over 1.5 mm in a single specimen. The tabulae were unequally spaced, averaging two per millimeter, and were usually horizontal and flat. The mural pores were biserially arranged, or with no obvious pattern, and averaged four in each intertabular space. In well-preserved specimens, septal spines could be seen, some reaching nearly to the corallite axis. The Newfoundland material appears to be sufficiently close to the Brassfield form to allow an assumption of conspecificity.

A specimen at the Yale Peabody Museum (YPM 25634) is attributed to Twenhofel as a type. It is from the Pike Arm Formation, in the outer coral zone, on Fossil Point, in Pike Arm, New World Island. This is doubtless one of the specimens studied by Shrock and Twenhofel (1939). The corallite diameters are about 0.5 to 0.6 mm, and long spines are present. This specimen is identical to my *F. hisingeri,* and that name appears on its label.

Northrop (1939) reported *F. hisingeri* from the Gaspé region of maritime Canada, from the following units: Clemville (Late Llandovery), La Vieille (Late Llandovery through Late Wenlock), Gascons, Bouleaux and West Point (Early Ludlow) and Indian Point (Middle Ludlow) Fms. Unfortunately, the only information he gave is the corallite diameters, usually not over 1 mm, rarely reaching 1.3 mm. While it is quite possible that Northrop's material is *Favosites hisingeri,* it is not possible to be certain, due to the lack of sufficient information.

Distribution. — England, Wenlock?; New York State, within the range of Late Wenlock through Middle Ludlow; northern Michi-

gan, within the range of Middle Llandovery through Middle Wenlock; Ohio and Indiana, mid-Llandovery (Brassfield Formation);
Kentucky: mid-Llandovery (Brassfield Formation), and Late Wenlock or Early Ludlow; Anticosti Island, Middle Llandovery through
Early Wenlock; Gotland, Late Llandovery?, Early Wenlock; northeastern Canadian Arctic, Silurian; Newfoundland, Silurian; Bohemia,
within the range of Wenlock to Lower Devonian (Pragian). (See
Kriz & Pojeta (1974) for discussion of the problems of Barrande's
stratigraphic designations.)

Brassfield occurrence. — In southwest Ohio, localities 9 (near
Lewisburg), 15 (near West Milton), and 55 (near Ludlow Falls, in
clay layer at top of Brassfield); in Kentucky (central), localities
1 (questionable) (near Panola) and 46 (near College Hill), both in
the Noland Member; in Indiana, locality 26 (Elkhorn Falls, near
Richmond).

Favosites discoideus (Roemer)   Pl. 8, fig. 7; Pl. 28, figs. 3, 4

? 1851. *Favosites forbesi* Milne-Edwards & Haime (*partim*), p. 238.
  1854. *Favosites forbesi* Milne-Edwards & Haime, Milne-Edwards & Haime
         (*partim*), pp. 258-259, pl. 60, figs. 2, 2a, 2b.
  1860. *Calamopora forbesi* var. *discoidea* Roemer, p. 19-20, pl. 2, figs. 10, 10a,
         10b.
  1928. *Favosites forbesi* Milne-Edwards & Haime, Twenhofel, pp. 127-128.
  1933. *Favosites forbesi* Milne-Edwards & Haime, Tripp, pl. 16, figs. 1a, b.
  1934. *Favosites* cfr. *forbesi* Milne-Edwards & Haime, Shimizu, Ozaki &
         Obata, p. 70, pl. 13, fig. 1.
  1939. *Favosites forbesi* Milne-Edwards & Haime, Shrock & Twenhofel, p.
         254.
? 1939. *Favosites forbesi* Milne-Edwards & Haime, Northrop, p. 150.
  1949. *Favosites discoideus* (Roemer), Amsden, p. 90, (*partim?*), pl. 18, figs.
         6, 7, 9; possibly(?) pl. 34, figs. 13, 15.
? 1964. *Favosites discoideus* (Roemer), Stumm, p. 62, pl. 58, fig. 10.

Type material. — Repository of the specimen figured by Roemer is unknown.

Type locality and horizon. — From the "Silurian" of western
Tennessee (possibly the Brownsport Formation, of Early Ludlow
age, as Amsden (1949) reported such corals from that unit).

Number of specimens examined. — Approximately 10.

Diagnosis. — *Favosites* in which the wider corallites (2.5 to 3.0
mm in diameter) are scattered in a matrix of smaller corallites,
down to less than 1.0 mm in diameter; corallum generally low, lens-
shaped, with pedunculate base; tabulae wide-spaced; in larger
coralla, corallite size-variation apparently becomes less marked.

Description. — The coralla occur as lenses of low relief, not attached to any solid object (assuming the sediment was not indurated). Apparently this form lived in a manner similar to that of *Siderastrea radians* in the Florida Keys, and could have been readily detached from the sea floor. A typical Brassfield specimen is 4 cm in diameter, with a height of 1.0 to 1.5 cm. The earliest portion of the corallum is in the center of the base, and growth was radial and centrifugal. No holotheca is preserved. The appearance of the base suggests that a peduncle was originally present.

The most distinctive feature of this species is the great variability of corallite diameter within a small area. In a space about 0.5 square centimeter may be found corallites ranging from 3 mm down to 1 mm across. The former is about maximum for this species, so far as my specimens show, and many corallites 1.0 mm and narrower occur on the same specimen.

The corallites are variable in cross-sectional shape, doubtless due to the requirements of packing, and three to six sided corallites occur in a single specimen. The corallite walls are thin, about 0.1 mm across.

The tabulae are relatively wide-spaced vertically, averaging three to four in 4 mm in the larger corallites, and possibly a bit closer in smaller ones. As a result, the surface of a corallum (Pl. 8, fig. 7) often appears to have rather deep calices. The tabulae are virtually all complete due to this wide spacing, and while most are flat, some may be convex or concave upward (though without much relief).

Spines were not clearly seen in my specimens. Mural pores also cannot be reported, though it seems likely they are present. The assignment of my specimens to the genus *Favosites* is based upon similarity to other reported species of that genus.

Discussion. — From the Silurian of western Tennessee, Roemer (1860) described *Calamopora forbesi* var. *discoidea*, as a new variety of *Favosites forbesi* Milne-Edwards and Haime (1851), a species from the Silurian and possibly Late Ordovician of England and the Silurian of Gotland. Roemer gave the following account of his corals (here paraphrased):

The corallum is small, round, disc-shaped, convex upward, and either flat or slightly convex on its lower surface. In most cases, the corallum diameter is only 6 to 10 "lines" [0.5-0.8 inch], seldom reaching 1 inch. Sometimes the outline is more elliptical than circular. In the center of the base is a small curved peduncle. The corallite apertures are characteristically quite unequal in diameter and of very irregular form.

On the inner surfaces of the corallites, indistinct longitudinal striae are reported, and indicated in Roemer's figures as well (though Amsden, 1949 could not detect such features in similar corals collected by him in this area). Roemer felt that as his form grew, the inequality of the corallite diameters would disappear, though he had no evidence of this. He reported collecting specimens at Visby, Gotland (probably from the Högklint beds, of Early Wenlock age according to Manten, 1971), which were identical to his own from Tennessee. Aside from the "indistinct" longitudinal striae reported in the calices, Roemer's material appears identical to mine, and they are regarded as conspecific.

Roemer identified his material with the specimens shown in plate 60, figs. 2, 2a and 2b of *Favosites forbesi* Milne-Edwards & Haime (1854). This species was reported by these authors in 1851, but first illustrated by them in 1854. Roemer noted a difference between these specimens (and his own) and what he referred to as the "typically-grown form of *C. Forbesi*" Milne-Edwards & Haime (pl. 60, fig. 2c). Consequently, he termed his specimens a variety (subspecies) of *Favosites* (or as he called it, *Calamopora*) *forbesi*. As Milne-Edwards and Haime had not assigned a type specimen for their species, this designation of fig. 2c as "typically-grown" is a valid selection of a lectotype for *Favosites forbesi* on the part of Roemer, whether or not he intended it so. His specimen, and those shown in Milne-Edwards and Haime's fig. 2, 2a and 2b apparently are distinct from the specimen chosen as lectotype. Further study may show that larger specimens of Roemer's species take on the form of Milne-Edwards & Haime's, which would make them synonymous under *F. forbesi*. Pending demonstration of this, however, I follow the practice of later authors (Amsden, 1949; Stumm, 1964) in regarding Roemer's "variety" as a distinct species, *Favosites discoideus*, to which I assign the Brassfield form. As may be seen in the synonymy, many authors have linked *F. discoideus* (Roemer) with *F. forbesi* Milne-Edwards and Haime, apparently because they were originally illustrated together under the same name.

Two other authors, Jones (1936) and Lecompte (1936), assigned lectotypes for *F. forbesi*. They did this independent of one another, and apparently unaware that Roemer had already made a selection. Their choosing of lectotypes was done more purposefully, though, as their wording leaves no doubt that this was their intention. Jones (p. 10, in a January paper, probably pre-dating Lecompte's) specifically stated that he was selecting Lonsdale's (1839) *Favosites gothlandicus* Lamarck (pl. 15 *bis*, fig. 4) as the lectotype. Lecompte (pp. 63-64) stated that Milne-Edwards and Haime based their species on the specimen shown by Goldfuss (1829) on plate 26, figure 4b, identified as *Calamopora basaltica*. Lecompte's decision was apparently based upon the fact that this was the first post-Linnaean reference included in Milne-Edwards and Haime's synonymy. In both these cases, the specimen chosen resembles Roemer's lectotype of *Favosites forbesi* far more than Roemer's specimen of *Calamopora forbesi* var. *discoidea*. Regardless, then, whose choice of a lectotype one feels is valid for *F. forbesi,* this species remains distinct from *F. discoideus.* Those of Milne-Edwards and Haime's specimens considered conspecific with *F. discoideus* are from the Wenlock Limestone (Late Wenlock in age, according to Ziegler, Rickards, and McKerrow, 1974) at Dudley, England.

Twenhofel (1928) reported *Favosites forbesi* from Anticosti Island. He noted that it was present in zone 4 of the Ellis Bay Formation (Early Llandovery), and that the species persisted through the Becscie and Gun River Formations, to the top of the Jupiter Formation (Late Llandovery). The coralla are larger than in Brassfield *F. discoideus,* to 20 centimeters in diameter, and are hemispherical in form. The central portion of the base is "conical" (pedunculate?), and the base shows concentric growth-lines on what Twenhofel considered a holotheca. The calices are distinctly unequal in size, and the larger corallites are above 3 mm across, while the smaller are 1.5 to 2.0 mm across with some narrower than 1 mm. In only a few specimens do the thin walls show evidence of spines. The tabulae are thin and horizontal, generally about 1 mm apart, and have "from 1 to 2 marginal fossulae to a side". The mural pores occur on the faces of the walls, in one or two rows, and are about 0.25 mm in diameter, spaced 1 mm apart. Each is encircled by a raised ring. Twenhofel's description fits my specimens, and the two are regarded as conspecific.

From Gotland's Högklint marl (Early Wenlock), Tripp (1933) illustrated a specimen he identified as *Favosites forbesi*. Included are a transverse and longitudinal section (pl. 16, figs. 1a, b). The former shows the inequality of corallite diameter, the maximum being 3 mm, the minimum less than 1 mm. In the longitudinal section, the tabulae appear zoned, and are spaced from less than 0.5 mm to about 1.5 mm apart. Their surfaces are flat, concave and convex, but rarely show much relief. Growth seems to have been radial, and the corallum was hemispherical. Tripp's material seems to be conspecific with the Brassfield specimens.

The presence of *Favosites discoideus* in the Silurian rocks of northwestern Korea was documented by Shimizu, Ozaki, and Obata (1934) who identified it as "*Favosites* cfr. *forbesi*." The single specimen found was hemispherical, with a few corallites 2.5 to 3 mm in diameter, scattered among smaller ones (down to less than 0.5 mm in diameter). The authors had no information on spines or mural pores. This specimen has the superficial features of the Brassfield material, and is tentatively regarded as conspecific with it.

Shrock and Twenhofel (1939) reported this species (identified by them as *Favosites forbesi*) from the Pike Arm Formation (outer coral zone) of northern Newfoundland. The age of this unit has not been determined with precision, and is here noted simply as Silurian. The coralla are "conical" (presumably meaning pedunculate), and characterized by a large variation in corallite diameter (from less than 1 mm up to 3.5 mm) that is typical of the present species. The corallite walls are thin and no mural pores were observed. No figures accompanied the description, but there appears to be sufficient agreement with the Brassfield form to warrant regarding them as conspecific.

Northrop (1939) reported specimens of corals from the Port Daniel-Black Cape region of Gaspé (maritime Canada) which he identified as *Favosites forbesi*. This material came from the following units: the Clemville and Anse Cascon Formations (Late Llandovery), La Vieille Formation (Late Llandovery and all of the Wenlock), probably the Gascons Formation (Early Ludlow), and the West Point and Indian Point Formations (Middle Ludlow). Northrop restricted himself to listing the size-ranges of the larger and smaller corallites of six specimens. The larger corallites were mostly about 2 to 3 mm in diameter, sometimes reaching 3.5 mm. The narrowest corallites were 1 mm or less in diameter. In the absence of further data, we can only note the possible occurrence of *Favosites discoideus* (which would fit these measurments) in Gaspé.

Amsden (1949) reported *Favosites discoideus* from the Brownsport Formation (Early Ludlow in age) of western Tennessee. The corallum is hemispherical with a convex to conical base. Attachment is by a narrow peduncle, and the rest of the base is reportedly covered by a holotheca. The largest specimen is 5.5 cm in diameter. Amsden noted that while there is some variation in corallite size, the average diameter is 2.5 mm. His two figures on plate 34 do not show the corallite size pattern associated with my specimens, but the figures of plate 18 do. The specimen illustrated in plate 18, figures 6 and 7 is the smallest Amsden collected, about 16 mm in maximum diameter. The larger corallites reach a maximum diameter of 2.5 mm, while the smallest are less than 1 mm. No septa are reported in these forms. The tabulae are apparently mostly flat or concave, spaced about 1 mm apart. Numerous mural pores, about 0.25 mm in diameter, are located on the wall faces of the corallites. They are irregularly-spaced, but generally about 1 mm apart. Amsden regarded the internal longitudinal elements reported by Roemer as non-existent. The agreement between the Brownsport and Brassfield specimens suggests they are probably conspecific (at least with regard to the specimens figured on Amsden's plate 18).

Stumm (1964) reported *Favosites discoideus* from the Louisville Limestone (Late Wenlock and Early Ludlow) in the vicinity of Louisville, Kentucky. The upper surface of the corallum is flat or convex with low relief. The lower surface is flat, with a concentrically-wrinkled "peritheca" (holotheca), and central point of attachment. The coralla are 2 to 7 cm across, and 0.5 to 2 cm thick. The corallites are bimodal in size, the larger being 3 to 4 mm across (slightly wider than in my specimens), and the smaller averaging 1.5 mm in diameter. The corallites of the several sizes are interspersed. The mural pores are uniserial or biserial. The tabulae are complete, horizontal, and spaced 1 to 3 mm apart. This wide spacing of the tabulae is in good agreement with that in my specimens. Stumm's material is similar to mine, though, the higher maximum corallite diameters and the smaller range in corallite diameter leaves his identification open to question.

There is good agreement on what smaller specimens of *Favosites discoideus* (Roemer) look like, but less unanimity on the characteristics of larger specimens. Much more work is needed before this form can be sufficiently understood.

Distribution. — Tennessee, Early Ludlow; England, Late Wenlock; Anticosti Island, Early to Late Llandovery; Gotland, Early Wenlock; Korea, Silurian; Newfoundland; Silurian; Ohio and Kentucky, mid-Llandovery (Brassfield Formation).

Brassfield occurrence. — Localities 7a (near Fairborn), 55 (clay layer at top of Brassfield near Ludlow Falls) and possibly 11 (near West Union), all in southwestern Ohio; locality 50 (near Colfax) in central Kentucky.

Favosites densitabulatus, sp. nov.   Pl. 9, figs. 6, 7; Pl. 30, figs. 1, 2

> 1906. *Favosites gothlandica* Foerste, pp. 298-299, pl. 2, figs. 1A, B (*non* Lamarck, 1816).
> Cf. 1937. *Favoites* cf. *F. favosus* (Goldfuss), Teichert, pp. 129-130, pl. 6, figs. 5, 6; pl. 7, fig. 2.

Type material. — Holotype: UCGM 41945.

Type locality and horizon. — Locality 1, near Panola, Kentucky, Waco bed, Noland Member, Brassfield Formation.

Number of specimens examined. — Eleven, all collected during this study.

Diagnosis. — *Favosites* with very broad corallites (diameter about 5 mm wall-to-wall, about 6 mm corner-to-corner); tabulae horizontal, often so closely spaced as to be almost touching; bases of tabulae with buttress-like features in margin, possibly reflecting marginal fossulae as in *Favosites favosus* (Goldfuss); tabulae without spines.

Description. — The coralla are expansive, the largest being 20 cm in maximum diameter (as preserved). This specimen indicates that corallum growth was centrifugally radial, and it may be reconstructed (assuming reasonable symmetry) as a low, upwardly-convex corallum, at least 24 cm in diameter and at least 4 cm high.

This species has the broadest corallites of all the Brassfield favositids; the wall-to-wall diameter is generally about 5 mm, and the corner-to-corner diameter about 6 mm. The corallites show no appreciable variation in diameter (as in *Favosites discoideus* (Roemer)), and are as a rule hexagonal.

Besides the great width of the corallites, the most striking aspect of *F. densitabulatus* is the close-spacing of its tabulae. These may be so densely-packed that they nearly touch, or may be as far apart as 1 mm. Usually they are 0.6 mm apart and display periodicity of spacing. The tabulae are generally nearly flat, though some show a slight up- or down-bowing in their axes. The margins of their undersides have buttress-like extensions that may reflect marginal fossulae, as in *Favosites favosus* (Goldfuss).

The walls are about 0.25 mm thick and undulose, with a distinct dark mid-line. Mural pores occur on the wall faces, though their pattern could not be determined, and each appears to have a thickened rim. Spines seem to be present on the corallite walls, but not on the tabulae.

The combination of broad corallites and close-spaced tabulae is quite striking, making this a distinctive species.

Discussion. — The only mention of this species in the literature is Foerste's (1906) report of a supposed *Favosites gothlandica* from the Waco bed, Noland Member of the Brassfield Formation, in central Kentucky. The localities given by Foerste are Irvine (our locality 53 or 53a), Panola (locality 1), Vienna, Indian Fields, and Tipton Ferry.

Foerste reported specimens up to 50 cm in diameter. The corallites were usually 4 to 5, and sometimes 6 mm across. The more-or-less flat tabulae were spaced 2 to 2.4 per millimeter. They reportedly have about 12 marginal depressions (apparently similar to those of *F. favosus*). Foerste noted the presence of rimmed mural pores (though not their arrangement), and was unable to clearly determine the presence of spines.

Teichert (1937) reported a favositid from the Silurian of the Brodeur Peninsula in the northeastern Canadian Arctic. This may be conspecific with another of the Brassfield species, but is mentioned here because of its similarity to *F. densitabulatus*. The tabulae are spaced quite closely, 2.4 per millimeter. The corallites are usually about 2.8 mm (though sometimes less) in diameter, and this does not seem to fit into the range of variation of *F. densitabulatus*.

Distribution. — Kentucky, mid-Llandovery (Brassfield Formation); possibly in same unit in Ohio.

Brassfield occurrence. — Localities 1 (near Panola) and 46 (near College Hill) in central Kentucky, in the Waco bed, Noland Member of the Brassfield. Possibly from locality 11, near West Union, Ohio. Foerste (1906) reported specimens from several other localities in central Kentucky (see discussion).

Favosites sp. A   Pl. 30, figs. 3, 4; Pl. 42, fig. 1

    1860. *Calamopora gothlandica* Roemer (*partim*), pp. 18-19, pl. 2, figs. 9, 9a, 9b (*non Favosites gothlandicus* Lamarck, 1816).
? 1902. *Favosites tachlowitzensis* Počta, pp. 231-233, pls. 72-75, 78-80, 86.
? 1902. *Favosites tachlowitzensis,* var. *delicata* Počta, p. 233, pl. 74.
Cf. 1934. *Favosites kennihoensis* Shimizu, Ozaki, & Obata, p. 71, pl. 12, fig. 7; pl. 13, figs. 2, 3.
    1937. *Favosites* cf. *F. favosus* (Goldfuss), Teichert, pp. 129-30, pl. 6, figs. 5, 6.
? 1937a. *Favosites kuklini* Chernyshev, p. 69, pl. 4, figs. 2a, b.
? 1949. *Favosites brownsportensis* Amsden, pp. 87-88, pl. 16, figs. 1-3.
Cf. 1957. *Favosites kuklini* (Chernyshev), Zhizhina & Smirnova, pp. 22-24, pl. 6, figs. 1-4; pl. 7, figs. 1, 2.
Cf. 1957. *Favosites crassimuralis* Zhizhina (*in* Zhizhina & Smirnova), pp. 29-30, pl. 11, figs. 1-4.
Cf. 1957. *Favosites undulatus* Tchernyshev (*sic*) var. *fulvus* Zhizhina, pp. 30-31, pl. 12, figs. 1, 2.
Cf. 1967. *Favosites gothlandicus* Stasínska, p. 80, pl. 23, figs. 3a-c (*non* Lamarck, 1816).

Type material. — Figured specimen (UCGM 41958).

Type locality and horizon. — Specimen collected at locality 46, near College Hill, Kentucky, in the top of the Noland Member, Brassfield Formation.

Number of specimens examined. — One.

Description. — The corallum appears to have been originally hemispherical. The portion preserved has a maximum width of 11 cm and a thickness of 7 cm. The corallites are generally uniform in size (Pl. 30, fig. 3), about 2.5 to 3.0 mm in diameter, and are as a rule hexagonal, with fairly equal sides. The tabulae are closely spaced, the closest nearly touching each other (sometimes to the point of becoming incomplete), and the most widely spaced about 1 mm apart. On average, there are about eight to ten in 5 mm. The tabulae tend to be weakly convex upward, and do not appear to have marginal pits, as are found in *F. favosus.* Their upper surfaces are free of spines.

The walls are about 1/7 mm thick, and have a dark mid-line (Pl. 39, fig. 1). Spines seem to be present, but recrystallization makes it difficult to ascertain their character and abundance. For the same reason, the distribution of the mural pores is obscure, but if the few breaks in the corallite walls that appear in section are mural pores, then it seems they are located on the wall faces, rather than in the corners. For this reason, the Brassfield form is placed in the genus *Favosites* Lamarck. The absence of thickened walls and of obvious squamulae precludes its being assigned to *Squameofavosites* Chernyshev.

Discussion. — Because of the lack of critical information on this specimen, especially the distribution of spines and mural pores, it is difficult to determine a satisfactory identification. It is clear, however, that it represents a form distinct from other Brassfield favositids. The combination of intermediate-sized corallites with very closely spaced, spineless tabulae, is distinctive. It could prove to be a sub-species of *Favosites densitabulatus*, characterized by a narrower corallite and no apparent marginal pits in the tabulae.

Roemer (1860) described favositids from the Brownsport Formation (Early Ludlow) of western Tennessee, which he assigned to *F. gothlandicus* Lamarck. They clearly differ from the neotype of that species, as designated by Jones (1936). The coralla were "flat- or cake-shaped", with mostly six-sided corallites fairly uniform in diameter, 2.5 to 3.0 mm. Four tabulae, apparently mostly horizontal, occurred in a vertical distance about equal to the diameter of the tube, representing a spacing about equivalent to that found in my specimen. The mural pores were irregularly distributed, sometimes in pairs, sometimes singly, even on the same corallite wall. In all characters observed in my specimen, Roemer's form agrees, and it is probable that these two forms are conspecific. Roemer also mentioned a variety of his form which he felt was distinguished solely by its narrower corallites (1.0 to 1.5 mm in diameter). This form, illustrated in his plate 2, figure 11, cannot confidently be placed in synonymy with the Brassfield form.

Počta (1902) described *Favosites tachlowitzensis*, a species which he ascribed to Barrande (apparently as an unpublished manuscript name), from Tachlowitz, Bohemia, level e2 (rocks representing discontinuous time periods from Wenlock to the base of the Lochkov, fide Kríz & Pojeta, 1974, p. 491). The corallites are mostly hexagonal, and fairly uniform in diameter, from 2.0 to 2.5 mm. The thin walls seem to thicken around the corallite apertures, and are invested with granules (spines?). Spines of varying length are reported in some coralla. Round or occasionally oval mural pores are distributed in two rows (sometimes in alternate positions) on the walls. On very broad walls, a third row of pores may be present. The pores have no thickened rims. The tabulae are horizontal and complete, spaced 14 to 16 per centimeter. *F. tachlowitzensis* is similar to the present form in the regularity of the corallite arrangement, most being six-sided and of fairly uniform size. As a rule they are narrower than mine, the broadest (2.5 mm) falling into the lower range of my corallites. The tabulae are not so closely spaced as in the Brassfield specimen, with density somewhat less than the average encountered in my corallites. Počta illustrated the ropy pattern in the corallite walls as seen on the surface of a weathered specimen, a pattern not seen in my specimen. Unfortunately, I have no data on the arrangement of mural pores in my form, so in this important character the two forms cannot be compared. My specimen may be conspecific with the Bohemian species, but it will be necessary to obtain more information on its morphologic features before a decision can be made.

Počta also described a variation of this species, *F. tachlowitzensis delicata* from level e2 at the same locality. The corallites of this form are nearly all hexagonal, of generally uniform diameter (averaging 2.5 mm). The walls are invested with granules that may form circles or squares around the mural pores. It is unclear if these granules represent septal spines. The pores are distinct, distributed in two rows on the wall faces. The tabulae are complete, horizontal, and evenly spaced, 16 in 1 cm, as in *F. tachlowitzensis* proper. The relationship between this form and my own is unclear, due to lack of information on the mural pores and spines of the latter.

Shimizu, Ozaki, and Obata (1934) described a coral fauna from the Ken-niho Limestone of northwestern Korea. The fauna suggested, by comparison with other parts of the world, an age of Late Valentian to Late Salopian (approximately, Early Wenlock to Early Ludlow). From this fauna, they described *Favosites kennihoensis*, which has some features in common with the present specimen. The massive corallum consists of corallites of fairly uniform size, about two mm in diameter. Their walls are thin and sometimes longitudinally-wrinkled. Mural pores are "irregularly-distributed". The tabulae show alternation of spacing. Where densely-packed, there are four to six per millimeter, and where more sparse, there are three or fewer per millimeter. The tabulae are spaced in a pattern comparable with the Brassfield specimen, but are not so distinctly convex. Further, my form does not show the longitudinally-wrinkled walls of the Korean form. The evidence suggests these two are distinct species.

Teichert (1937) described some specimens from the Silurian beds of the Brodeur Peninsula (Stanley Point) on Baffin Island, Canadian Arctic. The corallites are both regularly and irregularly hexagonal, with some pentagonal. Most were 2.8 mm in diameter, but smaller corallites are not uncommon. The wall thickness is 0.1 to 0.2 mm. The tabulae are spaced 12 in 5 mm (somewhat closer than my average). Spines were not observed, and the mural pores did not seem to be in the corners of the corallites. The interiors of the corallites appear to have been somewhat disrupted, but convex tabulae are not uncommon. This form and my own have many features in common, and are tentatively regarded as conspecific.

Favosites brownsportensis Amsden (1949), from the Brownsport Formation (Early Ludlow) of western Tennessee, is similar to my form. The corallites of the holotype (YPM 17616) are mostly 3 to 4 mm in diameter, though there are fields of smaller corallites where proliferation has apparently occurred. The corallites are as a rule six-sided, with the sides usually unequal. In a given region, they are generally uniform in size. The tabulae are horizontal or slightly convex and close-spaced, about 5.5 in 5 mm where most dense. The mural pores are up to 0.3 mm in diameter, and are located on the wall faces. No spines were observed in thin-section. Though the corallites are slightly larger than in my form, and the tabulae slightly sparser, the Brownsport specimen might be part of a continuum, including my form, and possibly *F. tachlowitzensis* of Bohemia.

Chernyshev (1937a) described *Favosites kuklini* from the Llandovery of Novaya Zemlya. His reference was not obtainable, but Zhizhina and Smirnova (1957) described specimens from the eastern Taimyrs, of Llandoverian and Wenlockian age, which they attributed to *F. kuklini*. The coralla are massive, with radially distributed corallites, five- to seven-sided, not homogeneous in diameter, ranging from 2 to 4 mm wide. The corallite walls are thin (0.08 to 0.3 mm thick) and weakly wavy. Thick, blunt spines are sparsely distributed. Mural pores 0.24 to 0.4 mm in diameter are distributed usually in two rows, but occasionally one or three rows. Most tabulae seem to be slightly convex upward, and depressed peripherally where they contact the walls (the same profile as in my corallites). About 6 to 10 tabulae occur in 5 mm vertically. This species (as described by Zhizhina & Smirnova) is similar to mine, but differs in the apparent greater range of corallite diameters, and more significantly, in the nature of the spines which might represent the squamulae of *Squameofavosites*.

In the same work (Zhizhina & Smirnova, 1957), Zhizhina described *Favosites crassimuralis* from the Llandovery of the eastern Taimyrs. The massive corallum consists of radially arranged corallites, usually five- or six-sided (except for very young corallites, which may be three- or four-sided). Their diameter is 2.5 to 3.5 mm. The walls are thick (0.25 to 0.4 mm), with a distinct mid-line. A wall may be weakly concave in one corallite and weakly convex where it faces on a neighboring corallite. The spines are stout, broad, and pointed, about 0.2 to 0.25 m long. The mural pores are round, 0.2 to 0.3 mm in diameter, and distributed alternately in two rows. The tabulae are close-spaced, sometimes wavy. Peripherally they often down-flex to join the wall (as in my form). There are 12 to 13 in 5 mm, rather denser than the average in the Brassfield specimen. The Taimyr form differs from mine in the thickness of the walls and the nature of the spines.

In the same work, Zhizhina described *Favosites undulatus fulvus*, a new variety of a species described by Chernyshev. Like the previous species, this form is from the Llandovery of the eastern Taimyrs. The corallum is massive, with corallites in radial or parallel patterns. The corallites are four to seven-sided, with rounded corners. The diameters are not uniform, mostly varying between 2.0 and 3.5 mm. The wall is 0.12 to 0.17 mm thick, and rarely as much as 0.2 mm. The spines are short, broad, and pointed. The pores are round or oval, and distributed on the walls in one or two rows. Their diameters can vary in a single specimen from 0.28 to 0.36 mm. The tabulae are closely spaced, with flat or concave surfaces, spaced 11 to 13 in 5 mm vertically. The Brassfield form differs from Zhizhina's in having more regularly convex tabulae and less-developed spines. Also, my tabulae are not generally as closely spaced as in the Taimyr form.

Stasínska (1967) described specimens from erratic boulders in Poland and from the Llandovery of Norway, that she identified as *Favosites gothlandicus* Lamarck. The coralla are massive, discoidal, with corallites 2.5 to 3.0 mm in diameter. The walls are straight, 0.1 to 0.2 mm thick, with a distinct mid-line. Mural pores, 0.2 mm in diameter, are arranged in from one to three, most usually two rows, with the rows 0.2 to 0.5 mm apart. The tabulae are closely spaced, thin, mostly horizontal, but also weakly concave or convex. They are wavy in profile. Where most closely spaced, there are 2.5 to 6 per mm. Elsewhere there are about 2 per mm. Short spines are present. This form is similar to mine, differing only in the absence of consistently convex tabulae and in the definite presence of spines on the walls. Also, the transverse section (Stasínska's figure 3a) suggests the presence of peripheral pits in the tabulae.

Distribution. — Kentucky, mid-Llandovery (Brassfield Formation); Tennessee, Early Ludlow; Baffin Island (Canadian Arctic), Silurian.

While several forms in various parts of the world (discussed above) bear similarities to mine, the Brassfield specimen is insufficiently understood to allow its confident identification with any of them. Further study will probably show it to be conspecific with some of these forms.

Brassfield occurrence. — Locality 46, near College Hill, central Kentucky, in the top of the Noland Member.

Genus PALEOFAVOSITES Twenhofel, 1914

1829. *Calamopora* Goldfuss, p. 77.
1914. *Paleofavosites* Twenhofel, p. 24.
1971. *Palaeofavosites* Oekentorp, pp. 158-160.

Type species. — (Original designation): *Favosites asper* d'Orbigny, 1850, p. 49 (= *Favosites alveolaris* Lonsdale [*non* Goldfuss] 1839, p. 681, Pl. 15 *bis*, fig. 1; *non* fig. 2).

Type locality and horizon. — "Silurian: Wenlock, Shropshire, England, and Island of Dagö, Estonia" (*fide* Lang, Smith, & Thomas, 1940, p. 94).

Diagnosis. — Identical to *Favosites*, except mural pores located in the corners of corallites, rather than on wall faces, often resulting in corallite corners being undulose.

Description. — The genus *Paleofavosites* encompasses those favositid species which consist of simple, polygonal tubes, traversed with tabulae, and bearing mural pores in the corners of the tubes only (none on the wall faces). The intercorallite walls contain a mid-line, which separates the wall portion belonging to each corallite from those of its neighbors.

Septal spines are variably prominent, short, and essentially normal to the corallite wall. They never form continuous sheets.

When three corallites meet along a single corner, there is an interdigitation of the mural pores. As an example: a pore will connect corallites *A* and *B*; above this a pore connects corallites *A* and *C*; above this, a tube again connects *A* and *B*, followed vertically by a tube connecting *A* and *C*, and so forth. The result is that the corallite corner does not define a straight line along the length of the corallite, as is the case in *Favosites*, but rather, a zigzag line. The prominence of this zigzag is presumably proportional to the vertical closeness of successive pores.

Discussion. — *Paleofavosites*, like *Favosites*, poses difficult taxonomic problems, as patterns of variation have confused the search for meaningful taxobases on the specific level. Sutton's work (1966) has been referred to in my discussion of *Favosites* Lamarck. In this work, he studied the variation of morphologic characters in *Paleofavosites asper* (d'Orbigny) (erroneously attributing this species to Twenhofel), and came to the same conclusions as for *Favosites gothlandicus* Lamarck: Abundance of spines and thickness of corallite walls vary greatly even within individual coralla, with greatest spinal abundance and wall thickness occurring rhythmically, corresponding to zones of close-spaced tabulae. The most stable character, he determined, is the mean corallite diameter (defined in the discussion of Sutton's work under *Favosites*), both within coralla from a single horizon, and from different horizons.

Paleofavosites Twenhofel has had a complicated taxonomic history, the problems of which have been only recently resolved. Following is a summary of the difficulties surrounding this name:

When Goldfuss (1829, p. 77) introduced the genus *Calamopora* he described several species under that generic name, including *C. alveolaris*. This was designated as the type species of *Calamopora* by King (1850, p. 26; *fide* Oekentorp, 1971). It is characterized in part by the mural pores being located in the corners of the corallite tubes.

Lonsdale (1839, p. 681) described English corals under the name of Goldfuss' species, but as he considered *Calamopora* to be a junior synonym of *Favosites* Lamarck (1816), he referred to his specimens as *Favosites alveolaris*. Of the two specimens Lonsdale illustrated, the one shown in his figure 2 (the "larger variety") was determined by Oekentorp (1971), after examining the type material, to be identical to Goldfuss' *C. alveolaris*. For the smaller specimen (Lonsdale's fig. 1), d'Orbigny (1850, p. 49) erected *Favosites asper* (spelling the trivial name "*aspera*"), a species in which the pores were also located in the corallite corners, as in *C. alveolaris*.

Twenhofel (1914, p. 24) recognized the distinctiveness of mural pores being restricted to corallite corners, and erected the genus *Paleofavosites* (often wrongly modified by later authors to *Palaeofavosites*), to accommodate *Favosites prolificus* Billings (1865b) and *F. capax* Billings (1866) [both from the Ordovician and Silurian of Anticosti Island], *Calamopora alveolaris* Goldfuss (1829), and *Favosites asper* d'Orbigny (1850). Twenhofel designated *F. asper* as the type species of *Paleofavosites*. Later (1928, p. 125) Twenhofel stated that he chose *F. asper* as the type species because he had believed it to be identical to Billings' later species, *F. prolificus*, but now he no longer felt they were synonymous. Consequently he sought to change the type species designation to *F. prolificus*. This latter assignment of type species is invalid.

Because the mural pores of both *C. alveolaris* and *F. asper* are restricted to the corallite corners, they are considered congeneric. Consequently, *Paleofavosites* Twenhofel is a junior subjective synonym of *Calamopora* Goldfuss. Arguing that the term *Calamopora* had not been used in the literature for half a century, and that *Palaeofavosites* [*sic*] had in the meantime become firmly established in the minds of paleontologists, Oekentorp (1971) requested that what he regarded as a potentially chaotic situation be avoided through the use of the plenary powers of the International Commission on Zoological Nomenclature to suppress *Calamopora* Goldfuss in favor of *Palaeofavosites* [*sic*] Twenhofel. To keep what was at that point a subjective synonymy from becoming an objective synonymy, Oekentorp further proposed that *F. asper* d'Orbigny be retained as the type species of *Palaeofavosites* [*sic*], rather than allowing Goldfuss' *C. alveolaris* to remain as the standard for the concept of this genus.

In accordance with Article 79(b) of the International Code of Zoological Nomenclature, which requires proof that the threatened name has been used by at least five different authors in at least ten different publications in the last fifty years, Oekentorp (1974) submitted a partial list of references in support of his appeal for the retention of *Palaeofavosites* [*sic*] Twenhofel and the suppression of *Calamopora* Goldfuss.

As noted above, the correct spelling of Twenhofel's generic name is *Paleofavosites*. Throughout his process of appealing to the ICZN, Oekentorp had used the spelling *Palaeofavosites*, and in 1975, Cara-manica (p. 1126) issued a reminder that the addition of an *a* after the first syllable of the name, in common practice to that time, was incorrect, differing as it did from Twenhofel's original spelling of the name as *Paleofavosites*. He was particularly concerned about the fact of Oekentorp's misspelling of the name in his application to the ICZN, as this might result in the improperly spelled name being placed on the Official List of Generic Names in Zoology, in the event that its continued use was authorized by the Commission.

In 1976, the International Commission on Zoological Nomen-clature ruled (Opinion 1059) in favor of suppressing *Calamopora* Goldfuss, 1829, using the plenary powers, "for the purposes of the Law of Priority, but not for those of the Law of Homonymy". In its place, the Commission accepted *Palaeofavosites* [*sic*] Twenhofel, 1914, with *Favosites asper* d'Orbigny, 1850 as the type species.

Following this ruling, Oekentorp (1977) noted that at the time he formulated his application to the ICZN he was unable to trace the reference in which the name *Paleofavosites* had been intro-duced by Twenhofel, and so he had used the "correct latinized form generally common in literature. . .", *Palaeofavosites*. He later realized that this was an emendation unjustified by the Code, acknowledging that *Paleofavosites* is the correct spelling (Oekentorp, 1976, *fide* Oekentorp, 1977). He went on to say (1977) that while he agreed with Caramanica (1975) on the correct original spelling of the generic name, the International Commission on Zoological Nomenclature's Opinion 1059 (1976), suppressing *Calamopora* Gold-fuss, had used the spelling *Palaeofavosites* for Twenhofel's generic name. Oekentorp felt that this firmly established the spelling *"Palaeofavosites"* again, and recommended that this spelling be retained.

Melville (1978) responded by saying that in Opinion 1059, the Commission "did not have the question of the spelling of the name under consideration", but rather the matter of suppression of a senior name in favor of a junior name that had allegedly become firmly established in the literature. Validation of the alternate spelling of *Paleofavosites* could have been accomplished only by use of the plenary powers by the Commission "after due advertisement of its intention to do so". As neither step had been taken, the supposed validation of the spelling *"Palaeofavosites"* had never, in fact, occurred. According to Melville, *"Paleofavosites"* is a correct original spelling under Article 32a of the Code, and is to be retained. *"Palaeofavosites"* is an intentional, unjustified change, and therefore an invalid junior objective synonym of *Paleofavosites*. Oekentorp's spelling error in his application had been simply repeated in the Commission's opinion, and a correction of this spelling error was published in the Corrigenda sheet on p. 264 of the same volume in which Opinion 1059 appeared (*fide* Melville, 1978).

The result of this long controversy is that *Paleofavosites* Twenhofel, 1914 (retaining its original spelling) replaces *Calamopora* Goldfuss, 1829 as the valid name for this genus, and the type species originally designated for *Paleofavosites* by Twenhofel, *Favosites asper* d'Orbigny, 1850, is the type of this genus.

Paleofavosites prolificus (Billings)   PI. 9, figs. 1-5; PI. 29, figs. 3, 4; PI. 39, figs. 1, 3

Cf. 1826. *Calamopora alveolaris* Goldfuss, p. 77, pl. 26, figs. 1a-c.
Cf. 1839. *Favosites alveolaris* (Goldfuss), Lonsdale, p. 681, pl. 15 *bis*, figs. 1, 1a, 1b, 2, 2a.
Cf. 1850. *Favosites aspera* [*sic*] d'Orbigny, p. 49.
  1865b. *Favosites prolificus* Billings, p. 429.
  1876. *Favosites niagarensis* Hall, Rominger, p. 23, pl. 5, fig. 1 (*non* Hall, 1852).
? 1890. *Favosites niagarensis* (Rominger), Foerste, p. 334.
  1899. *Favosites aspera* [*sic*] d'Orbigny, Lambe, p. 4-6, pl. 1, fig. 2.
  1919. *Paleofavosites asper* (d'Orbigny), Williams, pl. 6, figs. 1a, b; pl. 17, fig. 3.
  1928. *Paleofavosites prolificus* (Billings), Twenhofel, p. 126.
  1933. *Favosites aspera* [*sic*] d'Orbigny, Tripp, p. 96, pl. 8, fig. 3; pl. 9, fig. 1; text-fig. 20.
Cf. 1936. *Favosites alveolaris* (Goldfuss), Lecompte, pp. 66-68, pl. 11, fig. 4.
  1940. *Favosites asper* d'Orbigny, Sugiyama, pp. 126-127, pl. 19, fig. 11; pl. 20, figs. 2-4; pl. 21, figs. 9-11.

Non 1956. *Paleofavosites prolificus* (Billings), Stearn, p. 60, pl. 4, fig. 1; pl. 10, fig. 13.

     1963. *Palaeofavosites prolificus* (Billings), Nelson, p. 52, pl. 7, fig. 5.

? 1966. *Palaeofavosites asper* Bolton, pl. 1, figs. 25, 26.

? 1968. *Palaeofavosites asper* Bolton, pl. 12, fig. 20.

Type material. — Repository of type material unknown.

Type locality and horizon. — Anticosti Island, somewhere within the range of Late Ordovician to Middle Wenlock. The precise horizon is unknown in the case of Billings' specimens, but he (and Twenhofel, 1928) indicated the species is found through the above range.

Number of specimens examined. — Approximately 100, all collected during the present study.

Diagnosis. — *Paleofavosites* with corallites of fairly uniform size, generally 1.0 to 1.5 mm, rarely 2.0 mm in diameter; tabulae flat or concave; short spines on walls, but not on tabular surfaces; wall corners not strongly crenulate.

Description. — This is one of the most frequently encountered coral species in the Brassfield Formation. It often occurs as massive cobbles showing unmistakeable evidence of transport, but is also found in apparent growth-position, undisturbed after death. A maximum diameter of 30 cm is not unusual for a corallum of this species. The convexity of the upper surface of coralla varies, and there is no clear evidence of a peduncle or holotheca. It has never been observed to branch. Characteristically, the corallite diameter is 1.0 to 1.5 mm, rarely 2.0 mm. Smaller new corallites occur among the larger ones, but tend to be lost in the crowd of full-grown corallites, resulting in a fairly uniform appearance of the corallites on the surface of the corallum. This is in contrast to *Favosites discoideus,* where large corallites are often surrounded and isolated by numerous small ones. Neither do the small corallites occur in fields, as they do in *F. hisingeri.* These dissimilar size distributions doubtless reflect differences in off-setting patterns.

The corallite walls are about 0.2 mm thick. It is not clear if a mid-line is present. In some specimens, the corallite walls (in transverse section) appear undulose, while in others they are straight.

The few mural pores observed occur in the corners of the corallites, rather than on the wall faces. They are narrow, resembling pinholes. In one section (Pl. 29, fig. 3; Pl. 39, fig. 3) they appear to be about 1 mm apart vertically. The zigzag corallite junctures often seen in vertical sections of Ordovician specimens of *Paleofavosites* are not obvious in my material.

Very short spines, barely visible in whole specimens (Pl. 9, fig. 1) occur on the corallite walls. They are arranged in poorly-defined longitudinal rows. One count showed eight in a vertical space of 2 mm (UCGM 41926). Spines do not seem to be present on the tabulae.

The tabulae are usually concave upward, sometimes flat. They are 0.5 to 2 mm apart, and this spacing shows distinct vertical zonation (Pl. 9, figs. 1, 2; Pl. 29, fig. 3). This zonation results in layers of the corallum alternating in their resistance to erosion, so that it is common to find a worn specimen whose base or side is rugose.

In the best-preserved specimen (UCGM 41926, Pl. 9, fig. 3), the lower surfaces of the tabulae often have an axially-located protuberance, similar to those seen in the holotype of *Favosites favosus* (Goldfuss) (Pl. 8, figs. 4, 9). One corallite of the same specimen also has one tabula which appears to be radially-fluted, a feature also common in *F. favosus* (same figures), which represents marginal fossulae.

Discussion. — Goldfuss (1826) described *Calamopora alveolaris* from loose rock fragments from the Eifel and Groningen areas of Germany. The tubes are united in a bulbous mass (the corallum), and are five- and six-sided. They are fairly regular, straight, and of nearly equal "thickness" (diameter?). The tabulae are horizontal, and at each corner of the corallite, the tabula is depressed. Goldfuss believed these points might mark the apertures of mural pores. The mural pores are very numerous, and lie in the corners of the corallites. No indication is given of the corallite diameter, unless Goldfuss' figure 1a is natural size. If so, the corallites are rather broader than mine, commonly reaching 2.5, and often 3.0 mm. Goldfuss made a point of showing the marginal pits of the tabulae. As noted in my description, one corallite of a specimen had what appeared to be such depressions, and it is possible that they are environmentally induced features, rather than representing a specific taxobasis. They do appear in other species including some belonging to other genera (e.g., *Favosites favosus*). Goldfuss' figure 1c shows septal spines similar in size (proportional to the corallite) to mine. The mural pores seem similar in size and arrangement to mine. The possibly greater diameter of the corallites and the frequent occurrence of marginal pits in the German form, preclude identifying the Brassfield species with Goldfuss' material.

Lecompte (1936) described and figured the holotype of *Calamopora alveolaris* Goldfuss. His observations substantiate my interpretation of Goldfuss' description and figures. Lecompte noted that the probable age of Goldfuss' material is Silurian.

Lonsdale (1839) reported *Favosites alveolaris* from the Wenlock rocks of England. The diameters of his corallites are consistent with mine, about 1 mm, but the corners between corallites are strongly zigzagged, due to the presence of mural pores interlocking from several corallites. This longitudinal zigzagging is not found to such a degree in my specimens, and the English form is not regarded as conspecific with the Brassfield material.

D'Orbigny (1850) chose as the basis of his species, *Favosites asper*, Lonsdale's (1839) figure 1 of *F. alveolaris*. He recorded the date of original description of *F. asper* as 1847, but the referred-to publication has not been found, and 1850 is commonly cited as the original date (e.g., Bassler, 1915, p. 941). This trivial name (*asper*) has often been applied to forms similar to mine, but as noted above, Lonsdale's specimens are not considered conspecific with the Brassfield form.

Billings (1865b) introduced the name *Favosites prolificus* for specimens from Anticosti Island ("Hudson River group and throughout the Middle Silurian"). The original description is:

> Corallum forming large hemispheric or irregularly convex masses. Tubes about one line in diameter. Tabulae thin and either complete or imperfect, sometimes filling the tube with vesicular tissue. They are often very numerous, there being sometimes six or seven in one line. No septa or mural pores have yet been detected, and it may be that this species should be placed in another genus.

No figure accompanied this description. So far as it goes, the description agrees well enough with my specimens. Unfortunately, important information concerning the mural pores, corallite junctures, spines and topography of the tabular surfaces is lacking. Twenhofel (1914, p. 24) remarked that *F. prolificus* has pores in the corners. He later (1928, p. 125) expressed some doubt that *F. asper* and *F. prolificus* were (as was by then generally believed) identical. As the evidence discussed above shows, my form is not conspecific with either *C. alveolaris* or *F. asper*. As specimens described under the name of *Favosites* (later *Paleofavosites*) *pro-*

lificus are in general agreement with the Brassfield form, it seems best to use that name here, pending investigation of type material. From the "Niagara Group" of Point Detour, northern Michigan, Rominger (1876) described a form which he wrongly identified with *Favosites niagarensis* Hall. It is probable that this specimen (UMMP 8438) comes from the Manistique Group (Late Llandovery through Middle Wenlock). The corallites are polygonal, about equal in diameter (generally 1.5 mm). The tabulae are flat and spaced zonally. Delicate spines are present on the walls. The sparse mural pores lie close to the corners. In all important characters this specimen is identical with the Brassfield form and they are considered conspecific.

Foerste (1890) reported *Favosites niagarensis* (based on the similarity of his specimens to Rominger's material) from the "Clinton Group" (Brassfield Formation) in Ohio, noting it as the most common species in the unit. This abundance strongly suggests that Foerste was referring to *P. prolificus* but the paucity of data in his description (consisting of corallite diameters and tabula-spacing, with occasional references to tube-crenulation) leaves the true identity uncertain.

Lambe (1899) reported *Favosites asper* from Anticosti Island in the Canadian Maritime area, and from Stony Mountain and Stonewall, Manitoba. The age of Lambe's material is uncertain, but Bolton (1960, p. 30) noted Upper Ordovician units (Stony Mountain Formation and Stonewall Formation) at Stony Mountain and Stonewall quarry, both in Manitoba. It is likely, then, that the Manitoba material is of about this age.

The coralla are massive and large. Their form is convex upward, with a flat bottom. Lambe reported an "epitheca" covering the bottom, though this was not seen in any of my specimens. The corallites average 2 mm in diameter, and are of fairly uniform size in a given specimen. The tabulae are complete and flat, or less often, convex, or concave, and are about 0.5 to 1.0 mm apart, though sometimes more. Apparently the spacing shows vertical zonation. Small, usually inconspicuous marginal pits occur in some specimens. The pores are of "moderate" size, in or near the angles of the walls and numerous in some specimens. They are about 0.75 mm apart vertically, and some are enclosed by raised rims. Rarely the pores are located on the faces of the corallite walls.

Septal spines (generally short) occur on the corallite walls in longitudinal rows. The corallite diameter varies between specimens, ranging below 1 mm in some cases, while in others they range from 0.5 to 2 mm. The single figure accompanying the description shows an idealized corallite with large pores nearly as wide as the distance between them. The wall is slightly zigzagged. From the description and figures it seems that at least some of the specimens studied by Lambe are conspecific with our specimens. The greatest discrepancy is the apparent greater pore diameter in the Canadian specimens. The pore spacing, however, is comparable to mine. Recrystallization may have reduced the pore size in the Brassfield specimens, sometimes obliterating them completely. Lambe's specimens, then, may be considered conspecific with my own.

Williams (1919) illustrated specimens of what he referred to as *Paleofavosites asper* from the Manitoulin and Fossil Hill Formations (Early and Middle Llandovery, respectively) of Manitoulin Island. The corallite diameters are fairly uniform, mostly 1.0 to 1.5 mm, and the mural pores are apparently small and in the corners of the corallites. The resemblance to my specimens is such that they are regarded as conspecific.

Twenhofel (1928) reported *Paleofavosites prolificus* from the Late Ordovician through the Middle Wenlock ("throughout the Anticosti measures") of Anticosti Island. [This range is similar to that of *Propora conferta* on the island.] The specimens have corallite diameters of 1 to 1.5 mm, and flat, concave, and convex tabulae spaced 0.33 to 2 mm apart. Submarginal pits are rare and indistinct. The mural pores are small, less than 0.25 mm wide and spaced eight to nine in 5 mm. They are rarely visible, (as is the case with my specimens), and lie in the corallite corners. The walls marginal to each pore are "arched to make a depression about the pore". This presumably means that the pores are at the ends of lateral protuberances in the corners, a feature common in *Paleofavosites*, and apparently weakly developed in the Brassfield specimens. Spines are present. Twenhofel noted that this is one of the most common and wide-ranging species on Anticosti Island, extending from the base of the English Head Formation (Upper Ordovician) to the base of the Chicotte Formation (uppermost Llandovery). The Anticosti and Brassfield forms are probably conspecific.

Tripp (1933) recognized the lack of firm criteria for speciating favositids, and so resorted to using form-groups in his study of the Gotland favositids. His *Aspera*-forms are typified by the following (here paraphrased): The corallites are 1 mm in diameter, more or less constant in size throughout the colony. Spines occur only sporadically. The pores lie in the corners of the corallites, in small out-bayings of the walls, resulting in a waviness of the walls in longitudinal section. This waviness, however, is not always marked in his figures. The pore diameter is 0.1 mm, one-tenth the corallite diameter. Tripp noted that this relationship of pore to tube diameter holds for all of the members of the *Aspera*-form known to him. The tabulae are mostly gently concave upward, and spaced about 2 per mm (although the spacing is clearly zoned). The coralla are massive. Tripp's description and figures agree so well with my specimens that it is probable that the Gotland and Brassfield forms are conspecific. The specimen most similar to my material appears in Plate 8, figure 3 and comes from the Högklint beds (Early Wenlock).

Sugiyama (1940), in his study of the stratigraphy and paleontology of the Kitakami region of Japan, described several specimens similar to *P. prolificus* from the "*Clathrodictyon*" and "*Halysites*" Limestones. These horizons are assigned to the Salopian (Wenlock and Lower Ludlow) equivalents of England (Hamada, 1958, table 2, p. 95). The coralla are massive. In one of the three specimens described, the corallites are usually 1.0 to 1.5 mm in diameter, with thin, straight walls, about 0.06 to 0.1 mm thick. The tabulae are usually flat and complete, four to five occurring in 2 millimeters vertically. The mural pores are of "moderate" size, in two rows located very close to the angles, but sometimes on the wall faces. Rare septal spines are arranged in a regular manner.

A second specimen differed from the first in having smaller corallites (0.7 to 1.2 mm diameter), with normally concave tabulae.

A third specimen differed from the other two in having more numerous spines, and wavy walls.

Sugiyama regarded these specimens as conspecific. The available data indicate that these specimens are probably *P. prolificus*.

Stearn (1956) reported *Paleofavosites prolificus* from the Upper Ordovician of southern Manitoba (Interlake Group, Stonewall Formation, and especially the Stony Mountain Formation), but his corallites are too wide (uniformly 2.34 mm) to be considered *P. prolificus*, as that species is presently understood.

Nelson (1963) reported *P. prolificus* from the Chasm Creek Formation (Upper Ordovician) of Manitoba. The corallites are fairly uniform, 1.5 mm in diameter. Mural pores occur in the corners of the corallites, and are about 0.1 mm in diameter. The tabulae are flat and close-spaced, about four to six occurring in a vertical distance equal to the corallite diameter. In all features noted by Nelson, this form agrees with mine, and they are considered identical.

Bolton (1966, 1968) illustrated two specimens identified as *Palaeofavosites asper* from the Manitoulin Formation (Early Llandovery) of Manitoulin Island. Only surficial views are shown, and these are in good agreement with my specimens, showing massive coralla with a uniform corallite diameter of about 1.5 mm. If the original specimens (GSC 20481 and GSC 17961) have a zigzagging of the corallite corners at the mural pores, these specimens could be assigned to D'Orbigny's species, as Bolton has done. If the zigzagging is weak or absent it would be best to assign them to *P. prolificus*.

Our studies of *Paleofavosites prolificus* have served to emphasize the fact that as yet, taxonomic relations among the favositids are poorly understood. *P. prolificus* and *P. asper*, as understood here, differ mainly in the greater crenulation of corallite corners in the latter. Yet, in several cases, it appears that this crenulation may occur in some parts of a specimen and not in others. As with the marginal fossulae in *Favosites favosus*, and in some *Paleofavosites*, these features are sometimes ubiquitous, sometimes sporadic, and sometimes absent from specimens for whose taxonomic positions they are considered important taxobases.

It is possible that the development of crenulated sutures is related to the spacing of the mural pores. Where only two or three corallite corners meet at a suture, communication through pores can possibly be effected without formation of tubes. When four corallites meet along a narrow suture, they must alternate pore communication between opposing corallites. In this case, pore juncture between the east-west pair of corallites effectively forms the upper and lower walls of the pore junctures between the north-south pair. If this is so, then *P. prolificus* would be a junior synonym of *P.*

asper. Pending further study, however, they must be regarded as distinct.

Distribution. — Anticosti Island, Late Ordovician through Late Llandovery; northern Michigan, within the range of Late Llandovery to Middle Wenlock; Manitoba, Late Ordovician; Manitoulin Island, Early and Middle Llandovery; Gotland, Early Wenlock; Japan, Wenlock and Early Ludlow; Ohio (and possibly Indiana and Kentucky, mid-Llandovery (Brassfield Formation).

Brassfield occurrence. — Ohio localities: 2 (north of Manchester), 7a (near Fairborn), 11 (near West Union), 14 (Piqua), 37 (Turtle Creek near Sharpsville). Questionable occurrences include the following: Ohio localities 13 (Todd Fork Creek near Wilmington), 15 (near West Milton). Indiana locality 26 (Elkhorn Falls near Richmond). Kentucky localities 30 (near Seatonville) and 53a (near Irvine), the latter in the Noland Member.

Family **HALYSITIDAE** Milne-Edwards & Haime, 1850

Nom. transl. Duncan, 1872 (*ex* Halysitinae Milne-Edwards & Haime, 1850).

Cateniform Tabulata, with ranks only one corallite thick; corallites circular or elliptical in cross-section; corallite walls two-layered; longitudinal elements (septa), when present, are 12 acanthine septa; mural pores absent; space between neighboring corallites may contain an interstitial tube (smaller corallite?) which, like the larger corallites, is tabulate; tabulae of interstitial tubes, and the peripheral portions of the interiors of the corallites, may be vesiculose (*Cystihalysites*); balken may occur in margins of the small, intermediate corallites.

Genus **HALYSITES** Fischer von Waldheim, 1828

1813. *Alyssites* Fischer von Waldheim, p. 387.
1828. *Halysites* Fischer von Waldheim, p. 15.
1954. *Halysites* Thomas & Smith, p. 766 (*partim*).
1955. *Halysites* Buehler, pp. 21, 24-25.
1956. *Halysites* Duncan, pp. 222-223 (*partim*).
1957. *Schedohalysites* Hamada, pp. 397, 401.
1957. *Acanthohalysites* Hamada, pp. 397, 404.

Type species. — (By monotypy) *Tubipora catenularia* Linnaeus, 1767, p. 1270.

Type locality and horizon. — "Thrown up on the shores of the Baltic" (*fide* Buehler, 1955, p. 24). Neotype (see *Halysites catenularius*) from the Silurian of Gotland.

Diagnosis. — *Halysitidae* with circular or elliptical corallites, two or more in each rank; interstitial tubes present between some or all corallites; tabulae in both corallites and interstitial tubes complete.

Description. — These are corals of cateniform habit, with corallites arranged like the vertical poles of a palisade, to form the ranks of the corallum. The individual walls of the palisade branch and connect with other branches, producing, when viewed from above, what looks like an aggregation of interlinked chains. Each corallite is prostrate basally, except for those which arise later, through interstitial budding. Distally, the corallites flex upward, so that for most of its length, each corallite has a vertically-directed axis.

Two sorts of tubular structures constitute the corallum: the corallites, and the interstitial tubes (usually referred to as microcorallites or mesocorallites in the literature; see section on morphologic terms). The corallites are the larger and more obvious of these tubes, and the interstitial tubes alternate between all or some pairs of corallites in the ranks. Both kinds of tubes have two-layered walls, of which the outermost is thin and optically denser than the inner layer, and is continuous along both outer surfaces of the ranks. Sandwiched between the two sides of outer wall, the inner, coenosteum-like layer lines both the corallites and the interstitial tubes.

A distinctive feature called "balken" sometimes occurs in the form of vertical rods in the four corners of the interstitial tubes, parallel to the axis. These are of various forms, and are probably an important specific taxobasis for the genus. (Hamada, 1959, pp. 282-284 discusses this feature.)

The corallites and interstitial tubes contain tabulae; those of the latter are more closely spaced than those of the former. Spines (acanthine septa) occur in the corallites (not, apparently, in the interstitial tubes) of some species, and their development within a given specimen varies.

Discussion. — Fischer von Waldheim (1813, p. 387) erected the genus *Alyssites* (Greek for "chain rock") for *Tubipora catenularia* Linnaeus (1767), a species he determined not to be a true *Tubipora.* In a subsequent paper (1828, p. 15) he changed the spelling (though not the meaning) of the original generic name to *Halysites.* Fischer von Waldheim apparently considered this emendation insignificant, as he made no reference to it in his 1828 paper, and further stated there (p. 15) that, "Les Membres de notre Société, ainsi que mes élèves, connoissent ce genre sous le nom de *Halysites* depuis 1806." If this reference to long-standing usage was actually to the orthography *Halysites*, it suggests that the author considered the latinized transliterations of the Greek word to be as interchangeable taxonomically as they would be lexicographically.

Thomas and Smith (1954, p. 766) considered Fischer von Waldheim's emendation to be correct, quoting Lang, Smith, & Thomas (1940, p. 64) as judging this action a proper one, and noting its general acceptance. Buehler, however (1955, p. 21) considered such an emendation contrary to the rules of the International Code of Zoological Nomenclature but felt that a reversion to the use of *Alyssites* would cause undue confusion in the literature. He indicated his intention to petition the International Commission on Zoological Nomenclature to suppress *Alyssites*. Recently, however, Buehler informed me (personal communication) that such a petition never materialized, as he had become convinced that Fischer von Waldheim's emendation was legal. This does not seem to be the time or place to resolve the matter, and for the present at least, the subsequent and commonplace rendition for the generic name will be used.

Fischer von Waldheim (1828, p. 15) noted Lamarck's introduction of the generic name *Catenipora* (1816), but considered it a synonym of his *Halysites-Alyssites* (1813). Nonetheless, *Catenipora* continued in use, appearing in Goldfuss' work (1826) and that of Hall (e.g., 1852). The latter seems to have been the last author to employ *Catenipora* as a valid generic name until its resurrection by Buehler in 1955 (see *Catenipora* synonymy). Even in their taxonomic work on the genus *Halysites*, Thomas and Smith (1954) considered *Catenipora escharoides* Lamarck, 1816, the type species of *Catenipora* Lamarck, to be congeneric with *Halysites catenularius* (Linnaeus, 1767), placing both species in Fischer von Waldheim's genus.

As a result *Halysites* has long been used for *Catenipora* (see discussion of *Catenipora*). As noted above, Buehler resurrected the term *Catenipora* as a valid generic name, and Hamada, in his important work on the Halysitidae (1957) accepted this, while assigning spiniferous specimens, otherwise referable to *Halysites,* to *Acanthohalysites* Hamada (pp. 397, 404). It now seems that the development of spines in the halysitids may be gradational within a genus, species, or single specimen (as seems to be the case in *Favosites*), and consequently, *Acanthohalysites* is here considered a junior synonym of *Halysites.*

Duncan (1956), in her survey of Ordovician and Silurian corals of the western United States, used *Catenipora* (as well as *Cystihalysites* Chernyshev, 1941b, p. 70, based upon *Cystihalysites mirabilis* from the Upper Silurian of the southwestern Taimyrs in Siberia) as a subgenus of *Halysites.*

Hamada's (1957) *Schedohalysites* was intended to accommodate those halysitids in which interstitial tubes are present in certain portions of the corallum, but not in others. These structures, however, are not necessarily ubiquitous in a corallum of *Halysites,* either as a result of the natural living form of the colony, or the vagaries of preservation. Therefore a corallum is here assigned to *Halysites* if it has any interstitial tubes at all, provided that it does not fall into the realm of *Cystihalysites* (in having vesicular dissepiments in the interstitial tubes and the peripheries of the corallite lumens). Consequently, *Schedohalysites* is a junior synonym of *Halysites.*

Halysites catenularius (Linnaeus)　　　Pl. 32, figs. 2, 3; Pl. 37, figs. 1, 2; Pl. 41, fig. 2

1767. *Tubipora catenularia* Linnaeus, pp. 1270-1271.
? 1875. *Halysites catenularia* (Linnaeus), Nicholson, p. 227.
? 1876. *Halysites catenulata* [*sic*] (Linnaeus), Rominger (*partim*), pp. 78-79, pl. 29, figs. 1, 2 (*non* fig. 4).
? 1890. *Halysites catenulatus* [*sic*] (Linnaeus), Foerste, pp. 337-338.
1904. *Halysites cratus* Etheridge, pp. 27-29, pl. 1, fig. 1; pl. 4, figs. 3, 4; pl. 6, figs. 5, 6.
? 1909b. *Halysites catenularia* (Linnaeus), Foerste, p. 10.
? 1928. *Halysites catenularia* (Linnaeus), Twenhofel, p. 125.
? 1939. *Halysites catenularia* (Linnaeus), Shrock & Twenhofel, p. 255, pl. 29, fig. 18 (*possibly* 21, 23).
? 1939. *Halysites catenularia* (Linnaeus), Northrop, p. 153.
1954. *Halysites catenularius* (Linnaeus), Thomas & Smith, pp. 766-768, pl. 20, figs. 1a-c.
1955. "The Genotype of *Halysites*" Buehler, pp. 24-25.
1955. *Halysites catenularius* (Linnaeus), Buehler, pp. 28-29.
1958. *Halysites cratus* Etheridge, Hamada, p. 101, pl. 10, figs. 5, 6a, b.
1961a. *Halysites junior* Klaamann, pp. 93-95, pl. 12, figs. 1-5, text-fig. 6.
1962. *Halysites catenularius* (Linnaeus), Flügel, pp. 316-317, pl. 22, fig. 4.

Type material. — Neotype (of Thomas & Smith, 1954, p. 767): Specimen No. 1 of the Bromell Collection in the Paleontologiska Institution, Uppsala, Sweden; described and illustrated by Bromell (1728), p. 411, No. 5, and fig. 11 on the plate opposite p. 410. This specimen is figured by Thomas and Smith (pl. 20, figs. 1a-c), and in the present work (Pl. 37, figs. 1, 2; Pl. 41, fig. 2).

Type locality. — Silurian of Gotland.

Number of specimens examined. — One specimen found in the field.

Diagnosis. — *Halysites* with nearly ubiquitous interstitial tubes, lacking balken; corallites circular to elliptical in cross-section, measuring about 1.5 x 1.0 mm in inside diameter; interstitial tubes tetragonal in cross-section with longer side perpendicular to direction of rank extension, or equal-sided (square); spines absent, or weakly-developed; ranks frequently joined to form lacunae.

Description. — The original size and shape of the Brassfield corallum is difficult to deduce from the available material. The longest corallite is 3.5 cm in length, and seems to be based upon a very thin, apparently solid encrusting object. The length of the ranks varies considerably, most consisting of three to seven corallites. The lacunae may be long and narrow (26 x 2 mm) or more nearly equilateral and small (7 x 4 mm).

The corallites are elliptical to circular in cross-section. Their internal diameter parallel to the direction of rank extension is fairly uniform at 1.5 mm, while the diameter across the rank varies from about 1.0 mm (in elliptical tubes) to 1.5 mm (in circular ones).

Between almost every pair of corallites is an interstitial tube. These are tetragonal in cross-section, variable in size (but alwav⸗ much smaller than the corallites). In the direction of rank exten-sion, these are from 0.06 to 0.25 mm wide (mostly about 0.20 mm). Across the rank they are 0.06 to 0.3 mm wide. Characteristically, their greatest dimension is perpendicular to the direction of rank extension.

The corallites and interstitial tubes are enclosed on two sides bv a theca about 0.2 mm thick. They are not, however, separated from the immediately preceding or following corallite by this theca. Instead, surrounding each corallite, and separating it from its neigh-bors, is a deposit of skeletal material which is optically less dense than the theca. Within this deposit lies the interstitial tube. The above-mentioned theca appears to consist of two layers, the outer accounting for about 1/3 of the thickness, represented by a deposit optically denser than the inner layer.

The tabulae of both the corallites and interstitial tubes are shallowly concave upward. Those of the former are spaced vertically about 1.6 to 1.8 per mm, and are nearly all complete. Those of the latter are spaced about 4.8 per mm. There is a suggestion of zonation in the tabular spacing in the corallites, but not enough material was studied to determine the degree to which this occurs.

While no spines were observed on the walls of the corallites, vertical sections reveal longitudinal striations on these walls which may reflect sutures or planes of the intercorallite region (Pl. 32, fig. 3).

Discussion. — Linnaeus (1767) gave the name *Tubipora catenularia* to fossil corals "*projicitur frequens ad Littora* M. Balthici" ("frequently cast upon the shore of the Baltic Sea"), which he diagnosed as consisting of parallel tubes connected in contorted-plicated anastomosing laminae. He further noted the presence of a firm, erect skeleton, which in cross-section appears like simple series of chains. The tubes are equal (in diameter?), connected laterally. There is no mention of interstitial tubes.

While preparing his discourse on the halysitids, Buehler realized the necessity for an adequate description of the type species of *Halysites*. This need was fulfilled by Thomas and Smith (1954) who selected a neotype from the Bromell Collection. Bromell's work (1728) and specimens were referred to by Linnaeus in his synonymy for *T. catenularia,* and in some of his earlier works.

Thomas and Smith (1954, pp. 767-768) described the neotype, whose general characters are as follows (here paraphrased): There are about five corallites in 10 mm along the chains, and their calicular centers are about 1.9 mm apart. The internal diameters are about 1.52 x 1.18 mm, and the walls are about 0.19 mm thick. There are as a rule seven to eight corallites in a long rank, and two to three in a short one. The juncture of these ranks produces elongate lacunae, commonly only 1 to 2 mm wide and up to 30 mm long. An interstitial tube, square or rectangular in cross-section lies between each pair of corallites. Ranks always bifurcate from these interstitial tubes. The interstitial tubes, when not equal-sided, are elongate at right angles to the direction of rank extension, and average 0.31 x 0.19 mm in diameter.

The corallites contain complete tabulae that are either nearly flat or gently concave upward, spaced vertically about 2.2 to 2.4 per mm. The tabulae of the interstitial tubes are of similar topography, and average 2.4 to 2.8 per mm. Septa (including spines?) are reportedly absent. Sections of the neotype, loaned by the Paleontological Institution at Uppsala, Sweden, are shown on Plate 37, figures 1, 2 and Plate 41, figure 2.

There is good agreement between the neotype of Linnaeus' species and the Brassfield specimen assigned to it. The corallites and interstitial tubes are of the same size and shapes, and the tabulae have the same topography in both specimens. The lacunae tend to be elongate in both specimens, and the theca is of the same type. Balken are apparently absent in the neotype, as they are in the Brassfield specimen. The only difference is that the tabulae of the neotype corallites are somewhat more closely spaced, and those of its interstitial tubes more widely spaced, than in my specimen. The similarities seem strong enough, and the differences sufficiently trivial, to allow confident identification of the Brassfield form as *H. catenularius.*

Nicholson's (1875) reference to *Halysites catenularia* [*sic*] from the "Clinton Group" of southwest Ohio is based upon several specimens, and the information given is insufficient for determination of the appropriateness of the name. His observation that septa are present, the great range of the corallite diameters among his specimens (about 0.7 to 3.5 mm), and the absence of any reference to interstitial tubes, leaves Nicholson's identification open to question.

Similarly, Rominger's descriptions and figures of specimens from the Silurian of northern Michigan at Drummond Island and Point Detour, identified as *Halysites catenulata* [*sic*] Linnaeus, must remain in question. His material clearly represents several species, a probability he acknowledged, assigning the forms to Linnaeus' species only because he could find no break in their intergradation. Some of his specimens with particularly large corallite diameters of 2 mm or more may belong to *Halysites labyrinthicus* (Goldfuss), originally described from Drummond Island. Rominger's figure 4 has small corallites with nearly-parallel sides, and is a separate species, possibly conspecific with *Harmodites ramosa* Fischer von Waldheim (1828, fig. 5). The absence of mention of interstitial tubes or septal spines makes a species-level identification of Rominger's specimens difficult.

Foerste (1890) reported halysitids from southwest Ohio, apparently from Brassfield strata. His information covers only tube diameters, corallum sizes, and tabular spacing, none of which is sufficient to determine species.

Etheridge (1904) described *Halysites cratus* from New South Wales, Australia. The age is not given, but is probably Silurian. The corallum consists of chains from one to six corallites long, most commonly four. The lacunae are of variable form, and include equidimensional and very elongate examples. The corallites are elliptical to circular in cross-section, 1.5 to 2.0 x 1.0 to 1.5 mm in diameter, elongate parallel to the direction of rank extension. Etheridge did not indicate whether this was the inside or outside diameter. Between the corallites lie tetragonal interstitial tubes, transversely elongate, measuring 0.5 to 0.75 x 0.25 to 0.75 mm. The tabulae of these tubes are complete and horizontal, spaced four to five per mm. Those of the corallites are complete, and horizontal or concave, spaced two to three per mm. The sections in Etheridge's plate 6, and the description do not indicate the presence of spines or balken in the corallite interiors. Despite the greater diameter of the corallites, it seems reasonable to synonymize *H. cratus* with *H. catenularius*.

Foerste (1909b) reported *Halysites catenularius* from the West Union "horizon" (Late Llandovery or Early Wenlock) of Kentucky. The only information he gave is that the corallite diameters are 2 mm along their greatest width, and there are closely spaced, strongly arched (convex?) tabulae in the corallite lumens. It seems doubtful that Foerste was dealing with *H. catenularius*.

Twenhofel (1928) made reference to this species from the English Head Formation (Late Ordovician) through the Chicotte Formation (Early Wenlock) at Anticosti Island. Commonly, the corallites are about 1 mm in diameter, and are spaced four per cm. This agrees with the Brassfield specimen. Interstitial tubes are usually, but not always present. There is insufficient information to validate the identification of this species. The oldest of Twenhofel's specimens, from the English Head Formation, has no interstitial tubes, and the corallites are nearly square in outline. These features would certainly exclude it from Linnaeus' species as here understood.

Shrock and Twenhofel (1939) reported corals, some of which may belong to *H. catenularius*, from the Pike Arm and Natlins Cove Formations (Silurian) of northern Newfoundland. The corallites

range in length from 1 to 3 mm, with from 3 to 8 corallites in a distance of 10 mm, and in width from 0.5 to 1 mm. In most specimens there are tiny tubules between the corallites. These may be circular or quadrangular in transverse section. They range in length from 0.5 to 1 mm and in width from 0.3 to 0.75 mm. Both corallites and tubules are closely tabulate, with the tabulae in the former horizontal and spaced about 8 in 5 mm; in the latter, convex upward and spaced about 13 in 5 mm.

Figure 18 of these authors, an enlarged transverse section of part of a rank, strongly resembles my form in size and shape of the corallites and interstitial tubes, and in the apparent absence of septal spines. If, however, the tabulae of the interstitial tubes are convex upward as indicated in the description, then this specimen might be more reasonably assigned to *Halysites brownsportensis* Amsden (1949). It seems best to await more detailed data before accepting the presence of *H. catenularius* in Newfoundland.

Northrop (1939) reported the present species from Gaspé, maritime Canada, in beds ranging in age from Late Llandovery to Middle Ludlow (the Clemville, La Vieille, Gascons, Bouleaux, West Point, and Indian Point Formations). Unfortunately, the only morphologic data given is that interstitial tubes are present, and that there is considerable variety of form. This information is insufficient to validate identification with *H. catenularius* (Linnaeus).

Hamada (1958) reported *Halysites cratus* Etheridge from what he considered Lower Ludlovian beds in southern Japan. The chains consist of two to five corallites surrounding polygonal lacunae. The corallites are rounded-elliptical, about 1.6 x 1.4 mm in diameter (elongate in the direction of rank extension) and have a theca 0.1 mm thick. Between them lie interstitial tubes, tetragonal in cross-section (but Hamada describes them as triangular), that are elongate transverse to the direction of rank extension. These are about 0.3 x 0.8 mm on a side [*sic*]. The tabulae in both interstitial tubes and corallites are complete and flat, spaced two to three per mm in the corallites and three to four per mm in the interstitial tubes. Septal spines were not observed. Hamada considered that his specimens were similar to the neotype of *H. catenularius* (Linnaeus), but differed in lacking its labyrinthine lacunae and rather thick corallite walls. The difference in thecal thickness (0.1 mm, compared to 0.19 in the neotype) is not here considered critical, nor is the configuration of the lacunae. Etheridge's specimen, with which Hamada's is in good agreement, had both equidimensional and elongate lacunae. It seems reasonable to assign Hamada's form to *H. catenularius*, as was done with Etheridge's.

From the Yagarakhuskii horizon (Wenlock in age) near Saaremaa, Estonia, Klaamann (1961a) reported a new halysitid species, *Halysites junior*. This form consists of ranks with from two to 13 coralliates enclosing polygonal or elongate lacunae. The corallites are 1.8 to 2.3 mm in diameter. The theca is 0.25 to 0.30 mm thick, slightly less at the interstitial tubes, which appear to be nearly ubiquitous. These latter are tetragonal in cross-section, elongate at right angles to the rank extension, and as a rule are 0.25 to 0.3 mm wide, though they may be as narrow as 0.05 to 0.1 mm. The tabulae of the corallites are generally complete, and flat to slightly concave upward, spaced two to five per mm. Those of the interstitial tubes are usually complete, flat, and spaced four to five per mm. Spines are reportedly absent, and there is no indication of the presence of balken. *Halysites junior* is similar to the neotype of *H. catenularius*, and to my specimen of that species, except that its corallites are somewhat larger. Corallites whose diameters were of the lower limits reported by Klaamann (1.8 and 1.5 mm) would not be out-of-line with those of the neotype and the Brassfield specimen. Klaamann reported that his material consisted of twenty specimens, so it is unclear whether his range was among several coralla, or within one. It is likely, considering all the other similarities, that Klaamann's material belongs to *H. catenularius*, and represents a form with unusually large corallites.

From the Silurian of Ozbak-Kuh, northeastern Iran, Flügel (1962) reported *Halysites catenularius*. This form consists of chains of up to eight corallites enclosing what appear to be moderately elongate lacunae. The corallites have diameters of 1.75 x 1.125 mm, elongate in the direction of rank extension, with walls 0.2 to 0.225 mm thick. The walls have an outer layer 0.02 mm thick, which is optically-denser than the inner wall (as in the Brassfield form). There are five corallites per cm in the Iranian form (as compared with about four per cm in my specimen), and their centers lie 2.0 to 2.5 mm apart (as in the Brassfield form). Between the corallites are tetragonal interstitial tubes, ranging in size from 0.3 x 0.4 mm to 0.3 x 0.65 mm. Judging from Flügel's figure 4, these may be equidimensional, or elongate either parallel or perpendicular to the direction of rank extension. The tabulae of the corallites are flat, spaced two to three per mm. Those of the interstitial tubes are three to four per mm. There are no obvious septal spines, and no balken. As Flügel pointed out, his material agrees well with the neotype of *Halysites catenularius*, and there is no reason to doubt his identification.

Distribution. — Gotland, Silurian; Estonia, Wenlock; Ohio, mid-Llandovery (Brassfield Formation); Iran, Silurian (probably post-Middle Llandovery); Australia, Silurian; Japan, Early Ludlow.

Reported occurrences deserving further study include: Anticosti Island, Llandovery to Early Wenlock; Newfoundland, Silurian; Gaspé, Late Llandovery to Middle Ludlow.

Brassfield occurrence. — Locality 14, near Piqua, in southwest Ohio.

Halysites nitidus Lambe   Pl. 31, figs. 3, 4; Pl. 41, fig. 1

1899. *Halysites catenularia,* var. *nitida* Lambe, pp. 71, 76, 77, pl. 4, figs. 2, 2a, 2b.
? 1926. *Halysites pulchellus* Wilson, p. 15, pl. 3, figs. 8, 9.
1955. *Halysites nitida* Lambe, Buehler, p. 49, pl. 8, figs. 4-7; pl. 9, fig. 1.
? 1962a. *Halysites nitidia* Lambe, Norford, pl. 9, figs. 7, 12.
1964. *Halysites louisvillensis* Stumm, p. 79, pl. 80, figs. 8-10.
1971. *Acanthohalysites* sp. Scrutton, pp. 221-222, pl. 5, figs. 5, 6.

Type material. — Holotype: GSC 3035, the specimen figured by Buehler (1955, pl. 8, figs. 6, 7). Bolton (1960, p. 40) gave GSC 3035 as the catalogue number of the holotype, but Buehler gives the number as GSC 104761. His figures were made from photographs supplied by W. A. Bell of the Canadian Geological Survey, and it is not clear if the specimen or the number is in error.

Type locality and horizon. — According to Bolton (1960) the La Vieille Formation (Late Llandovery and Wenlock) at L'Anse à la Barbe, Chaleur Bay, in the Gaspé region of Quebec. (Lambe, 1899, p. 76, considered this unit to be of Lower Helderberg [lowest Devonian] stratigraphic position.)

Number of specimens examined. — Two, found in the field.

Diagnosis. — *Halysites* with corallite diameter 1.2 to 1.5 mm in direction of rank extension, and 0.7 to 1.0 mm at right angles to that direction; interstitial corallites tetragonal, not ubiquitous; corallites with 12 longitudinal rows of septal spines; ranks frequently join to form lacunae; balken absent.

Description. — Two coralla were found at different localities. The larger (UCGM 41651) was found in life position at locality 14, near Piqua, Ohio. Sections of this specimen are shown on Plates 31 and 41. About 5.5 cm of its original height remains, and the width of the upper portion of the corallum is 7 cm. The smaller specimen (UCGM 41997) is from locality 30, near Seatonville, Kentucky, and is 4.5 cm high and 3.5 cm across the top. The lacunae are of variable shape. In the Piqua specimen they include both equilateral (4 x 4 mm) and elongate forms (7 x 3 mm). There are usually three to five corallites to a rank, the latter more common than the former. The pattern in the Seatonville specimen is similar.

The corallites of the Brassfield specimens are of elliptical cross-section, longer in the direction of rank extension than at right angles to that direction. Those of the Piqua specimen are rather varied in size and shape, one having inside diameters of 2.0 x 0.7 mm, another having diameters of 1.3 x 0.8 mm. As in the Brassfield specimen of *Halysites catenularius* (Pl. 32, figs. 2, 3), the length of the corallites is more variable than the width, some being quite elongate, while others are nearly circular in cross-section. The corallites of the Seatonville specimen are more uniform, typically 1.3 x 0.7 mm internally. The theca of the Piqua specimen (the better-preserved of the two) is about 1.2 mm thick, and consists of two layers, the outer more optically dense than the inner.

Between some of the corallites are tetragonal interstitial tubes that are usually widest perpendicular to the direction of rank extension. Typically these are 0.3 x 0.1 mm but one equidimensional representative measures 0.1 mm in both directions. These features are set in skeletal deposits with about the optical density of the inner thecal layer, but are apparently structurally distinct. In this sense, the structure is similar to that in *H. catenularius*. There is no evidence of balken in these specimens.

The spines are well-developed, up to 0.2 mm long (Pl. 41, fig. 1), and are inclined upward toward the axis. They do not appear in every corallite, either because of differential preservation, or differential occurrence in life.

The tabulae of the corallites are concave upward, and spaced about two per mm. There are about six per mm in the interstitial tubes.

Discussion. — Lambe (1899, pp. 76, 77) gave the corallite diameters as 1.45 x 1.00 mm, and the average width of the interstitial tubes as 0.50 mm. The lacunae are reportedly long and narrow, 3 to 5 mm across, often with several parallel ranks. Spines are present, pointing upward in the figure (as noted in my section). The tabulae of the corallite tubes are described as flat or slightly convex (though in Lambe's figure they are nearly all flat or gently *concave*), spaced two to four per mm. Those of the interstitial tubes are reportedly straight (probably meaning flat) or slightly convex, at times vesicular (incomplete, also seen in my material).

Halysites pulchellus Wilson (1926), comes from the Beaverfoot Formation (Late Ordovician or Llandovery) of British Columbia. It is described as having corallites that were "oval" (elliptical?) in outline, 1 mm wide, with about three corallites in 5 mm (similar to my specimens). The walls are described as thicker than usual in so slender a species. The tabulae are complete, averaging 1.3 to 1.7 per mm. There is no mention of spines, and interstitial tubes are reportedly absent. Buehler (1955, p. 59) also reported the absence of spines and interstitial tubes. I have examined the much-recrystallized holotype however (GSC 6735), and found that it does have spines. I was unable to detect interstitial tubes, though, so that despite strong similarities between Wilson's specimen and mine, it cannot be confidently said that they are conspecific.

Norford (1962a) figured a specimen from the Upper Sandpile Group (probably Llandovery or Early Wenlock) of British Columbia, which he identified with *H. nitidus*. The corallites are about 1.5 mm in maximum diameter, and about 0.7 mm in minimum diameter, measured on the inside. Interstitial tubes are present. The tabulae of the corallites are mostly horizontal or concave, spaced about 3.5 to 5.0 per mm. Those of the interstitial tubes are horizontal or slightly convex, spaced five to six per mm. As no spines are clearly shown, it cannot be confidently said that this specimen is conspecific with Lambe's though the other features are in good agreement with *H. nitidus*.

Halysites louisvillensis Stumm from the Louisville Limestone (Late Wenlock and Early Ludlow) at Louisville, Kentucky appears to be conspecific with *H. nitidus*. The corallites are elliptical, slightly under 1 mm in maximum diameter, and about 0.8 mm in minimum diameter. Twelve septal spines are visible in cross-section. The interstitial tubes are tetragonal and small, averaging 0.1 mm across. The corallite tabulae are complete, flat or slightly convex, spaced about 2 to 4 (and reportedly, 6) per mm. The tabulae of the interstitial tubes are flat, and more closely spaced. There seems to be no significant difference between *H. louisvillensis* and *H. nitidus*, although it would be desirable to determine if the spines of the former project upward axially, as in the latter.

Scrutton (1971) reported an unidentified species of halysitid from sediments of what he regarded as Ludlow age in the Río Suripá region of Venezuela. The maximum inside diameters are 1.8 x 1.3 mm (in the largest corallites). There are as many as seven corallites to a rank. Between many of these are interstitial tubes, tetragonal or circular in cross-section, with maximum dimensions of 0.45 to 0.25 mm, and elongate perpendicular to the direction of rank extension. These are separated from the corallites by partitions which are enclosed, together with the corallites, in a theca 0.2 to 0.3 mm thick. The same structure is found in the Brassfield specimens. The tabulae of the corallites are flat or concave, spaced 2.6 to 3 per mm. They are rarely incomplete. The tabulae of the interstitial tubes also appear flat or concave, but their spacing was not described. Spines up to 0.25 mm long are reportedly present in some cases. Scrutton did not feel that this form could be referred to a known species, but hesitated to erect a new species, because his material was so poorly preserved. As described and figured, however, this form is quite similar to *H. nitidus* as I understand it, and it is assigned to that species.

Distribution. — Gaspé, Late Llandovery or Wenlock; Venezuela, Ludlow; Kentucky, mid-Llandovery (Brassfield Formation), and Late Wenlock or Early Ludlow; Ohio, mid-Llandovery (Brassfield Formation).

Possible occurrence in British Columbia (Llandovery or Early Wenlock).

Brassfield occurrence. — Localities 14 (near Piqua, southwest Ohio), and 30 (near Seatonville, central Kentucky).

Halysites(?) meandrinus (Troost) ?   Pl. 32, fig. 1

? 1840. *Catenipora meandrina* Troost, pp. 68-69.
Cf. 1955. *Halysites meandrina* Buehler, pp. 34-35, pl. 4, fig. 1.
Cf. 1967. *Catenipora jarviki* Stasínska, pp. 47-48, pl. 2, figs. 1a, b.
Cf. 1969. *Halysites praecedens* Webby & Semeniuk, pp. 349-353, figs. 2-6.

Type material. — Neotype (of *C. meandrina* Troost) is YPM 19237, selected by Buehler (1955, p. 35) and figured in his plate 4, fig. 1.

Type locality. — Lobelville Formation (now part of the Brownsport Formation), of Early Ludlow age, at Short Creek in Linden, Perry County, western Tennessee.

Number of specimens studied. — One, found in the field.

Diagnosis (of *H. meandrinus*). — Species of *Halysites* with long ranks (commonly 10 or more corallites to a rank) rarely forming closed lacunae; interstitial tubes not ubiquitous; corallites generally 1.7 mm in diameter (parallel to rank extension) by 1.0 mm across the rank (measured inside corallite), possibly ranging downward from this size.

Description (of Brassfield specimen). — The corallum is composed of long ranks of up to 10 corallites linked together by shorter ranks of one or two corallites. The ranks are seldom observed to close around a lacuna. The area occupied by the colony thus included a relatively sparse population of polyps.

The corallites typically have inside diameters of 1.0 x 0.5 mm, widest in the direction of rank extension. The theca is about 0.3 mm thick. Each corallite is consistently separated from its daughter by a distance equivalent to its long diameter. While the intervening spaces are broad enough to accommodate interstitial tubes, no unequivocal examples were observed in the single specimen studied. The specimen is so poorly preserved that the relatively fine structure of an interstitial tube would probably be obscured.

There is some suggestion of spines in the corallites, but again, there has been too much alteration of the specimen to be certain. It was not possible to make longitudinal sections, so nothing is known of the tabulae.

Discussion. — The Brassfield specimen is fragmentary and poorly-preserved, so that the identification must necessarily remain tentative. The similarities to *H. meandrinus*, however, are distinct.

Troost's original description of this species is quoted *in toto* below, because of the relative difficulty of obtaining his reference:

The tubes and apertures are in size and form like those of *C. labyrinthica,* [size of this species was not given in Troost's work, but he did refer to its tubes as oval, rather than lanceolate] but they do not contortuously anastomose as in the latter, and meandering in an irregular manner in every direction, they form no meshes. They seem to have formed large masses. I have a mass of compact limestone from the vicinity of Nashville of the size of five by four inches, which is water worn, and the Catenipora, being less soluble, and having thereby come to light, shows these lamellae running in every direction over the whole surface of the stone, sometimes appearing as chain-form apertures, sometimes showing their sides.

It is only when worn down in that manner, that they are perceptible, but are invisible in the fractures of the rock.

I discovered them, not only near Nashville but I found the same, though not in such large masses, in a more granular limestone, associated with Cyathophillum [sic] gracile and its associates near Brownsport, Perry county.

The only critical morphologic data is the oval shape of the corallite cross-sections, and the infrequency with which the ranks close to form lacunae. The Brassfield specimen displays these traits, but there is insufficient information for reconstruction of the specimens Troost described.

Buehler (1955) told of finding two specimens (he appears to have worked with three) in the collection of the Yale Peabody Museum with coralla displaying features described by Troost, and corallites with the size and form of those found in *H. labyrinthicus* (Goldfuss), from the same locality as Troost's specimens. From these he selected a neotype for *Catenipora meandrina* Troost.

The neotype, YPM 19237, is viewed from the bottom in Buehler's figure, and is silicified, as are the two other specimens associated with it in the YPM collection. It is larger than the Brassfield specimen, having corallite diameters (inside measurements) of 1.7 x 1.0 mm, and a rugose theca about 0.4 mm thick. The corallites are situated about 1.0 mm apart, and these intervening spaces occasionally contain an interstitial tube. These latter are tetragonal in cross-section, either equidimensional or with the longer side perpendicular to the direction of rank extension.

Four ranks are preserved whole in the neotype: two of ten corallites, one of six, and one of three. Two of the broken ranks contained at least nine corallites. In this respect, the corallum is similar in form to the Brassfield specimen.

No septal spines were observed, and only a glimpse of the tabulae was visible when the neotype was examined. These tabulae appear to be gently concave upward, not convex as reported by Buehler, and are insufficiently clear to permit determination of their spacing.

Two specimens identified as *H. meandrina* are associated in the YPM collection with Buehler's neotype, though neither has a catalogue number. One of them agrees closely with the neotype in all respects. The other is basically similar, but some of the corallites are unusually narrow (0.9 mm). On the same rock as the latter specimen, however, is a corallum with the basic form of *H. mean-drinus*, but much more delicate in form (internal corallite diameters of 0.9 x 0.3 mm). The smaller corallum is not clearly attached to the larger, but lies between two of its ranks. If these two corals are conspecific, then the Brassfield form lies on a morphologic continuum between them.

Stasínska (1967) described *Catenipora jarviki* from the Wenlock of Gotland near Visby (probably the Early Wenlock Högklint beds), a form that shares several characters with my own. It is characterized by corallites with long diameters of 0.6 to 0.9 mm and short diameters of 0.4 to 0.5 mm, arranged in long ranks of up to ten corallites. These ranks enclose long, narrow lacunae. A transverse section of the holotype, exposing many corallites and ranks, shows few closed lacunae, and might approximate the pattern in the Brassfield specimen. The corallite walls are thin (0.06 mm), whereas those of the Brassfield specimen are comparatively thick (0.3 mm). The "intercorallite walls" (presumably the spaces separating adjacent corallites) are 0.2 mm thick (in my specimen the corallites are separated by 1 mm of skeletal deposits). Spines are present in the Gotland specimen, while their presence in mine is uncertain. Finally, the tabulae of *C. jarviki* are horizontal or weakly convex, while the nature of the tabulae in my specimen is unknown. The true relationship between the Gotland and Brassfield specimens cannot be ascertained at this time, but the similarities show that further comparative study is warranted.

Webby and Semeniuk (1969) introduced a new species, *Halysites praecedens*, from the Bowen Park Limestone and Canomodine Limestone (both Caradoc in age, Late Ordovician) of New South Wales, Australia. The corallum consists of long ranks of corallites with infrequent T-shaped junctions between them. In one rank were counted 31 corallites. There are no enclosed lacunae. Corallites and interstitial tubes are distinct from one-another. The corallites are "oval" (elliptical) in cross-section, measuring 1.2 to 2.0 mm along the ranks, and 1.2 to 1.6 mm across the ranks. The interstitial tubes are square in cross-section and 0.2 to 0.5 mm on a side. The wall is 0.2 to 0.3 mm thick, consisting of a thin, discontinuous "epitheca", a middle fibrous layer with the fibers perpendicular to the outer wall, and an inner, non-fibrous layer. The last of these layers forms the barrier between corallites and interstitial tubes. Tabulae are thin, mostly horizontal (though in fig. 6a, these authors showed longitudinal sections of corallites having obliquely directed tabulae, in what they referred to as "deformed" corallites), and spaced from six to seven in 5 mm in the corallites, and nine to 11 in 5 mm in the interstitial tubes.

Webby and Semeniuk marked the similarity of *H. praecedens* and *H. meandrinus*, noting that the two forms have similar dimensions, but that rank junctions occur more frequently in the latter than in the former. The corallites are somewhat wider than those of the Brassfield specimen, but are in good agreement with the neotype of *C. meandrina*. The fact that *H. meandrinus* is poorly understood makes it inadvisable to equate it with *H. praecedens*, especially in light of the considerable time separating them. Nevertheless, the many features shared by these two forms should encourage closer scrutiny in future studies.

Distribution. — Kentucky, mid-Llandovery (Brassfield Formation); similar forms in the Early Ludlow of Tennessee, Early Wenlock of Gotland, and Caradoc of Australia.

Brassfield occurrence. — Locality 53a, near Irvine, central Kentucky.

Genus **CATENIPORA** Lamarck, 1816

1816. *Catenipora* Lamarck, p. 206.
? 1826. *Catenipora* Goldfuss, p. 74 (*partim*).
1843. *Catenipora* Hall, illustration No. 22 (*partim, non* fig. 2).
1850. *Halysites* Milne-Edwards & Haime, p. lxii (*partim*).
1851. *Halysites* Milne-Edwards & Haime, p. 281 (*partim*).
1852. *Catenipora* Hall, pp. 127-130 (*partim, non* pp. 129-130).
1879. *Halysites* Nicholson, pp. 226-231 (*partim*).
1880. *Halysites* Nicholson & Etheridge, Jr., pp. 274-276 (*partim*).
1899. *Halysites* Lambe, pp. 64-68 (*partim*).
? 1904. *Halysites* Etheridge, pp. 15-22 (*partim?*).
1940. *Catenipora* Lang, Smith & Thomas, p. 33 (as synonym of *Halysites*).
1940. *Halysites* Lang, Smith & Thomas, pp. 64-65 (*partim*), p. 33.
1941a. *Palaeohalysites* Chernyshev, p. 36.
1954. *Halysites* Thomas & Smith, p. 766 (*partim*).
1955. *Catenipora* Buehler, pp. 21-22.
1955. *Quepora* Sinclair, p. 96.
1956. *Catenipora* Hill & Stumm, p. 469.

1956. *Catenipora* Duncan, pp. 222-223 (as a subgenus of *Halysites*).
? 1957. *Eocatenipora* Hamada, pp. 396, 398.
1957. *Quepora* Hamada, pp. 396, 398-399.
1957. *Catenipora* Hamada, pp. 396, 399-401.
1961. *Catenipora* Flower, pp. 47-49.
1962. *Catenipora* Sokolov, pp. 255-256.

Type species. — (SD, Lang, Smith, and Thomas, 1940, p. 33): *Catenipora escharoides* Lamarck, 1816, p. 207.

Type locality and horizon. — Lamarck's original account (*fide* Buehler, 1955, p. 25): "Washed up on the shores of the Baltic". The neotype (Thomas & Smith, 1954, p. 25) is from the Silurian of Gotland.

Diagnosis. — Similar to *Halysites* but lacking interstitial tubes.

Description. — Except for the absence of interstitial tubes, this genus is basically identical in appearance to *Halysites*. The presence of balken in *Catenipora* appears to be uncertain. Hamada (1959, p. 283) claimed that this structure of the intercorallite region is absent in *Catenipora* (and common among *Halysites*-like chain corals), with the single exception of a supposed specimen of *Catenipora escharoides* figured by Fischer-Benzon, a specimen Hamada considered anomalous. On the other hand, Flower (1961, p. 49) reported balken in his species, *C. workmanae,* though his figures are not clear enough to be convincing.

The tabulae of the corallites are complete, and there is a double wall. The spines, as in *Halysites,* vary in development from species to species, though *C. escharoides,* the type of *Catenipora,* has long spines, while the type species of *Halysites, H. catenularius,* has no noticeable spines.

Discussion. — The extensive synonymy for this genus serves to show that a considerable period of time passed during which the name *Catenipora* was discarded in favor of *Halysites* and other generic names. Hall (1852) was the last author to use the name *Catenipora* in its currently-accepted sense (applying it to *C. escharoides* with illustrations showing no interstitial tubes, though he incorrectly applied the generic name to *C. agglomerata* Hall, which proves, in thin-section, to be a *Cystihalysites*), prior to an interregnum which ended with Buehler's resurrection of the name in 1955.

Chernyshev (1941a) erected the generic name *Palaeohalysites* for *Halysites gotlandicus* Yabe (1915) from the Silurian of Gotland.

His purpose was to separate from *Halysites* Fischer von Waldheim, this and related species which lack interstitial tubes. His taxonomic criterion is the same as that for *Catenipora*, which predates his new name, by virtue of the designation of *C. escharoides* as the type species (Lang, Smith, and Thomas, 1940).

The neotype of *Catenipora escharoides* (selected by Thomas & Smith, 1954, p. 25) bears prominent septal spines. For a *Catenipora*-like form from the Ordovician of Quebec, which lacked spines, Sinclair (1955) created the generic name *Quepora*. The variation of development that spines so commonly show in tabulates (*e.g.*, *Favosites*), suggests that this alone is not a sound generic taxobasis; hence the suppression shown in the synonymy.

Hamada (1957) introduced the name *Eocatenipora* for *Halysites cylindricus* Wilson (1926) from the Upper Ordovician of British Columbia (see discussion under *Catenipora favositomima*). This is a halysitoid tabulate coral, lacking interstitial tubes and spines, with occasional free corallites (corallites unattached laterally to neighbors over portions of their length). The relationship between this form and *Catenipora* Lamarck cannot be determined with confidence until careful study is made of Wilson's specimens. As Sinclair (1955, p. 95) pointed out, halysitoid corals are likely a polythetic group, and it is not even certain that *Halysites cylindricus* is co-familial with *Halysites, Catenipora,* etc.

Catenipora gotlandica (Yabe)   Pl. 9, fig. 8; Pl. 33, figs. 1, 2; Pl. 42, fig. 2

Cf. 1829. *Catenipora approximata* Eichwald, pl. 2, fig. 9 (*fide* Klaamann, 1966, p. 46).
Cf. 1843. *Catenipora agglomerata* Hall, illustration 22, fig. 2.
Cf. 1852. *Catenipora agglomerata* Hall, Hall, p. 129, pl. 35, figs. 2a-g.
? 1876. *Halysites catenulata* [*sic*] (Linnaeus), Rominger (*partim*), pp. 78-79, pl. 29, fig. 1 (*non* fig. 2, 4).
    1915. *Halysites gotlandicus* Yabe, pp. 34-35, pl. 7, figs. 1, 2.
? 1919. *Halysites catenularia* (Linnaeus), Williams, pl. 19, fig. 2.
? 1938a. *Halysites gotlandicus* Yabe, Chernyshev, p. 128, pl. 4, figs. 3a, b.
    1955. *Halysites gotlandicus* Yabe, Buehler, p. 57.
    1966. *Catenipora* cf. *gotlandica* (Yabe), Klaamann, pp. 41-42, pl. 6, fig. 6; text-fig. 18.
Cf. 1966. *Catenipora approximata* Eichwald, Klaamann, pp. 46-48, pl. 11, figs. 1-4; pl. 12, figs. 1, 2; text-figs. 22, 23.
    1972. *Catenipora gotlandica* (Yabe), Leleshus, pp. 45-46, pl. 21, figs. 3, 4.
Cf. 1972. *Catenipora approximata* Eichwald, Leleshus, pp. 46-47, pl. 22, figs. 1-4; pl. 23, figs. 1, 2.

Type material. — Lectotype (of Leleshus, 1972, p. 45): The specimen of Yabe (1915, pl. 7, figs. 1, 2).

Type locality and horizon. — Korpklint, north of Visby on the Island of Gotland. The horizon is probably the Högklint beds, which are of Early Wenlock age, according to Manten.

Number of specimens examined. — Three, all found in the field.

Diagnosis. — *Catenipora* lacking balken; close-packed ranks of circular to broadly elliptical corallites with interior diameter generally 1.0 to 1.5 mm; spines prominent, apparently in 12 rows.

Description. — This species occurs in the Brassfield as broadly-convex lenses. The most rewarding source is a bed of grey-green silty clay at locality 7a (the "clay-coral bed"; see section on Brassfield Lithosome), where the coralla occur as worn cobbles piled one upon the other, often upside-down. As a result of this wear, it is difficult to estimate the original size of the coralla. A typical cobble has a maximum diameter of 13 cm, a minimum diameter of 8 cm, and a thickness of 4.5 cm.

The corallites of these specimens are so densely packed that it is necessary to look closely at them to be certain they are not favositid coralla.

The ranks are generally fairly short, containing a maximum of four corallites. The lacunae tend to be narrower than the corallites. The short chains may appear superficially longer than they really are because parallel chains tend to be linked by a very short chain (often a single corallite), while a new chain, coextensive with the first, continues beyond the branching to make what looks like a single long chain, parallel to its neighbors. At other times, however, the lacunae may be polygonal, even equilateral, but are always small (Pl. 9, fig. 8; Pl. 33, fig. 2).

The corallites are from 1 to 2 mm (most commonly 1.5 mm) in the inside diameter parallel to the direction of rank extension, and about 1.2 mm at right angles to this direction. They are sandwiched on two sides by the theca, which seems to consist of a single layer 0.2 to 0.25 mm thick. In the intercorallite zones, and also sandwiched by the theca, is a solid carbonate deposit, with an apparent suture running through it lengthwise. In cross-section, this deposit usually is less than 1 mm across (parallel to the rank extension) and about 1 mm wide perpendicular to that direction. Apparently, it is the same sort of structure found in *C. favositomima*. No balken were observed in the Brassfield specimens.

The corallites usually display well-developed spines (Pl. 33, figs. 1, 2; Pl. 42, fig. 2), arranged in vertical rows with about six spines per millimeter vertically, and extending up to two-fifths the distance to the axis of the corallite. The maximum number of spines seen in transverse section is 12. Commonly, a spine projects directly from the portion of the wall consisting of the inter-corallite deposits.

The tabulae are usually somewhat convex upward, spaced about four per millimeter, though some instances of denser and sparser spacing may be found.

Discussion. — The neotype of *Catenipora approximata* Eichwald, 1829 (chosen and described by Klaamann, 1966) bears a strong resemblance to this species. It is discussed under *Catenipora favositomima* in the present work. *C. approximata* differs from *C. gotlandica* in having larger, rounder corallites, which are even more close-packed than in the latter.

Hall (1843, 1852) described *Catenipora agglomerata* from the "Niagara Limestone" in Monroe County, New York. Buehler (1955), who chose as lectotype AMNH 1690/2, and figured it on his plate 4, fig. 4, noted that a search of this locality revealed no outcrops. He supposed the specimen must have come from the till boulders of local streams, and that the lithology most closely resembles that of the Guelph Dolomite (of Ludlow age). Both the lectotype and paralectotype closely resemble the Brassfield specimens in having rounded corallites of similar size, forming closely spaced and often sinuous ranks. Closer examination of these specimens, however, shows no spines (or none preserved), and the inter-corallite spaces are vesiculose, indicating that this form belongs to *Cystihalysites* Chernyshev (1941a).

Rominger's (1876) specimen of *Halysites catenulata* [*sic*] from the "Niagara group" at Point Detour, Michigan, (UMMP 8541, his pl. 29, fig. 1) resembles my form in having rounded-elliptical corallites with well-developed spines. The corallites are commonly 2 mm in inside diameter, more commonly than in my specimens, and the ranks are not nearly as close-packed as in my form. The inter-corallite spaces, however, appear to be occupied by interstitial tubes, suggesting that this specimen belongs to the genus *Halysites*, possibly the species *H. labyrinthicus* (Goldfuss). Unfortunately, it was not possible to section the specimen, so that the structure is not definitely known.

Halysites gotlandicus Yabe is the form most similar to mine. A holotype does not appear to have been originally assigned, but Leleshus (1972, p. 45) stated that the holotype is the specimen figured by Yabe (pl. 7, figs. 1, 2). This comes from Korpklint, north of Visby, Gotland, and is probably derived (assuming it was in place) from the Early Wenlock Högklint beds (age indicated by Manten, 1971). The corallites are rounded-elliptical in cross-section, with diameters of 1.9 x 1.7 to 2.1 x 1.6 mm (no indication if this is inside or outside measurement). There are no interstitial tubes, though Buehler (1955, p. 57) believed some could be seen in Yabe's figure; short spines occupy the lumens of the corallites. The tabulae may be flat or concave upward, and are spaced on average three per mm. Yabe's figure 1 suggests a somewhat sparser distribution of ranks than in my specimens, but some relatively broad lacunae do occur in my coralla. There is no significant difference between the specimen described by Yabe and my specimens, and they are regarded as conspecific.

Williams (1919) identified as *Halysites catenularia* a specimen (GSC 5124) from the Fossil Hill Formation (Middle Llandovery through Middle Wenlock) of Manitoulin Island. This specimen has corallites with the same size and shape as my specimens, in a somewhat looser pattern of ranks, some of which contain up to seven corallites. The specimen appears to be recrystallized, and shows no evidence of spines or interstitial tubes. Without sectioning the specimen, it is not possible to be certain of its identification.

Chernyshev (1938a) reported *Halysites gotlandicus* from the Silurian of Vaygach Island, northern Siberia. The corallum consists of chains of from one to eight corallites. The lacunae are broad, or elongate and narrow, with a maximum breadth of 7 mm, and a maximum length of 15 mm. The corallites range from elliptical to nearly circular in cross-section, 1.5 to 2.0 mm wide and 2.0 to 2.3 mm long. Their tabulae are complete, and flat to concave upward, spaced 2 to 2.4 per mm. Interstitial tubes are reportedly absent, and the apparent lack of spines was ascribed by Chernyshev to strong recrystallization. These specimens are in generally good agreement with Yabe's material, but their corallites are somewhat larger than in either the Gotland or Brassfield forms. Chernyshev's inability to locate spines in the lumens leaves his identification questionable, so *C. gotlandica* can only tentatively be reported from this region.

Buehler's reference to the present species (1955, p. 57) consists of a quotation of the original description, with the type occurrence, and the remark (as noted above) that, while Yabe stated there were no interstitial tubes in his material, his figures suggest their presence. Study of Yabe's figures convinces me that there is no evidence of interstitial tubes, and that this species should be placed in *Catenipora* Lamarck.

Klaamann (1966) reported "*Catenipora* cf. *gotlandica*" from the Early Llandovery Juuru Stage of Estonia. It consists of circular to elliptical corallites grouped in winding chains which form loops containing 15 to 20 corallites. From Klaamann's figures, these loops appear to be free from connection with more than one outside rank of corallites, but only very small portions of the coralla are shown, so that the pattern of the lacunae is unclear. The long diameter of the corallites is 1.7 to 2.0 mm, and the short diameter is 1.3 to 1.5 mm, similar to the diameters of my form. Wall thickness is 0.2 to 0.25 mm. The tabulae appear flat, with occasional weakly concave or convex examples, some incomplete, and are spaced about 2.5 to 5 per mm. Spines in the corallite lumens, apparently spaced vertically about six per mm in longitudinal rows, have about the same arrangement as in my specimens. Klaamann felt his corallites were somewhat narrower than those of Yabe's specimen, but Klaamann's figures do not seem to bear this out. The corallite variation in this species seems to encompass both the Gotland and Estonian forms.

Leleshus (1972) reported *Catenipora gotlandica* from the Late Llandovery (horizon *G*) and Early Wenlock (horizon *K*) at Mount Daurich, Tadzhikstan. The coralla are discoidal or hemispherical, usually 100 to 250 mm in diameter (though in layer *K* coralla up to 1 meter in diameter and 40 cm thick were encountered). The corallites are rounded-elliptical in cross-section, with representative diameters of 2.3 x 1.4 mm, 1.8 x 1.6 mm, and 2.1 x 1.5 mm (in a single corallum). It is not indicated whether these are inside or outside diameters. The corallites are commonly arranged in groups of one to five, and occasionally up to ten in a rank. The lacunae reportedly tend to be elongate, about 3 to 30 mm long, and 1 to 6 mm wide. Thecal thickness is about 0.15 to 0.2 mm. There are no interstitial tubes. Short, sharp spines are rare in most corallites, but abundant in a few. The tabulae are thin and flat, spaced about 3 to 3.5 per mm. Leleshus makes a curious statement

> In some coralla in the wall between adjacent corallites is occasionally encountered an opening (diameter from 0.2 to 0.6 mm), through which tabulae from one corallite often cross into the cavity of the other.

As Leleshus pointed out, this is a rare occurrence in halysitids. His suggestion that this is the result of some accident, cannot be substantiated without further study. Despite the somewhat greater size of the corallites, and sparseness of spines, this form is probably conspecific with mine.

In the same work, Leleshus reported *Catenipora approximata* Eichwald from layer *G* (Late Llandovery) at the same locality as his specimens of *C. gotlandica*. His specimens seem identical in general form with Klaamann's neotype of Eichwald's species (see discussion under *Catenipora favositomima*), and they are referred to here only to point out that this form, so similar to *C. favositomima*, occurs in the same locality and horizon as *C. gotlandica*, just as these two species occur at about the same locality and horizon in the Brassfield.

In terms of the close-packing of corallites, the specimens of *Catenipora gotlandica* found in the Brassfield resemble *Catenipora favositomima*, sp. nov. (which see) from the Brassfield at locality 7, near Fairborn, Ohio. This latter species has larger, rounder, even more closely-packed corallites, and lacks spines, thus clearly differing from *C. gotlandica*.

Distribution. — Gotland, Early Wenlock; Estonia, Early Llandovery; Tadzhikstan, Late Llandovery; Ohio, mid-Llandovery (Brassfield Formation).

This species possibly occurs in the Silurian of Vaygach Island, off the north Siberian coast (Chernyshev, 1938a).

Brassfield distribution. — Locality 7a, near Fairborn, southwestern Ohio. Specimens were found in the talus, and in the clay-coral layer at this locality (discussed under Brassfield Lithosome).

Catenipora favositomima, sp. nov.   Pl. 33, figs. 3, 4; Pl. 40, fig. 2

Cf. 1829. *Catenipora approximata* Eichwald, pl. 2, fig. 9 (*fide* Klaamann, 1966, p. 46).
Cf. 1876. *Halysites compactus* Rominger, p. 79, pl. 29, fig. 3
Cf. 1915. *Halysites gotlandicus* Yabe, pp. 34-35, pl. 7, figs. 1, 2.
Cf. 1926. *Halysites cylindricus* Wilson, p. 15, pl. 2, figs. 6, 7.
Cf. 1966. *Catenipora approximata* Eichwald, Klaamann, pp. 46-48, pl. 11, figs. 1-4; pl. 12, figs. 1, 2; text-figs. 22, 23 (transl. of diagnosis on p. 91).

Type material. — UCGM 41908, specimen figured on Plates 33, 40.

Type locality and horizon. — Brassfield Formation (mid-Llandovery) at locality 7, near Fairborn Ohio; from upper part of this unit.

Number of specimens examined. — One, collected in the field.

Diagnosis. — *Catenipora* apparently without balken; corallites circular in cross-section, large (1.8 to 2.0 mm inside diameter), usually arranged in long, parallel ranks so crowded as to often touch each other across the narrow lacunae; intercorallite space occupied by solid carbonate deposit with a lengthwise suture; spines apparently absent.

Description. — The corallum was apparently a large, densely-populated hemisphere. The preserved portion has corallites 7 mm long, and is about 15 cm across the top. It forms about 120° of what originally was probably a 180° expanse. In surface aspect, the corallites are so large (1.8 to 2.0 mm across the inside diameter, with a circular cross-section) and closely-packed as to have a favositid appearance (whence the trivial name). The lacunae are typically much longer than wide, one measuring 15.0 x 1.5 mm. It appears that as many as nine corallites may comprise a rank, but this is difficult to determine, as corallites of parallel ranks sometimes touch due to the narrowness of the lacunae.

Spines were not observed in thin-section, though enough corallites were studied that, were they present, they should have been detected. This is in marked contrast to the stout spines found in *Catenipora gotlandica* from the same vicinity. Longitudinal lines often observed when longitudinal sections cut part of the intercorallite walls are not septal structures, but rather reflect the structure of the skeletal deposits in this region. The tabulae may be flat, convex, or concave, and are often incomplete. In areas where they *Catenipora gotlandica* from the same vicinity. Longitudinal lines are evenly distributed, they are spaced 2.0 to 2.2 per mm. The spacing may occur in rhythmic zones: six per mm in denser regions, and two per mm in sparser regions.

The corallites are sandwiched between thecal deposits on two sides. The thecal wall is about 0.2 mm thick, and does not appear to consist of two layers, as it does in *Halysites catenularius*.

Also enclosed on two sides by this theca, and situated in the intercorallite spaces, are zones of solid carbonate deposits. These are about 0.3 mm wide (across the rank) and 0.4 to 0.7 mm long. They are divided by a lengthwise suture into two equal halves, and it is probably this suture and the boundaries with the theca that appear as longitudinal lines in vertical sections of the corallite walls.

Discussion. — This form of *Catenipora* is distinctive by virtue of its large, round, and densely packed corallites. Yet there are species from various parts of the world that closely resemble it. When *C. favositomima* becomes better known, its variations may prove to include some known species, but for the present, it seems best to regard it as a separate entity.

Halysites compactus Rominger closely resembles *C. favositomima* in several details. It was described by Rominger (1876) from the Silurian beds (probably Llandovery or Wenlock) at Epoufette Point, Lake Michigan. The corallites are the same size and shape as in the Brassfield specimen, and have about the same wall thickness. Also, they appear to lack spines. Examination of the lectotype (UMMP 8543, selected by Buehler, 1955, p. 42, and figured by him on pl. 4, figs. 5, 6; pl. 5, figs. 2, 3) showed intercorallite spaces filled with vesiculose deposits, indicating that this form belongs to the genus *Cystihalysites* Chernyshev, 1941b.

Yabe (1915) described *Halysites gotlandicus* from Korpklint, north of Visby, Gotland (probably from the Högklint beds of Early Wenlock age, according to Manten, 1971). (Yabe's description appears here under *Catenipora gotlandica*). The most important difference between Yabe's species and *C. favositomima* is the presence of spines in the former.

Wilson (1926) described *Halysites cylindricus* from British Columbia (Upper Ordovician, unit unknown). The corallites, as in *C. favositomima*, are circular in cross-section, but are considerably narrower, about 1 mm in diameter. Interstitial tubes are reportedly absent. There are 2.0 to 2.3 tabulae per mm. Wilson was impressed by the fact that some of the corallites appeared to be free, unattached laterally to other corallites in the plane of transverse section. The significance of this is unclear. Wilson did not comment on the presence or absence of spines. The narrowness of the tubes, and the apparent free habit of some of the corallites set this species apart from *C. favositomima*, and as Buehler (1955, p. 59) pointed out, the latter character suggests that placing this species in the *Halysitidae* may be inappropriate.

Catenipora approximata was described by Eichwald (1829) from the alluvial deposits in the vicinity of Vilnus (Vilna?), Estonia. It was redescribed by Klaamann (1966), and a neotype was designated. The location and horizon are the Tamsalu Stage (Middle Llandovery, according to Manten, 1971) of the Estonian island of Hiyumaa. According to Klaamann, the corallum consists of large (diameter = 1.7 to 2.0 x 1.7 to 2.4 mm), circular, densely-packed corallites, commonly four to six, rarely as many as nine in a rank. The ranks are closely spaced and often parallel. The thecal diameter is 0.15 to 0.20 mm. The tabulae are flat or slightly concave, and spaced about two to three per mm. Spines are arranged in 12 vertical rows within the corallites. Of all species encountered, this one is most like *C. favositomima*, differing solely in having septal spines. Klaamann did not present a clear picture of the nature of the intercorallite spaces, but they do not seem obviously dissimilar to those in the Brassfield specimen. Should *C. favositomima* prove to have spines, there would be strong grounds for assigning it to *C. approximata* Eichwald.

Distribution. — Ohio, mid-Llandovery (Brassfield Formation).

A species closely resembling *C. favositomima* occurs in Estonia, in rocks of Middle Llandovery age.

Brassfield occurrence. — Locality 7, near Fairborn, in southwestern Ohio.

Family **MONILOPORIDAE** Grabau, 1899

Tabulate corals in which the corallum consists of corallites shaped like smoking-pipes; each new corallite arises at the base of the "bowl" of the parent corallite, to which it remains attached by the "mouth-end" of its "stem"; proximal portion of corallum grows attached to and prostrate on a firm substrate (though not in a reticulate pattern), and may have a free-growing, unattached, and shrub-like distal portion; theca thick, consisting of two layers; septal elements, when present, consisting of widely spaced spines; vesicular dissepiments sometimes present, but apparently the living space is never cut off by tabulae from continuity with the lumen of the immediate ancestor and descendent of the corallite; each corallite is attached to the parent only at its proximal end, without transverse stolons connecting adjacent corallites laterally.

Genus CLADOCHONUS M'Coy, 1847

1847. *Cladochonus* M'Coy, p. 227.
1851. *Pyrgia* Milne-Edwards & Haime, pp. 159, 310.
1879. *Monilopora* Nicholson & Etheridge, p. 293.
1885a. *Aulocystis* Schlüter, pp. 148-151.
1887. *Drymopora* Davis, explanations to pls. 70, 71, 72, 74.
1899. *Ceratopora* Grabau, pp. 414-416 (*non* Hagenow, 1851, *nec* Hickson, 1911 [*fide* Sokolov, 1962, p. 246]).
1940. *Monilipora* Lang, Smith, & Thomas, p. 86 (*nom. van. pro Monilopora* Nicholson & Etheridge, 1879).
1946. *Plexituba* Stainbrook, pp. 424-426.
1962. *Grabaulites* Sokolov, p. 246 (*pro Ceratopora* Grabau, 1899; *non* Hagenow, 1851, *nec* Hickson, 1911).
1972. *Cladochonus* M'Coy, Laub, pp. 364-370, pl. 1, text-figs. 1-2.

Type species. — Cladochonus tenuicollis M'Coy, 1847, p. 227 (SD Milne-Edwards & Haime, 1850, p. lxxvi).

Type locality and horizon. — Lower Carboniferous (Burindi), Dunvegan Shale, Dunvegan (on the Patterson River), New South Wales, Australia.

Diagnosis. — Moniloporidae with corallum consisting of a proximal attached or unattached prostrate phrase, from which may arise a distal, freely-branching phase; attached phase never truly reticulate; corallites not normally in lateral contact, except when crowded; no more than three corallites arising from a parent.

Description. — (from Laub, 1972)

The corallum consists of a proximal prostrate region (attached or unattached, dendroid; or, when encrusting a pelmatozoan column, annular), and distal free region of trumpet- or pipe-shaped corallites with their interiors communicating only at point of origin. Corallites of the free-growing region may alternate the direction in which the aperture faces, producing a zig-zag branch (as in M'Coy's figure of *C. tenuicollis*), or the apertures may all face in the same general direction. The initial portion of the corallum consists of two corallites with apertures at opposite ends of a common stolon. Offsets occur at the base of the calice theca, or higher up towards the aperture, and occasionally in both positions in a single corallite. Offsets may be single, bifurcating, or rarely trifurcating. The theca is typically thick and in two layers. Thecal lamellae may be bowed to produce vesicular dissepiments. Pits are sometimes observed in the interior side of the theca. Rod-like structures are frequently embedded in the theca in longitudinal rows, with their tips usually protruding into the interior of the corallite to form acanthine septa.

Discussion. — The morphology and taxonomic status of *Clado-chonus* were discussed by Laub (1972); hence only a summary appears here, along with a consideration of *Grabaulites* Sokolov (1962), a generic name that has recently come to my attention.

To 1972, *Cladochonus* was regarded as being free of spines and dissepiments, the corallum attached only at its base, and branching freely, like a low shrub. *Pyrgia* was erected by Milne-Edwards and Haime (1851, p. 310) for two European Carboniferous species which restudy clearly showed to be congeneric with *Cladochonus*. In 1879, Nicholson and Etheridge (p. 293) proposed *Monilopora,* to accommodate *Cladochonus crassus* (M'Coy, 1844) from the Mountain Limestone (Lower Carboniferous) of Ireland. This species they considered distinct from *Cladochonus* by virtue of the habit of encircling pelmatozoan columns, and by the possession of a "reticulate" structure in the theca, described by Rofe (1869). Hill and Smyth (1938) demonstrated that *Monilopora crassa* is the attached portion of the proximal end of a *Cladochonus* colony, and that a phase of freely branching corallites extended from one of the attached corallites. The meaning of the reticulate structure is not yet resolved. Despite photographs offered by Hill and Smyth (1938) its nature remains unclear. Hill and Smyth's suppression of *Monilopora* under *Cladochonus* seems warranted.

Schlüter (1885a) introduced the genus *Aulocystis* for his *A. cornigera* from the Middle Devonian of Germany. With *Cladochonus* it shared a freely-branching growth-habit and a thick theca. It differed from *Cladochonus* in possessing internal vesicular dissepiments and spines. Two genera subsequently proposed in the literature, *Drymopora* Davis (1887) and *Ceratopora* Grabau (1899) bear all the critical features of *Aulocystis,* and are here regarded as subjective junior synonyms of Schlüter's genus. *Grabaulites* was proposed by Sokolov (1962, p. 246) to replace *Ceratopora* Grabau, which he noted was a name already in use.

Specimens from the Devonian of New York (see Laub, 1972) show that corallites with the characters of *Cladochonus* and *Aulocystis* can coexist in a single corallum, and separate corallites from the same locality show a gradation of morphology from those bearing abundant dissepiments and spines, to those without dissepiments and obvious spines. Hence, it is clear that the definition of *Clado-chonus* must be expanded to include the features of *Aulocystis*.

Stainbrook (1946) introduced *Plexituba*, from the Upper De-vonian of Iowa. These corals, he said, have an adnate growth habit (that is, they are attached-prostrate, but do not form a reticulate pattern as does *Aulopora*), have thick walls, abundant dissepiments, and lack spines. Study of New York Devonian specimens, as stated above, shows a considerable variation in *Cladochonus* with respect to spines and dissepiments. Further, as Hill and Smyth showed, *Cladochonus* does possess a proximal attached phase of several corallites. Grabau (1899, pl. 3, fig. 12; pl. 4, figs. 17, 18) figured specimens of his *Ceratopora* (= *Cladochonus*) attached-prostrate on other fossils. Consequently, there remain no differences between *Cladochonus* and *Plexituba*, and the latter must be regarded as a junior synonym of the former.

Cladochonus had the appearance, in life, of a low shrub, with a basal attached portion ("*Plexituba*"), from which arose freely-branching corallites, unattached except at the base ("*Cladochonus*" and "*Aulocystis*").

Cladochonus sp. A   Pl. 10, fig. 1

Type material. — Figured specimen of *Cladochonus* sp. A is UCGM 41987.

Type locality and horizon. — Locality 46 (near College Hill, Kentucky) in the top of the Noland Member, Brassfield Formation.

Number of specimens examined. — 1.

Description. — The specimen studied comes from the type sec-tion of the Noland Member, Brassfield Formation, and lay within an inch of a budded specimen of *Tryplasma* sp. (Pl. 6, fig. 5). The *Cladochonus* specimen consists of a complete corallite with two offsets, one represented by a complete corallite and the other by only the proximal stolon (see Laub, 1972, p. 367 for *Cladochonus* morphology).

The parent corallite is pipe-shaped, with a preserved length of 4 mm. The proximal end of the stolon is 0.5 mm in diameter. At its juncture with the calice, the lumen is about 1 mm in diameter. The axis of the calice lies at about 115° to the axis of the stolon. The calice itself has a diameter of about 2 mm. The theca of the calice is thin, about 0.1 mm or less.

This group of corallites appears to have been damaged at the time of final deposition, so that the offsets are slightly disjointed from the parent. Viewed from above, within the vertical plane of symmetry of the parent corallite, the offsets make an angle of about 90° with each other. They arose at the antero-lateral portion of the parent's calice, as is usual in *Cladochonus*, but while one originated at the base of the calice, its sibling arose slightly above this level on the other side. This suggests that these corallites came from the free distal portion of a colony, probably not in contact with the substrate.

Discussion. — This specimen and those categorized in this study as *Cladochonus* sp. B are small, and were not sectioned to determine the internal structure. These two forms consist of corallites having quite different shapes and forming clusters of different appearance. Consequently, until further study is possible, and additional material has been collected, they are tentatively placed in two separate, unnamed species. The corallites of different parts of a single corallum among the auloporid corals can have strikingly diverse forms (Laub, 1972), and it is conceivable that *Cladochonus* sp. A and *Cladochonus* sp. B could prove to be a single species.

Cladochonus sp. A differs from *Cladochonus* sp. B primarily in having dichotomous offsets (rather than single offsets), a stolon that expands distally, and a calice that is appreciably wider than the stolon.

Cladochonus sp. B   Pl. 9, figs. 11, 12

Type material. — Figured specimen of *Cladochonus* sp. B is one of several on a slab labelled UCGM 41859.

Type locality and horizon. — Locality 13 (near Wilmington, Ohio) near the top of the Brassfield Formation in Todd Fork Creek.

Number of specimens examined. — 5, collected in the field.

Description. — Five fragmentary coralla on a single slab of rock from near the top of the Brassfield Formation were studied. On the same slab are two fragments of *Favosites favosus* (Goldfuss) (one overturned), and a *Calostylis* corallite. Both of these species, interestingly, are characteristic of the Brassfield fauna in the southeast portion of the Cincinnati Arch region, some distance from this locality.

All offsets in this form occur singly. The largest fragment consists of a line of six to eight corallites (Pl. 9, figs. 11, 12) whose stolons describe a portion of a circle 1 cm in radius. Both the stolons and the calices are about 1 mm in diameter. The calices all occur on the convex side of the line of stolons, though they are not all in quite the same plane. The stolons are only slightly narrower at their proximal end than at their distal end. The distance from offset to offset is about 4 mm.

The other specimens of *Cladochonus* sp. B on this slab have the same general form as the one just described.. In one specimen, a corallite is rather long (6.5 mm) but has the same width and shape. A corallite in another cluster was worn on the side, and it was possible to clean the sediment from the stolon lumen. There appear to be no vesicular dissepiments or spines. While there has been recrystallization of the corallite, and it is possible that spines were present in life, the absence of dissepiments is probably not due to alteration. The theca is just over 0.1 mm thick.

The fragments of *Cladochonus* sp. B from this locality almost certainly came from free, erect portions of shrub-like *Cladochonus* colonies.

Discussion. — See "Discussion" section for *Cladochonus* sp. A.

Cladochonus(?) sp. C   Pl. 9, figs. 9, 10; Pl. 31, fig. 7

Type material. — Figured specimens are UCGM 41692 and 41693.

Type locality and horizon. — Figured specimens are from the Brassfield Formation at locality 15, near West Milton, Ohio.

Number of specimens examined. — Five.

Description. — Specimens of a species possibly belonging to *Cladochonus* M'Coy were recovered at two localities in the Brassfield Formation. The coralla consist of short branching cylinders, about 0.75 mm in diameter, and up to 7 mm long (though usually rather shorter), with fairly uniform diameter throughout their length. They occur in thick clusters in the grey-green silty clay near West Milton (locality 15) and Ludlow Falls (locality 55) in southwest Ohio. One of those from the Ludlow Falls locality occurs on the same small rock as *Striatopora flexuosa* Hall. It is uncertain if there was an extensive attached prostrate phase, most of the corallites appearing to have been part of a small, erect, shrub-like colony.

Some corallites arise as offsets with their axes proximally perpendicular to the parent corallite, distally bending until parallel with the parent (Text-fig. 11a). In other cases, offsets arise from a parent corallite at a point where the parent's axis has bent from its more proximal direction (Text-fig. 11b; cf. *Cladochonus* sp. A, Plate 10, fig. 1). In this case, the axes of the offsets are proximally co-planar with the proximal axis of the parent, while distally their axes bend until about parallel with the distal axis of the parent. Both these offset patterns occur in *Cladochonus* M'Coy (see Laub, 1972, text-fig. 1b).

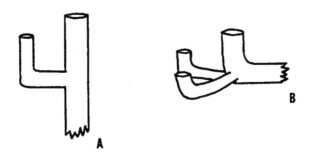

Text-figure 11. — Corallite offset patterns in *Cladochonus* (?) sp. C.

The internal structure is not well-preserved, unfortunately (Pl. 31, fig. 7). The walls are about 1/6 mm thick, so that the lumen occupies more than half the total diameter of the corallite. Significantly, the dissepiments appear to be tabulae, separating the lumen into compartments. This contrasts with the typical vesicular dissepiments of *Cladochonus,* and might call into question the propriety of placing this species in that genus. No spines were detected in the corallites.

Many corallites parallel one another, so that on many surfaces of the specimen, much of the field occupied by the corallum consists of transverse sections of these parallel tubes. The pattern is similar to that often seen in *Syringopora* (cf. *S.* (?) *reteformis* here), but in no case has a stolon been observed connecting the sides of adjacent corallites.

Discussion. — A remarkable specimen (UCGM 41782) was found at the old DeBoldt Quarry in southeastern Richmond, Indiana. This lay loose on an outcrop of Richmondian rocks (Whitewater Fm. overlain by Saluda Fm.), and its lithology suggests it came from one of these units, probably the Whitewater. It appears to be conspecific with *Cladochonus* (?) sp. C. The corallites are prostrate on what may be a bryozoan colony, and in places they bend upward, apparently initiating an erect phase. They are of the same diameter (0.75 mm) as the Brassfield form and show no change in diameter throughout their length, which appears as much as 7 mm. Offsets seem to form both singly and doubly, though insufficient material is exposed to be certain of this. The internal structure was not observed.

Of the other specimens of *Cladochonus* found in the Brassfield (Pl. 10, fig. 1; Pl. 9, figs. 11, 12), *Cladochonus* (?) sp. C most resembles *Cladochonus* sp. B from the top of this unit at locality 13 (Todd Fork), which shows little change in diameter through the corallite lengths. This is in contrast with the gradually broadening corallites (1 mm proximally, 2 mm at the calice) of *Cladochonus* sp. A from locality 46. The Todd Fork specimens are thicker than the present form (about 1 mm throughout), stouter in general appearance, and more curved. They probably represent a separate species.

Distribution. — Ohio, mid-Llandovery (Brassfield Formation); Indiana, Late Ordovician.

Brassfield occurrence. — Localities 15 (near West Milton) and 55 (near Ludlow Falls) in southwestern Ohio.

Family **SYRINGOPORIDAE** Nicholson, 1879

Tabulate corals with corallum dendroid or phaceloid; corallites connected laterally by cross-tubes dispersed throughout the corallum; basal portion of corallum auloporoid in habit, each corallite flexing upward to extend into the vertical, branching phase; theca consists of two layers; septal spines may be present inside corallites; tabulae horizontal and complete, or incomplete, sometimes inclined at large angles to produce infundibuliform dissepiments with axial canal.

Genus **SYRINGOPORA** Goldfuss, 1826

1826. *Syringopora* Goldfuss, p. 75.
1828. *Harmodites* Fischer von Waldheim, p. 19.
1841. *Caunopora* Phillips, p. 18.
1933. *Kueichowpora* Chi, p. 22.

Type species. — (SD Milne-Edwards & Haime, 1850, p. lxii): *Syringopora ramulosa* Goldfuss, 1826, p. 76.

Type locality and horizon. — Carboniferous, "aus dem Ueber-gangskalk von Olne im Limburgischen," Germany.

Diagnosis. — Corallum phaceloid, consisting of a proximal portion of prostrate-unattached corallites, which flex upward to form a phase of more vertically directed corallites; adjacent corallites connected by stolons; corallite interiors contain infundibuliform dissepiments that meet to form an axial tube; development of spines variable; mural pores absent.

Description. — This genus consists of species which are essentially like *Cladochonus* colonies, except that the distal, up-flexed portion of the corallite is longer, and generally parallel to its fellows, and there are cross-tubes (stolons) connecting these distal portions of neighboring corallites.

The proximal portion of the corallum is prostrate-attached, but distally progresses through a prostrate-unattached to a free growth habit. The proximal portion of each corallite is subhorizontal, and the distal portion flexes upward continuing this upward growth for the greater portion of the length of the corallite. The corallites of the distal portions of the corallum are sub-parallel. Corallum increase is by offset of the corallites.

The corallites are separate from the point of offset, except where they are joined laterally to a neighbor by a tubular stolon. The stolons are important features in distinguishing the syringoporids from other corals.

Internally, the corallite wall consists of two layers, the outer optically denser than the inner. Embedded in the inner layer, normal to the wall, are septal spines aligned longitudinally into a series of rows (reportedly 12). These rows are acanthine septa and never form solid sheets.

The tabulae are infundibuliform (funnel-shaped). They commonly consist of complete sheets of skeletal matter in the corallite lumens, each depressed axially to join with the depressed axial portion of the next lower tabula, and thus forming a hollow axial tube.

Discussion. — Milne-Edwards and Haime (1850, p. lxii) indirectly made *Syringopora ramulosa* Goldfuss the type of this genus by declaring *Syringopora* Goldfuss a synonym of the younger generic name, *Harmodites* Fischer von Waldheim (1828), and assigning *Syringopora ramulosa* as the type species of *Harmodites*. Subsequently, Lang, Smith, and Thomas (1940, p. 130) designated *Harmodites distans* Fischer von Waldheim (1828, pp. 19-20, fig. 1) as the type of *Harmodites*. (These authors gave the locality and horizon of the latter species as Carboniferous of Russia.) The original illustration and description of *H. distans* leave little doubt that it is congeneric with *S. ramulosa*.

As type species of *Caunopora* Phillips (1841), Lang, Smith, and Thomas (1940, p. 33) selected *C. placenta* from the Middle Devonian of England. Although Phillips (p. 18) ascribed this species to Lonsdale (as *Coscinopora placenta*), Lang, Smith, and Thomas (p. 33) indicated that Lonsdale obtained the specific name from Goldfuss (1826, p. 31). These authors also found that Lonsdale's specimen was not conspecific with Goldfuss'. They therefore assigned the species to Phillips, who was the first to use it without reference to *Coscinopora*. Lang, et al. stated that *Caunopora placenta* Phillips was the type species of *Caunopora* Phillips by monotypy. As Phillips described two species under this generic name, *C. placenta* is more properly considered the type species by subsequent designation of Lang, Smith, and Thomas (1940).

Lang, et al. noted that Phillips' specimen of *Caunopora placenta* is actually a stromatoporoid enclosing a specimen of *Syringopora*, and Phillips' figure (pl. 10, fig. 29), while unclear, makes this conclusion credible. Lang, et al. took the *Syringopora* corallum as the lectotype of *C. placenta*, thus making *Caunopora* a junior synonym of *Syringopora*.

Chi (1933) introduced the genus *Kueichowpora* from the Carboniferous of China. His figures of the type species clearly show the stolons and infundibuliform dissepiments characteristic of *Syringopora*, of which *Kueichowpora* is a junior synonym. The bifurcation of corallites found in *Kueichowpora* also occurs in the type species of *Syringopora* Goldfuss.

Syringopora (?) reteformis Billings   Pl. 10, figs. 2-4; Pl. 11, figs. 1, 2;
                                            Pl. 32, fig. 4; Pl. 41, fig. 4

1858. *Syringopora reteformis* Billings, p. 170.
? 1876. *Syringopora tenella* Rominger, p. 81, pl. 30, fig. 4
? 1890. *Syringopora (Drymopora) fascicularis* (Davis) Foerste, pp. 338-339 (*non Drymopora fascicularis* Davis, 1887).
1899. *Syringopora retiformis* [*sic*] Billings, Lambe, p. 52, pl. 2, fig. 3.
1919. *Syringopora retiformis* [*sic*] Billings, Williams, pl. 17, fig. 1.
Cf. 1926. *Syringopora columbiana* Wilson (*partim*), p. 17, pl. 3, fig. 6 (*non* fig. 5).
? 1931. *Syringopora* cf. *retiformis* [*sic*] Billings, Foerste, p. 185, pl. 19, fig. 13.
? 1939. *Syringopora retiformis* [*sic*] Billings, Northrop, p. 155.
? 1972. *Syringopora* sp. cf. *S. retiformis* [*sic*] Billings, Bolton & Copeland, pl. 4, fig. 7.

Type material. — Lectotype (Williams, 1919, explanation of pl. 17, fig. 1): GSC 2617 (the specimen figured by Williams, 1919, pl. 17, fig. 1). (Pl. 10, figs. 3, 4 herein).

Type locality and horizon. — According to Billings (1858, p. 170), "Upper Silurian. Isthmus Bay, Lake Huron". Bolton (1960, p. 45) gives the locality as Owen Sound, Ontario, and the horizon as the Fossil Hill Formation, which is Late Llandovery to Middle Wenlock in age.

Number of specimens examined. — Nine, including the holotype and eight from the Brassfield Formation.

Diagnosis. — *Syringopora* with corallites about 1 mm in diameter, usually with geniculate axes; cross-tubes sparse, corallites usually in mutual contact through genicular approximation; spines extremely abundant in lumen; tabulae appear complete, possibly because of narrow axial canal.

Description. — The corallites are uniformly about 1 mm in diameter and characteristically geniculate. Contact between them most often occurs through the bending together of neighboring corallites, followed by sharp separation, or less frequently, *via* cross-tubes (stolons) with diameters equal to or slightly less than those of the corallites (Pl. 10, fig. 3; Pl. 11, fig. 2). Sparseness of stolons prevented a meaningful determination of their spacing.

The corallite walls are about 0.3 mm thick, and the lumen has a diameter of about 0.4 mm. The inner surfaces of the corallite walls are richly-invested with spines, spaced vertically about six per mm in longitudinal rows.

The nature of the internal dissepiments is difficult to discern. Usually they appear as complete tabulae (Pl. 32, fig. 4; Pl. 41, fig.

4), but sometimes they may be seen as vesicular dissepiments, similar to those of *Cladochonus*. If this is a true *Syringopora* (a fact not yet established with certainty), it is probable that the dissepiments are infundibuliform (funnel-shaped) and concave upward, but possibly with an axial tube so narrow that it seldom appears in longitudinal section. There are about five or six tabular plates per mm.

Discussion. — Billings' original description of the present species says:

Forming large masses; corallites much geniculated, frequently anastomosing or connecting by stout processes; diameter of corallites about two-thirds of a line [about 1.4 mm], distant from each other from half-a-line to a line and a-half [about 1 mm to 3.1 mm]; distance of connecting processes about one to three lines [about 2.1 to 6.3 mm], usually about 2 lines [about 4.2 mm].

This agrees well with the lectotype figured by Williams (1919), and with the Brassfield specimens. The lectotype (Pl. 10, figs. 3, 4) is silicified, and partly free of the matrix. It was not possible to section it, but in the openings of some of the corallites are apparent spines (rather obscured by recrystallization). The wall thickness and lumen diameter are in the same proportion as in my specimens, but it was not possible to study the dissepiments.

Syringopora tenella Rominger is similar to *S. reteformis*. The specimen shown by Rominger is from the Silurian of Drummond's Island, Lake Huron and was found in the "drift" (presumably glacial or alluvial deposits). The corallites are reportedly 1 mm or less in diameter and branching. They interconnect by "approximation" (bending toward each other) rather than through distinct stolons. There is a faint longitudinal striation on the outer walls (not noted on corallites of the lectotype of *S. reteformis* Billings, nor on the Brassfield specimens, possibly due to poor preservation) and the lumens of *S. tenella* show 12 spinose crests. The dissepiments, described as funnel-shaped, are not always developed, so that the lumens are often totally clear. My specimens commonly show the chaotic aggregate of corallites shown in Rominger's figure, and Lambe (1899) reported this condition especially in the basal region of specimens that he identified with the present species. The close similarities make it likely that *S. tenella* is a junior synonym of *S.* (?) *reteformis*.

From the Clinton beds (probably Brassfield or Dayton Formation) of southwest Ohio, Foerste (1890) reported *Syringopora (Drymopora) fascicularis* (Davis, 1887). Different coralla had rugose-sided corallites with diameters ranging from 0.7 to 1.1 mm. The corallites connected by approximation, rather than through stolons, and no spines or dissepiments were noted (Foerste indicating that no careful search was made for them). While the description suggests *S. reteformis* in most respects, the lack of any report of stolons makes it unwise to equate these two species.

Wilson (1926) reported a new species, *Syringopora columbiana*, from the Late Ordovician Beaverfoot Formation of British Columbia. This species does not appear to be conspecific with the Brassfield material. The corallites are slender (0.5 to 0.75 mm in diameter), giving the corallum a much more delicate aspect than in my specimens of *S.* (?) *reteformis*. It should be noted, however, that Wilson's figure 6, a twice-natural-size view of a polished section (of the holotype?), strongly resembles the present species in general form. The corallites are geniculate and, while stolons are present, much of the connection between corallites is by approximation. *S. columbiana* may prove to lie on an evolutionary line ancestral to *S.* (?) *reteformis*.

Foerste (1931) figured a specimen identified as *Syringopora* cf. *retiformis* [sic] from what he considered to be the Dayton Formation (Late Llandovery) of Adams County, Ohio. This specimen has all the outward characters of the species, the corallites connecting more often through approximation than through stolons. Foerste probably was correct to associate his form with Billings' species, but the figure alone does not provide sufficient information for certainty.

Northrop (1939) reported *Syringopora retiformis* [sic] from the Bouleaux Formation (Early Ludlow) of the Gaspé region of Maritime Canada. The only pertinent facts supplied by his unillustrated description are that the corallite diameter ranges from 0.8 to 1.2 mm, and that lateral contact between corallites appears to be due more to approximation than stolons. Northrop, in fact, noted that it was uncertain if any stolons were present. In the absence of information on the internal structure, it is possible to say only that his specimen is probably (but not certainly) identical to the present species.

Bolton and Copeland (1972) described *Syringopora* sp. cf. *S. retiformis* [*sic*] from the Thornloe Formation of Late Llandovery to Early Wenlock age, in the Lake Timiskaming region of eastern Canada. Their material bears a general resemblance to my specimens. The corallites are geniculate, and about 1 mm in diameter. The absence of prominent spines in the corallites of this specimen, however, leaves the identification open to question.

Distribution. — Ontario, Late Llandovery or Early or Middle Wenlock; Michigan, Silurian; Ohio, mid-Llandovery (Brassfield Formation) and probably Late Llandovery (Dayton Formation); Kentucky, mid-Llandovery (Brassfield Formation).

This species possibly occurs also in the Early Ludlow of the Gaspé region, Maritime Canada.

Brassfield occurrence. — Localities 7a (near Fairborn, Ohio), 11 (near West Union, Ohio), 23 (at the Highland County - Adams County border in Ohio), and 53 (near Irvine, Kentucky, in the Noland Member).

Family **SYRINGOLITIDAE** Waagen & Wentzel, 1886

(*Nom. transl.* Sokolov, 1950, *ex* Syringolitinae Waagen & Wentzel, 1886)

Tabulate corals with massive corallum; corallites prismatic, pressed together (cerioid) or slightly diverging (fasciculate); cerioid corallites connected by mural pores, while cross-tubes connect corallites that diverge and lose contact; tabulae strongly concave axially, some forming a distinct axial tube; septal elements spinose where present.

Genus **SYRINGOLITES** Hinde, 1879

1879. *Syringolites H*inde, p. 244.

Type species. — (By monotypy) *Syringolites huronensis* Hinde, 1879, p. 246.

Type locality and horizon. — Hinde (1879, p. 246): "Not uncommon in the Niagara dolomite (Wenlock), near Manitouwaning, Great Manitoulin Island, Lake Huron", Canada. Hill and Jell (1970, p. 173) give the horizon of Hinde's specimens as the Manitoulin Dolomite (Llandovery).

Diagnosis. — Corallum similar to *Favosites* in general plan, but having infundibuliform dissepiments resulting in the develop-

ment of an axial tube; mural pores located on corallite walls; spines may be located on corallite walls and on upper surfaces of dissepiments.

Description. — This is a coral of cerioid form, superficially resembling *Favosites*. The striking features of this genus lie in the infundibuliform tabulae (apparently both complete and incomplete), as in *Syringopora.* They are rather wide-spaced and plunge in a biologically proximal direction in the axis, so that each tabula forms a funnel, with its tube fitting into that of its proximal neighbor. In some instances, the continuity of this axial tube is broken by a transverse tabula within it.

The upper surfaces of the tabulae bear radial ridges. Rows of spines often appear to cap these ridges, one row to each ridge, but the relationship between them is not always clear. Hinde (1879, p. 244) reported that in some specimens the calice walls are crenulated longitudinally (presumably as extensions of the radial ridges of the tabular surfaces). In addition, spines sometimes occur within the axial tubes.

As in favositids, the corallite wall faces bear mural pores.

Discussion. — In all descriptions and figures of members of the genus *Syringolites* Hinde, the tabulae are indicated as being complete. That is, the walls of each funnel are a single sheet (e.g., Hinde, 1879; Hill & Jell, 1970). Only in the specimen reported as *S. huronensis* by Foerste (1906, pp. 301-302, described below under *Syringolites vesiculosus,* sp. nov.) are the tabulae incomplete. In this instance, a transverse section (Pl. 31, fig. 2) reveals that the funnels are made up of several imbricating (vesicular) dissepiments. This genus is insufficiently understood to allow a meaningful judgement on the taxonomic importance of this difference, and so incomplete tabulae are here considered acceptable in *Syringolites.*

Syringolites vesiculosus sp. nov.   Pl. 10, fig. 5; Pl. 31, figs. 1, 2

Cf. 1879. *Syringolites huronensis* *H*inde, p. 246, text-fig. A-D on p. 245.
Cf. 1896. *Roemeria kunthiana* Lindström, pp. 14-17, pl. 2, figs. 19-25; pl. 3, figs. 26-29.
    1906. *Syringolites huronensis* *H*inde, Foerste, pp. 301-303, pl. 2, fig. 3; pl. 4, fig. 2 (*non* Hinde, 1879).

Type material. — Holotype: USNM 84893 (Pl. 10, fig. 5; Pl. 31, figs. 1, 2) described and figured by Foerste (1906).

Type locality and horizon. — Waco bed, Noland Member of the Brassfield Formation (mid-Llandovery), "two miles southwest of Clay City, along the road immediately north of Tipton Ferry" in east-central Kentucky.

Number of specimens examined. — All data based on holotype.

Diagnosis. — A species of *Syringolites* Hinde in which the axial tubes are formed not by the down-flexing of complete tabulae, but of individual vesicular dissepiments; in all other aspects similar to *S. huronensis* Hinde.

Description. — The specimen studied is a thin (2 to 11 mm), flat fragment of a corallum of which the original diameter was probably about 7 to 8 cm, judging by the disposition of the basal corallites. Those corallites exposed in the base of the corallum have their axes in a horizontal position, and are distinctly rugose. The axes of the corallites visible on the upper surface of the corallum are vertical (perpendicular to the surface). Corallum growth was centrifugally radial.

The corallites range in diameter from 2 to 3 mm, though most are 2 to 2.5 mm. Adjacent corallites share walls whose thickness is about 0.20 to 0.25 mm. In the axis of each corallite is a raised circular ring, 1 mm or less in diameter, with its axis parallel to and coinciding with that of the corallite. The calice floor slopes down gently from the rim of the corallite wall to form a shallow dish about 0.5 mm deep, at the foot of the central ring. The calices appear to have radial, rounded ridges and valleys reaching from the corallite wall, which they corrugate, to the foot of the axial ring. Additionally, the calice floors are studded with short spines which in a minority of specimens appear to be aligned radially. Their relationship to the radial undulation of the calice floor is unclear.

The corallites contain vesicular dissepiments which are downflexed axially to produce axial tubes, apparently continuous for the length of the corallite. Clearly, the axial ring visible in the calices reflects the juncture of a segment of the central tube with the upper surfaces of the dissepiments immediately proximal to it. This juncture appears in all cases to be centered in the calice. The inner surface of each tube is spinose.

The mural pores typical of *Syringolites* were not observed in the Brassfield specimen, perhaps due to its imperfect state of preservation.

Discussion. — It is not surprising that Foerste identified the Brassfield form of *Syringolites* as *S. huronensis,* which was described by Hinde (1879, p. 246) from Silurian rocks at Manitouwaning, Manitoulin Island in Lake Huron. Hill and Jell (1970, p. 173) designated as lectotype of this species BMNH R19949 which they took to be the original of Hinde's fig. A. "Niagara dolomite (Wenlock)", the stratigraphic designation given by Hinde for his specimens, is interpreted as Manitoulin Dolomite (Llandovery) by Hill and Jell. The descriptions of Hinde, and of Hill and Jell indicate that *S. huronensis* strongly resembles the Brassfield species. The corallite diameter (1.8 to 2.5 mm), the fluting of the calice floor and resulting corrugation of the corallite walls, the presence of spines on the calice floor (apparently better aligned than in Foerste's specimen) and in the axial tube, and the centered position of the tube axis on the calice floor are all attributes held in common by *S. huronensis* and the Brassfield form.

The major difference between these two forms is in the nature of the axial tube. In *S. huronensis,* this structure is commonly formed by the down-flexing of complete tabulae in the axis. In the Brassfield specimen, it is formed by similar axial down-flexing of vesicular dissepiments. Mural pores were not observed in Foerste's specimen, probably due to poor preservation, and so cannot be compared with those reported in *S. huronensis.* Also, the "horizontal or gently saucered tabellae" crossing the tubes at wide intervals, as reported by Hill and Jell (p. 175) were not noted in the Brassfield form, but this, again, could be due to the state of its preservation.

The only other species commonly assigned to *Syringolites* is *Roemeria kunthiana* Lindström (1896) from the Silurian of Gotland. The lack of evidence of distal divergence of the corallites or of their lateral connection with syringoporoidal tubes makes it more reasonable to assign it to *Syringolites* than to *Roemeria.* Basically, this species has the same axial tube structure as *S. huronensis* (thereby differing from the Brassfield form), and is distinguished from the present species and from *S. huronensis* Hinde by the presence of spines on the walls of the corallites between the tabulae. Hill and Jell (p. 175) reported that the axial tubes of *R. kunthiana* often lie in an eccentric position.

The genus *Syringolites* is known from relatively few places (Canada, Kentucky, Gotland, and Estonia; see Hill & Jell, 1970, pp. 173-174), and apparently only a small number of specimens have been studied. It is possible, then, that as more information becomes available, the structure of the axial tube will not be considered sufficient grounds for separating *S. vesiculosus* from *S. huronensis*. For the moment, however, there is no evidence of a continuum between them.

Distribution. — Kentucky, mid-Llandovery (Brassfield Formation).

Brassfield occurrence. — Foerste gave the following localities, all in the general vicinity of Irvine and Waco, Kentucky: "Two miles southwest of Clay City, along the road immediately north of Tipton Ferry" (the type locality, and source of the only specimen studied); "a mile southeast of Indian Fields, east of the home of Brownlow Bruner, where the road from Kiddville joins the road from Indian Fields to Clay City; half a mile east of Waco, where the road to Cobb Ferry starts off from the Pike to Irvine; along the road north of Estill Springs, north of Irvine" (localities 53 and 53a of this work). All reported occurrences are in the Waco bed of the Noland Member, Brassfield Formation.

Family **ALVEOLITIDAE** Duncan, 1872

Tabulata with massive coralla of hemispherical, encrusting or ramose form; corallites open oblique to corallum surface, with compressed, crescentic apertures; mural pores large; spines sometimes present, often restricted to one wall of the corallite; tabulae complete.

Remarks. — Sokolov (1962, p. 346 in English translation) reported that "septal scales" (presumably squamulae) may be present in this family.

Genus **ALVEOLITES** Lamarck, 1801

1801. *Alveolites* Lamarck, p. 375 (*non* Defrance, 1816).
1887. *Platyaxum* Davis, explanations of pls. 60, 61, 63.

Type species. — (by subsequent designation of Nicholson & Etheridge, 1877, p. 356) *Alveolites suborbicularis* Lamarck (1801, p. 376).

Type locality and horizon. — Upper Devonian (Frasnian) near Düsseldorf, Germany.

Diagnosis. — *Alveolitidae* with corallum usually encrusting or ramose; corallite axes at about 45° to corallum surface at aperture, more nearly parallel to surface proximally; spines variably developed, in some specimens a single spine or ridge in middle of lower wall of corallite; no squamulae present; corallite walls thick.

Description. — These are massive tabulate corals that are generally flat and encrusting, though in some specimens the upper surface may be uneven, or even ramose.

Typically, the corallites have a vaulted upper wall and a comparatively flat lower wall. Proximally, they are subparallel with the upper surface of the corallum, but distally they flex upward at about 45° (as compared with about 90° for *Crassialveolites* Sokolov, *fide* Sokolov, 1962). Their compressed cross-section produces a crescentic or semi-circular corallite aperture on the corallum surface.

The corallite walls are rather thick, but usually narrower than the lumen. The walls are perforate, but this is not always obvious in thin section.

Septal structures are commonly restricted to the middle of the lower wall of the corallite, where they may take the form of a single longitudinal ridge, though spines may be present in other specimens.

The tabulae are thin and complete.

Discussion. — Nicholson and Etheridge (1877) presented a lengthy discussion of the validity and criteria for recognition of *Alveolites* Lamarck. They considered six groups of alveolitid corals, including the type species of the genus. Of the type species, *A. suborbicularis* Lamarck, they stressed the following characters as being of particular diagnostic usefulness: corallite obliquity, presence of mural pores and complete tabulae, the general shortness and layered arrangement of the corallites, and the presence of a single septal ridge in each corallite on the ventral wall. They noted that only the last of these traits might possibly serve as a criterion for distinguishing this genus from related genera. They concluded (pp. 360-361) that

> With our present knowledge it seems unwise to abandon the genus *Alveolites* altogether; but it is at the same time perhaps improbable that future investigations will justify its permanent retention.

Nicholson and Etheridge may have underestimated the importance of corallite obliquity and compression of the apertures. In stressing the fact that such obliquity is observed in clearly non-alveolitid forms under exceptional circumstances, they overlooked the fact that it is universal among species of *Alveolites*. Nicholson and Etheridge were most concerned with finding consistent means of distinguishing *Alveolites* from *Favosites*. As the latter is generally understood, obliquity and corallite compression are certainly not typical features. As noted above, other characters are needed to distinguish *Alveolites* from other alveolitids, but its position as an entity distinct from favositids, seems secure.

Platyaxum Davis (1887) is based (by subsequent designation of Lang, Smith, and Thomas, 1940, p. 102) upon *P. turgidum* Davis, probably from the Silurian beds near Louisville, Kentucky. Lang, *et al.* noted that the prior selection, as type species, of *Pachypora frondosa* Nicholson from the Middle Devonian of Ontario by Bassler (1915), is invalid, as this species is not among those included by Davis under *Platyaxum*. This genus appears to be a junior synonym of *Alveolites* Lamarck. It is distinctive for its erect, palmate growth habit, with corallite apertures on both sides of the frond, and the corallites diverging from the midplane of the frond. The general corallite morphology is enough like *Alveolites* that, pending further study, it seems best to regard these two forms as congeneric.

Alveolites labrosus (Milne-Edwards & Haime)   Pl. 10, figs. 8, 9; Pl. 24, figs. 1, 2

1851. *Coenites labrosus* Milne-Edwards & Haime, p. 302.
1852. *Limaria laminata* Hall (*partim*), p. 143, pl. 39, fig. 6c (*possibly* figs. 6a, b, d).
1854. *Coenites labrosus* Milne-Edwards & Haime, Milne-Edwards & Haime, p. 277, pl. 65, figs. 6, 6a.
1874. *Coenites lunata* Nicholson & Hinde, p. 151, figs. 2a-c.
1875. *Coenites lunata* Nicholson & Hinde, Nicholson, p. 55, figs. 25 a-c.
1875. *Coenites laminata* (Hall), Nicholson, p. 55, figs. 25d, e.
Cf. 1876. *Limaria laminata* Hall, Rominger, p. 45, pl. 18, fig. 2.
Cf. 1876. *Limaria crassa* Rominger, pp. 45-46, pl. 18, fig. 1 (*non Coenites crassa* Davis, 1887).
? 1887. *Coenites laminata* (Hall), Davis, pl. 4, fig. 5.
1899. *Coenites lunata* Nicholson and Hinde, Lambe, p. 28.
? 1928. *Coenites labrosus* Milne-Edwards & Haime, Twenhofel, p. 131.
? 1939. *Coenites labrosus* Milne-Edwards & Haime, Shrock & Twenhofel, pp. 256-257, pl. 27, fig. 16.

Type material. — Repository unknown.

Type locality and horizon. — Silurian of Dudley, England, probably from the Upper Wenlock Dudley Limestone.

Number of specimens examined. — One, collected in the field.

Diagnosis. — *Alveolites* with crescentic calices about 0.6 to 0.7 mm long, generally separated from their nearest neighbors by this distance; calice opening distinctly indented by a lip which is sometimes bifurcated, projecting from its convex wall; concave ventral wall of corallite with broad median longitudinal ridge that does not appear in the aperture.

Description. — The specimen collected is a laminar fragment of a corallum, with maximum width about 1.5 cm and thickness about 1 mm. The corallites are shaped rather like quarter-moons, one side of the aperture being invested with a broadly-jutting lip. The apertures lie in depressions, about 0.2 mm deep, and are usually about 0.6 to 0.7 mm long, and 0.3 mm across. These corallite apertures are widely spaced, each distant from its nearest neighbor by a space about equal to its width.

The corallite interiors show no dissepiments. Running longitudinally along the lower (concave) wall is a single ridge which appears to die out at a point where the subhorizontal proximal portion of the lumen flexes in an upward direction distally, to reach the surface of the corallum. Thus, this basal "septal" ridge is not visible in the aperture of the corallite. No other longitudinal structures, including spines, were observed in the interiors.

Discussion. — Many reports of specimens with nearly all the characters of this rather distinctive species have appeared in the literature, but in some of these the lip projecting into the aperture from the convex wall was medially emarginate (bilobate). This apparently varies in its development: The specimen figured by Milne-Edwards and Haime (1854, pl. 65, fig. 6a) shows some apertures with a faintly emarginate projecting lip, whereas others are like the singly lobate lip of my specimen. The configuration of the lip, then, is not regarded here as a specific criterion.

Milne-Edwards and Haime's original description of *A. labrosus* (1851), and a subsequent description with figures (1854), present a form with lunate apertures about 0.7 mm long, having a singly-lobate or faintly bilobate "tooth" projecting from the convex corallite wall into the aperture. The spacing of these apertures is similar to that in my specimen.

Hall (1852) described *Limaria laminata* from the lower part of the "Niagara Limestone" at Lockport, New York (probably of approximately Wenlockian age). His description gives little information, stressing only the fact that the corallum consists of a series of laminae with extraneous deposits between them. His figures show what are probably more than one species, but his figure 6c is most like the Brassfield specimen. It has a crescentic aperture, with lateral spacing as in my specimen, but with a distinctly bilobate apertural "tooth". If his other figures can serve as an index, the corallite openings are about the same size as in mine, and it seems safe to consider his material conspecific with mine.

Coenites lunata was described in 1874 by Nicholson and Hinde, from Silurian rocks at Owen Sound, Ontario. The original reference was unfortunately not available. In the following year Nicholson described and figured this species. The corallum is a thin crust, about 1.4 mm thick, containing lunate or crescentic calices about 0.5 to 0.6 mm long, and about 0.3 mm across. Their spacing is as in my specimen. The convex wall of the aperture contains a "single strong rounded tooth", giving the crescentic appearance. Nicholson apparently repeated his and Hinde's original description of the species, as a later treatment of this form by Lambe (1899) includes the same description, attributing it to Nicholson and Hinde.

Limaria crassa Rominger (1876), is based upon material from the Silurian of Michigan. It represents a form similar, but not identical with the Brassfield specimen. The coralla are laminar growths, often with several superposed upon one another. The corallites tend to be broadly kidney-shaped, though apparently not so distinctly lunate as in my specimen. A lip is sometimes present on the dorsal (in my specimen, convex) margin, but this is absent in other corallites, leaving them asymmetrically oval in outline. Corallites are about the same size as in my form, but their spacing varies from wide to quite narrow. The ventral surface of the corallite contains a longitudinal ridge, apparently like that found in my specimen. On the dorsal side, two others of similar size project, bifurcating above the vental ridge. Large pores are numerous in the corallites, and the thinner-walled corallites have dissepiments. *L. crassa* is clearly distinct from *A. labrosus*. These pores, dissepiments, and dorsal internal ridges are absent from *A. labrosus*. In addition, *L. crassa* has more closely spaced corallites than mine does. It more closely resembles the specimens shown by Hall (1852) in fig. 6a, b, and d of his description of *L. laminata*.

Another species, from the Silurian rocks at Point Detour, northern Michigan, is *Limaria laminata* Hall (1876). This has a laminar corallum with stoutly-walled corallites, sometimes more than a diameter apart, and less than a millimeter wide. The convex side of the apertures has two rather obscure tooth-like crests, while the concave side has a more conspicuous crest on its median line. Judging from the figure and description, this may be a small variety of Rominger's specimen of *Limaria crassa* Rominger. If so, it would be a junior synonym of that species.

Davis' (1887) figure of *Coenites laminata* from the "white clay" near Louisville, Kentucky (according to Stumm, 1964, p. 71, from the Louisville Limestone, Late Wenlock and Early Ludlow in age) is not clear enough to determine its relationship to *L. laminata* Hall. Davis' specimen was not illustrated by Stumm in his work on the corals of the Falls of the Ohio (1964), but Stumm placed Davis' concept into junior synonymy with another Davis species, *Alveolites fibrosus*. Stumm's figures of *A. fibrosus*, including the holotype, do not resemble Hall's figures of *L. laminata*.

Twenhofel (1928) reported *Coenites labrosus* from the Jupiter and Chicotte Formations (Late Llandovery and Early Wenlock) on Anticosti Island. Unfortunately, he gave no diagnostic morphologic information, and so the presence of *A. labrosus* on Anticosti Island must remain in question.

Shrock and Twenhofel (1939) reported *Coenites labrosus* from the Pike Arm Formation (probably Llandovery, though possibly extending into the Wenlock, or even Ludlow) of northern Newfoundland. The encrusting corallum takes the form of sheets about 1 mm thick. The lunate or crescentic corallite apertures are 0.65 to 0.85 mm long, and about a third as much across. A lip projects into the aperture from the convex side, but no tooth-like projection was observed on the concave side. The apertures are widely spaced, though apparently not to the extent found in my specimen. The corallite lumens are prostrate through most of their length, as mine are, and distally flex upward to appear at the surface. Between them, the deposits are reportedly flattened vesicles. These were not observed in my specimen, but this may be due to the recrystallization to which it appears to have been subjected. The interspaces in the Brassfield specimen appear to be solid. The Newfoundland form has no obvious mural pores, and a few poorly-preserved transverse tabulae were reported. Shrock and Twenhofel's material can only questionably be identified with *A. labrosus.*

Distribution. — England, Late Wenlock; New York, Middle Silurian; Ontario, Silurian (possibly Fossil Hill Formation, of Late Llandovery to Middle Wenlock age); Kentucky, mid-Llandovery (Brassfield Formation).

This species is possibly present in the Late Llandovery and Early Wenlock of Anticosti Island, and in the Silurian of northern Newfoundland.

Brassfield occurrence. — Locality 1 (near Panola), in central Kentucky, in the top of the Noland Member.

Alveolites labechii Milne-Edwards & Haime   Pl. 31, figs. 5, 6;
                                                      Pl. 41, fig. 3

? 1839. *Favosites spongites* (Goldfuss) Lonsdale (*partim*), pp. 683-684, pl. 15 bis, figs. 8, 8a, 8b (*Non* figs. 8c, 8d, 8e; *non Calamopora spongites* Goldfuss, 1826).
   1851. *Alveolites labechii* Milne-Edwards & Haime, p. 257.
   1854. *Alveolites labechei* Milne-Edwards & Haime, Milne-Edwards & Haime, p. 262, pl. 61, figs. 6, 6a, 6b.
Cf. 1876. *Alveolites niagarensis* Rominger, pp. 40-41, pl. 16, figs. 1, 2 (*non* Nicholson & Hinde, 1874).
Cf. 1877. *Alveolites undosus* Miller, p. 262 (*pro Alveolites niagarensis* Rominger, 1876, *non* Nicholson & Hinde, 1874).
? 1890. *Alveolites niagarensis* Rominger, Foerste, pp. 336-337.
Cf. 1909a. *Alveolites inornatus* Foerste, p. 103, pl. 3, fig. 56.
? 1919. *Alveolites labechei* Milne-Edwards & Haime, Williams, pl. 14, fig. 3.
   1928. *Alveolites labechi* Milne-Edwards & Haime, Twenhofel, pp. 130-131.
? 1939. *Alveolites labechei* Milne-Edwards & Haime, Northrop, p. 152.
? 1966. *Alveolites undosus* Miller, Bolton, pl. 8, fig. 6.

Type material. — Repository of types unknown. Milne-Edwards and Haime (1854, p. 262) reported specimens at the "Parisian Museum" (Jardin des Plantes?), and the Museum of Practical Geology in London.

Type locality and horizon. — Milne-Edwards and Haime's material is from Wenlock and Benthall Edge, England (probably of Wenlock age).

Number of specimens examined. — Three, collected in the field.

Diagnosis. — *Alveolites* with apertures forming pattern reminiscent of pillow-lava on surface of encrusting corallum; apertures about 1.5 mm long (on interior) and roughly one-third as wide; tabulae present in lumens, complete; spines probably present to varying degrees, on all sides of lumen.

Description. — The coralla are encrusting and expansive, the largest being 11 cm in maximum diameter and 2 cm in maximum thickness. The corallite apertures are separated only by the thickness of their walls, about 1/8 mm. They open obliquely on the surface in a pattern reminiscent of pillow-lava, and have an inside length (the maximum dimension) of about 1.3 to 1.8 mm, and a breadth (the minimum dimension) of about 0.3 to 0.8 mm (most commonly 0.5 mm). Sparse spines apparently were present on all walls of the corallites, but this is not completely clear in the specimen sectioned. The tabulae of the lumens are horizontal, concave or convex, and are spaced about 1.7 to 2.0 per mm. No mural pores were observed.

Discussion. — Milne-Edwards and Haime (1851, 1854) described this species from the Wenlock of England. The apertures are reportedly not very prominent on the surface, with the maximum dimension being between 0.7 and 1.0 mm, and the minimum dimension one-third as long. The walls are thin, and the aperture has a pillow-lava-like configuration like my form. The inner process (presumably a septum or spine) is indistinct. Some of the apertures of their figure 6a show a process or spine on the convex (lower) wall, not quite centered, and in some calices of my specimens a thickening occurs in this position (as well as elsewhere), which may be a spinal remnant. Their figure 6b shows tabulae shaped and arranged similarly to those of the Brassfield specimens.

Milne-Edwards and Haime listed as a synonym of their species *Favosites spongites* of Lonsdale (1839, *non* Goldfuss, 1826, which has a basal medial longitudinal ridge), from Benthall Edge, England (probably Wenlockian). It is difficult to interpret this specimen. It is low and expansive, with narrow apertures and internal tabulae similar to those in my form, and that described by Milne-Edwards and Haime. This is the specimen of Lonsdale's figures 8, 8a, and 8b. The specimen of Lonsdale's figures 8 c-e is from Wenlock, England, and appears to have more rounded apertures. No information is available on spines.

A form similar in apertural outline to the second specimen of Lonsdale is *Alveolites niagarensis,* described by Rominger (1876) from what is probably the Llandovery or Wenlock of Drummond's Island at Point Detour, northern Michigan (lectotype, Rominger's fig. 2), and from the drift in the lower peninsula of Michigan. As this name was preoccupied by a species of Nicholson and Hinde (1874), it was renamed *A. undosus* by Miller (1877). The apertures appear to be unthickened, and are rounder than in my form. Rominger reported spines around the walls of the lumen, apparently in longitudinal rows. In some cases, the spines do not appear in apertures, but Rominger attributed this to the vagaries of preservation. The tubes measure 0.5 to 1.0 mm in the longer direction (diameter), and contain tabulae which are "somewhat distant and oblique". Large mural pores produce pouch-like dilations "of the tube wall at the spot where situated". The degree of compression of the apertures reportedly varies, but in Rominger's figures they are considerably rounder than in my specimens. This fact, and the presence of large mural pores, which were not seen in the Brassfield specimens, suggest that *A. niagarensis* is distinct from the Brassfield material.

Foerste (1890) reported *A. niagarensis* Rominger from probable Brassfield beds at Ludlow Falls, Ohio (near the source of my specimens, locality 55). The average maximum dimension of his apertures is 1.2 mm, similar to mine. The shorter diameter varies from 0.4 mm to nearly the same length as the longer diameter, so that some of his apertures are reportedly as round as in Rominger's specimens. The corallite walls of Foerste's specimens are crenulate, a feature not seen in my specimens (unless Foerste was referring to spinal remnants). The tabulae are horizontal, spaced about 2.5 per mm. It is likely that Foerste's form is the same as mine, and that the roundness of some of the apertures reflects variability inherer' in *A. labechii*. Without further study, however, this cannot be certain.

From the Brownsport Formation of western Tennessee (Early Ludlow in age), Foerste (1909a) described *Alveolites inornatus*. Foerste's account of the aperture size (4 to 5 apertures in a width of 5 mm) is unclear, as he doesn't specify whether he speaks of the elongate or short dimension of the aperture. Judging from his figure, they are comparable to mine in size and shape. They vary in the degree of roundedness they display, but appear to be of the same pattern as mine, or within that range. Spines were not observed on the lumen walls, but Foerste noted that a longitudinal striation is sometimes seen along the median portion of the lower wall, and less often on the upper wall (apparently as in my specimens of *Alveolites labrosus,* which see). Foerste reported no marginal pores in the walls. This form seems quite similar to mine, except for the reported median striations. If these exist, it is unlikely that Foerste's specimen is conspecific with mine.

In at least two instances, *A. labechii* was reported by workers who did not supply sufficient information to substantiate their identifications. Williams (1919) reported it from the "Lockport Formation" (probably Middle Silurian) of Manitoulin Island, Lake Huron. He supplied only a figure of a specimen (GSC 2642a) which appears to have apertures with the size and roundness of *A. niagarensis* Rominger (= *A. undosus* Miller) from the same area.

Northrop (1939) reported the same species from the Gaspé region of Maritime Canada, but gave no morphologic data or figures. His material is from the La Vieille (Late Llandovery), the Gascons (Early Ludlow), the Bouleaux (Early Ludlow), and the Indian Point (Middle Ludlow) Formations.

From the Jupiter and Chicotte Formations of Anticosti Island (Late Llandovery to Early Wenlock), Twenhofel (1928) reported the present species. The apertures are compressed, the long diameter being 0.5 to 1.0 mm, and the shorter usually no more than a third that length. The tabulae are oblique, about two per mm, complete as well as incomplete. Pores occur at the angular edges of the corallites. Spines appear to be present, but are poorly-preserved. Judging from Twenhofel's description, it appears that *A. labechii* does indeed occur at Anticosti Island.

Bolton (1966) figured a specimen identified as *Alveolites undosus* from the Fossil Hill Formation (Middle Llandovery through Middle Wenlock) of Manitoulin Island. His specimen (GSC 20546) does have the general features of the present species, with apertures less rounded than in *A. undosus,* and about the same size as mine. Unfortunately, no other information than a figure of the upper surface is available, and so identification must await further study.

Of some interest is a specimen (UCGM 41781) found in the upper Whitewater beds (Richmondian, Late Ordovician) at Richmond, southeast Indiana. This is an *Alveolites* remarkably similar to the Brassfield form. The apertures are slightly smaller (maximum dimension = about 1.1 mm, minimum = 0.5 mm), the walls are of the same thickness (1/8 mm), and the tabulae are of the same sort, mostly concave upward, and spaced about 2.5 to 4.0 per mm. No mural pores were observed in this specimen. The major difference between the Ordovician specimen and those from the Brassfield is that the former appears to be more spinose. Spines are present on all walls of the corallites (though not seen in every corallite). If the thickenings on the walls of the Brassfield specimens are spinal remnants, it is likely that the present species occurs in the Late Ordovician and the Early Silurian, on either side of an unconformity. It is possibly, on the other hand, related to *A. undosus*, in which case also, the same species occurs on both sides of the Ordovician-Silurian boundary in eastern North America. Rominger's report of abundant spines in his specimen agrees with the features of the Ordovician specimen, but the apertures of the latter do not seem to be quite so round.

Distribution. — England, Wenlock; Ohio, mid-Llandovery (Brassfield Formation), and possibly Late Ordovician (Whitewater Formation); Anticosti Island, within the interval of Late Llandovery to Early Wenlock.

This species may occur on Manitoulin Island, in the interval from Middle Llandovery through Middle Wenlock (Bolton, 1966).

Brassfield occurrence. — Locality 55 (near Ludlow Falls, southwestern Ohio), in the grey-green clay at the top of the Brassfield Formation.

Family **PACHYPORIDAE** Gerth, 1921

Tabulate corals with corallites arranged in cerioid pattern in ramose coralla; corallites open approximately perpendicular to the corallum surface; corallite walls thickened by stereome, so that the mural pores are tunnel-like; septa spinose or represented by low ridges; tabulae complete.

Genus **STRIATOPORA** Hall, 1851

1844. *Cyathopora* Owen, p. 69 (*nom. null. pro Cyathophora* Michelin, 1843, p. 104).
1851. *Striatopora* Hall, p. 400.
1852. *Striatopora* Hall, p. 156.

Type species. — (By monotypy): *Striatopora flexuosa* Hall, 1851, p. 400; 1852, p. 156.

Type locality and horizon. — Lockport Formation (of Wenlock and Ludlow age) of New York.

Diagnosis. — *Pachyporidae* in which proximal portions of the corallites lie parallel to the axis of each corallum branch, but distally flex away from axis to open on surface; corallite walls thicken distally; tabulae complete, flat; mural pores present; septal crests sometimes present in calices.

Description. — *Striatopora* includes ramose tabulate corals in which the proximal portions of the corallites parallel the axis of the branch, while the distal portions curve from the axis to open obliquely on the surface of the branch. The corallite aperture is always somewhat inclined so as to face toward the more distal end of the branch.

The corallite lumen contains complete tabulae, generally widely spaced. Spines are present in some specimens. The walls are perforated by mural pores. In the corallite aperture, septal ridges are sometimes present.

The corallites, and their walls, broaden distally.

Discussion. — The taxonomic history of this genus has involved some matters of judgement that introduce considerable subjectivity into the status of the name *Striatopora* Hall. Michelin (1843) erected the name *Cyathophora* for a coralline, cerioid genus that (according to Lang, Smith, and Thomas, 1940, p. 43) is a Mesozoic scleractinian. In 1844, Owen introduced a new species, *Cyathopora iowensis* (p. 69), from what he called the "carboniferous limestone" of Iowa (but which Lang, *et al.* claimed is from Devonian rocks of that state). Judging from Owen's description and figure, it appears likely that this is a species belonging to *Striatopora* Hall (1851).

The difficulties begin in trying to determine whether Owen was introducing a new genus (this being the first use of the generic spelling *Cyathopora*), or was simply misspelling *Cyathophora* Michelin. Lang, Smith, and Thomas (1940, p. 44) argued that

although it might be argued that *Cyathopora* Owen is not a mistake for *Cyathophora* but was erected as a new genus by him, nevertheless, in view of the absence of any definitive evidence to that effect, and as Owen was not in the habit of creating new genera, we prefer to regard Owen's term as an error for *Cyathophora* (a Jurassic hexacoral genus).

Lang, *et al.* pointed out that other authors, including Lindström, Meek and Worthen, and Bassler concurred with this conclusion. To give this view its due, Owen indeed did not indicate that he intended to make a new genus by introducing this species. On the other hand, he nowhere mentioned Michelin (though neither did he mention Goldfuss in describing *Cyathophyllum caliculare*), and his figure of *Cyathopora* bears no resemblance to *Cyathophora* Michelin.

If one chooses to accept the historical view, that Owen's name was an invalid emendation (conscious or unconscious) of *Cyathophora* Michelin, then Hall's *Striatopora* must be the first valid name applied to the genus in question. On the other hand, if *Cyathopora* is valid as a generic name, it must displace *Striatopora* in usage. If Lang, *et al.* have erred, it could only be in their judgement of an entirely subjective matter, as the International Code of Zoological Nomenclature remains neutral in this affair. The long-established usage of Hall's generic name also argues in favor of upholding the historical view.

Hall first used the name *Striatopora* in 1851, in a brief paper whose purpose appears to have been to affix his name to genera which were to be described and figured in greater detail in his 1852 volume of *Palaeontology of New York*. In fact, in the former paper, Hall gave the page references for the latter publication.

Hall (1851) described *Striatopora* as follows:

> Ramose; corallum solid; stems composed of angular cells; apertures of the cells opening upon the surface into expanded angular cup-like depressions; interior of the cell rayed or striated, striae extending beyond the aperture of the cell.

With this description, he gave as an example of this genus the species *S. flexuosa*. His description of *S. flexuosa*, however, did not appear until 1852, with his more formal description of the genus *Striatopora*.

One possible interpretation would make *Striatopora* in 1851 a *nomen nudum*, as at that date it did not have a properly described type species. In this case, the valid date of *Striatopora* Hall would be 1852.

I choose a second interpretation: As *S. flexuosa* is the only species given by Hall (1851) as an example of *Striatopora*, the description of *Striatopora* provided in that paper applies also to *S. flexuosa*. *Striatopora*, then, was originally described in 1851, with a validly described single species, *S. flexuosa* Hall, 1851, which becomes the type species by monotypy.

Striatopora flexuosa Hall   Pl. 11, figs. 3, 4; Pl. 24, figs. 3-5; Pl. 39, fig. 2

1851. *Striatopora flexuosa* Hall, p. 400.
1852. *Striatopora flexuosa* Hall, p. 156, pl. 40B, figs. 1a-e.
? 1875. *Striatopora flexuosa* Hall, Nicholson, pp. 55-56, fig. 26.
1890. *Striatopora (Cladopora?) proboscidalis* (Davis), Foerste, p. 337 (*non*
        Cladopora proboscidalis Davis, 1887).
? 1899. *Striatopora flexuosa* Hall, Lambe, p. 40.
1901a. *Striatopora flexuosa* Hall, Grabau, p. 148, fig. 45.
1901b. *Striatopora flexuosa* Hall, Grabau, p. 148, fig. 45.
1944. *Striatopora flexuosa* Hall, Wells, p. 259, 260, pl. 40, figs. 1, 2.

Type material. — Lectotype (Wells, 1944, p. 260) — AMNH
1685, the original of Hall (1852, pl. 40B, figs. 1a, d(?)); shown by
Wells (pl. 40, figs. 1, 2).

Type locality and horizon. — Hall cited the locality of this
species as the shale of the Niagara Group at Lockport, New York.
This is probably the Rochester Shale, which spans most of the Wenlock (Berry and Boucot, 1970).

Number of specimens examined. — Two (one containing many
fragments), found in the field.

Diagnosis. — *Striatopora* with corallite apertures opening only
slightly oblique to the surface of the corallum branches, where they
are positioned in shallow, polygonal depressions; striae running from
aperture across floor of depression toward wall of depression; spines
not in evidence; tabulae generally farther apart than diameter of
lumen.

Description. — This species was found in the form of disjointed
branching twigs at two localities. The specimen with best internal
preservation (UCGM 41962) is a small fragment, apparently about
8 mm in original diameter, from locality 37. Full-grown corallites
in the interior of the cylindrical fragment are about 1.1 mm in
diameter, with a polygonal outline. Smaller, presumably younger
ones occur interstitially among the larger corallites. Apparently the
corallites extend along the axis of the corallum branch, and eventually flex outward, to open on the surface like the zooecia of a trepo-
stome bryozoan.

The corallite walls are less than 0.1 mm thick, with distinct,
dark midlines. The tabulae are usually spaced farther apart than the
diameter of the lumens, about one per mm. Occasional breaks in
the corallite walls of this specimen may represent mural pores. No
spines were observed in any of my specimens.

Another specimen (UCGM 41994), from locality 55, has similar internal features, and corallites of equivalent size to UCGM 41962. Its diameter, however, and the diameters of most of the corallum fragments associated with it in a small rock, is smaller, about 2.5 mm. Occasional breaks appear in the walls of some corallites where they appear to bud from a neighbor. While preservation of the specimen from this locality is poor, some surface details can be seen that were not visible in UCGM 41962. The corallite apertures open slightly oblique to the surface, facing upward. Each opens into a depressed area, up to 1.5 mm across, that rises gradually outward to a low bounding wall, giving it a polygonal outline. In most cases the aperture could not be clearly seen in the bottom of the depression, but in others, there is a distinct deepening where the aperture would be expected. Rarely, grooves radiate from the inferred aperture outward toward the wall of the depressed area.

Discussion. — See section on *Striatopora* Hall for detailed discussion of the original date for the description of *S. flexuosa* Hall. Hall's specimens consist of ramose, terete branches, with round apertures opening into polygonal depressions. The apertures open somewhat obliquely, facing upward in the depression, and striae radiate from their rims, the longest directed toward the upper walls of the depression. The figures and description agree closely with my specimens, and I have little hesitation in identifying the Brassfield form with *S. flexuosa.*

Nicholson's reference to *S. flexusoa* (1875) consists of a reillustration of some of Hall's figures, with the statement that the form is not uncommon in the Niagara Limestone of Thorold, Ontario. It is probable that the unit to which Nicholson was referring is the Gasport Member of the Lockport Formation (Upper Wenlock), as I have collected what appears to be *S. flexuosa* from this horizon.

Foerste (1890) reported a form of *Striatopora* from the Ludlow Falls region of southwestern Ohio, in the vicinity of the source of some of my material. The ramose coralla had an average diameter of 3 mm and tapered to about 2.2 mm within about an inch in some specimens. The thicker coralla seemed to branch more frequently than the more slender ones. The apertures were directed obliquely, outward and upward, opening into polygonal areas which were about five to six in a length of 5 mm (approximately comparable to my specimens). The floors of these depressions were often marked by striae, radiating from the aperture, that were more obvious in large calicular regions. In some specimens, Foerste reported that distinct pores could be seen connecting adjacent corallites. His description, and the provenance of his material suggest that Foerste was dealing with S. *flexuosa*.

Foerste identified his specimens with *Cladopora proboscidalis* Davis (1887, pl. 97, fig. 21) from the Louisville Limestone (Late Wenlock and Early Ludlow) near Louisville, Kentucky. Davis' figure is unclear, but Stumm's figure of his type (1964, pl. 69, fig. 7) indicates that the apertures open into upward-directed grooves, rather than polygonal depressions.

Lambe (1899) repeated Hall's description and noted that the Canadian Geological Survey had a specimen of this species from the Niagaran of Grey County, Ontario. As no figure or description accompanied this report, it is impossible to verify Lambe's identification.

Grabau (1901a, 1901b) reillustrated two of Hall's figures and gave a diagnosis of S. *flexuosa*, agreeing with Hall's concept of the species and with mine. He reported it as common in the bryozoan bed of the Rochester Shale at Niagara and at Lockport, western New York.

Distribution. — New York, Wenlock; Ohio, mid-Llandovery (Brassfield Formation).

Possible occurrences are those reported from Ontario, Canada by Lambe and by Nicholson, the latter from the Upper Wenlock.

Brassfield occurrence. — Localities 37 (near Sharpsville) and 55 (near Ludlow Falls) in southwestern Ohio.

HELIOLITIDA

Family **PROPORIDAE** Sokolov, 1950

Massive, encrusting or hemispherical heliolitids with distinct corallites lying in matrix of vesiculose coenosteum; corallite boundaries marked by the vesicles of the coenosteum (i.e., having no distinct, integrated walls), or else formed by lateral fusion of 12 longitudinal ridges representing thickened rows of spines; spines may be present even where thickened walls are lacking, being based upon the sides of coenosteal dissepiments bordering the corallites; tabulae usually complete, occasionally incomplete; spines sometimes present on upper surfaces of coenosteal vesicles.

Genus **PROPORA** Milne-Edwards & Haime, 1849

1849. *Propora* Milne-Edwards & Haime, p. 262.
1851. *Lyellia* Milne-Edwards & Haime, pp. 150, 226.
1878. *Pinacopora* Nicholson & Etheridge, p. 52.
? 1913. *Calvinia* Savage, p. 65 (*non* Nutting, 1900).
? 1922. *Cavella* Stechow, p. 152 (*pro Calvinia* Savage, 1913, *non* Nutting, 1900).
1934. *Koreanopora* Ozaki, p. 68 (*in* Shimizu, Ozaki, & Obata, 1934).

Type species. — (By monotypy) *Porites tubulatus* Lonsdale, 1839, p. 687 (*partim*), pl. 16, figs. 3, 3a, 3b (other figures not included).

Type locality and horizon. — Wenlock Limestone (Late Wenlock) at various localities in Shropshire, Herefordshire, Gloucestershire and Worcestershire, England.

Diagnosis. — Corallum of cylindrical corallites separated laterally by vesiculose coenosteum; corallites with distinct walls; corallite walls thickened at 12 points around the circumference, forming 12 septal ridges, parallel to the corallite axis, and at the surface forming the 12 nodes of a thickened exsert aperture; corallite tabulae complete; coenosteum may contain trabeculae; no mural pores.

Description. — Coralla in species of *Propora* consist of cylindrical corallites separated from one another by a vesiculose coenosteum. The corallites contain complete tabulae, and each corallite has its own wall, separating it sharply from the coenosteum. The wall contains 12 longitudinal rows of spines or nodes, and these rows are usually thickened to varying degrees to form 12 septal ridges. In transverse view, these ridges appear as 12 nodes around the periphery of the corallite. The nodes project above the level of the upper surface of the coenosteum, though it is not clear if this was the condition in life, or simply the result of a greater resistance to erosion on the part of the thickened parts of the wall than was possessed by the coenosteum. These 12 projections constitute the exsert margin of the corallites, noted by Milne-Edwards and Haime (1849, p. 262).

Corallites appear to arise directly from the coenosteum, rather than as offsets from older corallites. Vertical spines project from the dissepimental surfaces of the coenosteum in some forms.

Discussion. — Milne-Edwards and Haime (1849) first described *Propora* based upon *Porites tubulatus* Lonsdale (1839) from the Silurian of England. They were struck particularly by the protruding apertures of the corallites, marked by a surficial prolongation of each septal ridge into a knob at the corallum surface. It is not clear from the original description of this genus, nor from Lonsdale's description of the type species, if the coenosteum is vesicular. Hill and Stumm (1956, pp. F460, F461) described the genus and figured the type species, showing a vesiculose coenosteum, and so this genus is here accepted as having the structure usually attributed to it.

In 1851, Milne-Edwards and Haime erected *Lyellia*, based upon the heliolitid species *L. americana* from the Silurian of Drummond Island, Lake Huron. In all respects, this appears identical to *Propora*, and is regarded as a junior synonym thereof.

Nicholson and Etheridge (1878) introduced the genus *Pinacopora* for Silurian heliolitids from near Girvan, Scotland. Though their description of the genus and its type species, *P. grayi*, is replete with references to tabulate tubules between the corallites, the figures (e.g., pl. 3, fig. 3d) clearly show a vesiculose coenosteum, often so narrow due to close approximation of the corallites, as to be about one dissepiment broad. In this, and in all other discernible features, *Pinacopora* is identical to, and a junior synonym of *Propora*.

From the Lower Silurian of Illinois, Savage (1913) described the heliolitid genus *Calvinia*, distinguished by its vesiculose coenosteum, through which small tubules reportedly run parallel to the corallites. Stechow (1922) noted that the name *Calvinia* was preoccupied, and renamed the genus *Cavella*. Except for the presence of the tubes, Savage's specimen would clearly belong to *Propora*. Though Savage's plate 2, figure 15 suggests that these tubes are real features, it will be necessary to study the type material before a final determination of the relationships of this genus can be made.

Ozaki (*in* Shimizu, Ozaki, & Obata, 1934) described the genus *Koreanopora* from Silurian beds of northwest Korea (North Korea). This genus differs from the typical form of *Propora* solely in having a slight axial elevation of the tabulae in the corallites, a feature that seems more likely to be a normal variation for *Propora*. The two genera are here considered synonymous.

Propora conferta Milne-Edwards & Haime　　　　Pl. 34, figs. 2, 4

1851. *Propora conferta* Milne-Edwards & Haime, p. 225.
1865b. *Heliolites affinis* Billings, p. 427.
1899. *Propora conferta* Milne-Edwards & Haime, Lindström, p. 93-94, pl. 8, figs. 32-39; pl. 9, figs. 1-23, 31, 32, 35.

1909a. *Lyellia thebesensis* Foerste, p. 95, figs. 69a, b.
? 1913. *Lyellia thebesensis* Foerste, Savage, pp. 66-67, pl. 3, figs. 6, 7.
1928. *Lyellia affinis* (Billings), Twenhofel, p. 133.
1934. *Propora* cfr. *magnifica* Počta, Shimizu, Ozaki, & Obata, p. 67, pl. 10, figs. 9, 10; pl. 11, fig. 1 (*non Propora magnifica* Počta, 1902).
1934. *Propora yabei* Shimizu, Ozaki, & Obata, pp. 67-68, pl. 11, figs. 2, 3.
1937. *Plasmopora lambei* Schuchert, Teichert, pp. 53-54, pl. 4, fig. 13; pl. 5, figs. 1, 2 (*non Plasmopora lambii* Schuchert, 1900).
1939. *Lyellia affinis* (Billings), Northrop, p. 148.
1939. *Lyellia affinis* (Billings), Shrock & Twenhofel, p. 253, pl. 28, figs. 10, 11.
? 1940. *Propora conferta* Milne-Edwards & Haime, Jones & Hill, p. 209, pl. 11, fig. 5a, b (*caet. excl.*).
1940. *Propora affinis* (Billings), Sugiyama, pp. 141-142, pl. 24, figs. 1-3; pl. 32, figs. 1, 2.
1957. *Lyellia thebesensis* Foerste *paucivesiculosa* Bolton, pp. 64-65, pl. 6, figs. 1-7.
1957. *Propora conferta* Milne-Edwards & Haime, Iskyul, pp. 97-98, pl. 7, figs. 3, 4.
1962a. *Lyellia affinis* (Billings), Norford, pl. 8, fig. 17.
1962. *Propora conferta* Milne-Edwards & Haime, Sokolov, p. 425 (English translation), fig. 4.
? 1972. Heliolitid undet. Bolton & Copeland, p. 30, pl. 4, figs. 1, 2.
1972. *Propora* sp. Bolton, p. 26, pl. 3, figs. 5, 8; p. 28, pl. 4, figs. 8, 13; p. 32, pl. 6, figs. 5, 9, 10, 11; p. 34, pl. 7, figs. 6, 17; p. 40, pl. 10, figs. 2, 3.

Type material. — Lindström (1899, p. 93) described two syntypes in the de Verneuil Collection, École des Mines, Paris.

Type locality and horizon. — Late Ordovician of Estonia (Windau and Borkholm).

Number of specimens examined. — Twelve, found in the field.

Diagnosis. — *Propora* with corallites about one to two mm in diameter, separated from the nearest neighbor by less than their diameter; corallite walls thickened into 12 longitudinal "septal" ridges; tabulae concave upward or flat, occasionally incomplete.

Description. — The coralla are gently to strongly convex upward. Within a single specimen, corallite diameters may range from 1.0 to 1.5 mm, though descriptions in the literature indicate an upward continuum to 2.0 mm. Characteristically, each corallite is separated from its nearest neighbor by a distance less than its diameter, and sometimes the coenosteum is almost absent between a pair of corallites. Where the upper surface of the corallum is exposed, the corallite rims are exsert, a character described by Milne-Edwards and Haime (1850, p. lix) as typical of *Propora*. While this condition may have existed in life, it may also be a result of the thickened corallite wall having a greater resistance to erosion than has the vesiculose coenosteum. None of the specimens so far encountered can be unequivocally said to possess a holotheca.

Well-preserved corallites have the wall thickened at 12 points around the perimeter, producing a cogwheel-like appearance, with the teeth pointing axially. This results in a corallite wall about 1/9 mm thick. All 12 of these "septal" thickenings appear concurrently in a given transverse section of a corallite. This suggests they are continuous longitudinal ridges, as does the absence of any discontinuity in wall thickness as seen in longitudinal section (see Pl. 34, figs. 2, 4). The wall deposits are optically discontinuous from the rest of the corallum.

The corallites are contained in a matrix (coenosteum) of vesicular dissepiments. They appear to originate by differentiating directly from this "common skeletal material", rather than by budding from one another. Strangely, they also appear to disintegrate distally, merging once again with the vesiculose tissue without being bounded above by the wall deposits as would be expected if they were merely passing from the plane of the section. (This is not the case seen in Pl. 34, fig. 4, where the extension of the septal ridges upward after the disappearance of the rest of the corallite indicates that the corallite has merely passed out of the plane of section.) The spaces between neighboring corallites vary so that there are as few as one, and as many as three or more dissepiments ranged laterally between them, depending upon the size of the space. These dissepiments vary in size and convexity. A large one may be 1 mm across and 0.3 mm high. A rather small one may be 0.3 mm across and 0.1 mm high. In some specimens (Pl. 34, fig. 4) these size-variations are grouped, giving the coenosteum a vertical zoning. Dissepiments cannot be definitely said to bear the spines on their upper surfaces that occur in other species of *Propora*.

The tabulae of the corallites are as a rule concave upward, occasionally flat, and rarely convex. They are spaced from 1/2 to 1/7 mm apart vertically, in bands correlating with those of the coenosteum. The tabulae are mostly complete, but some incomplete examples occur as well.

Discussion. — *Propora conferta* is one of the most widespread of all Early Paleozoic coral species. In addition, it had a strong superficial resemblance to at least one much older species (discussed below).

P. conferta was first described by Milne-Edwards and Haime (1851), without illustration, based upon material from Estonia.

Their assignment of a Wenlockian age to the specimens is apparently wrong, a Late Ordovician age being more likely (see discussion of Lindström, 1899 below). The important characters follow: Coralla poorly-preserved hemispherical masses; corallite diameter 2 mm, with calices close-spaced, making the zones of vesiculose "coenenchyme" (coenosteum) ill-developed. The corallite walls are serrated, with 12 projections. The tabulae are close-spaced. The description, while quantitatively vague, agrees well with my specimens.

Billings (1865b) reported as new, *Heliolites affinis* from Anticosti Island. The syntype in the collection of the Canadian Geological Survey (GSC 2340) is from the Chicotte Formation (Late Llandovery through Early Wenlock), but I have not examined it. Lindström (1899, p. 93) examined type *P. conferta* and typical specimens of *H. affinis*, and judged them conspecific. The corallites of *H. affinis* are "from half a line to little less than one line in diameter, the more common width being about two-thirds of a line. . ." A line is 1/12 inch (just over 2 mm), so the corallite diameter ranges from slightly over 1 mm to about 2 mm, with most about 1.3 mm across. The corallites are in a vesiculose matrix (which, together with other features, indicates that *H. affinis* Billings belongs to *Propora*, and not *Heliolites*). The corallites are separated from one another by about half their diameter, so that the intervening coenosteum is quite narrow. Sometimes the corallites are so close that their usually round profile becomes polygonal. As is common in *Propora*, the rims of the corallites are thin and exsert, and well-preserved apertures display 12 small tubercles around the opening. Septa in some specimens appear as striations of the corallite walls. The tabulae are usually flat, with three or four in 2 mm vertically. In every respect, this description agrees with my material, and *H. affinis* is probably a junior synonym of *P. conferta*.

Lindström (1899) took the important step of searching out and describing the unfigured specimens on which Milne-Edwards and Haime's (1851) description of *P. conferta* was based. These were located in the École des Mines in Paris, in the care of Douvillé, who assured Lindström that the writing on their labels was in the hand of Jules Haime. The two specimens involved were collected by E. deVerneuil during his expedition in Russia (1844) with Murchison. One was from Borkholm, and the other from "Chavli

Canal de Windau", both in Estonia. Lindström believed the second specimen was not in place, as there were no Silurian strata at that locality.

The coralla are in discoidal or spheroidal masses, with corallites about 1.5 mm across. These tubes are so closely spaced that little room remains for the vesiculose coenosteum. The apertural rims are only slightly exsert, and are crenulated into 12 small nodules. No spines were seen in the corallites, but the walls were fluted. The tabulae are variably spaced: those of the Borkholm specimen (Lindström's pl. 8, fig. 34) average about 4 per mm, while those of the Windau specimen (fig. 39) are about 2 to 3 per mm. The tabulae are mostly flat, or slightly concave. A very few are incomplete. The coenosteum is free of spines.

Lindström (p. 130) indicated that the Borkholm specimen is from the F-1 horizon, (Late Ordovician in age, according to Manten, 1971). His description and figures indicate that Milne-Edwards and Haime's material is probably conspecific with mine, as these authors' less detailed description suggested.

Foerste (1909a) described *Lyellia thebesensis* from the vicinity of Thebes, Illinois. Stratigraphically, it lay three feet above the "Cape Girardeau" Limestone (now called the Girardeau Limestone) in the Edgewood Formation, of Early Llandovery age. Foerste's description is vague, but the important points are that the coralla are massive, with corallites slightly over 1 mm in diameter nearly in contact with one another, with narrow interspaces of coenosteum. This intermediary tissue is not distinctly vesiculose. The corallite walls are crenulate and longitudinally striate. The corallites have eight to nine tabulae in 5 mm vertically. This description agrees with all the characters of the Brassfield specimens except the nature of the coenosteum. In my specimens, however, where corallites are closely spaced, the interstitial deposits may lose some of their vesicular aspect, so that it seems likely that *L. thebesensis* is a junior synonym of *P. conferta*.

Savage, (1913, pp 66-67), described several more specimens from the Edgewood Formation near Thebes, Illinois, and Pike County, Missouri, which he assigned to *L. thebesensis* Foerste. The description and figures, while poor, agree with Foerste's, and are tentatively assigned to *P. conferta*. Closer study of Savage's material would be desirable before the identification can be confirmed.

Bolton (1957) described *Lyellia thebesensis paucivesiculosa,* a new variety of Foerste's species, from the Manitoulin Formation (Early Llandovery) of Manitoulin Island, Lake Huron. The type specimen (ROM 783cl) has corallites 1.0 to 1.25 mm in diameter, with flat or concave tabulae spaced one to three per mm that show some vertical zonation. The septa are solid ridges. This form is characterized by the extreme narrowness of the coenosteum between the corallites. Where corallites are especially closely spaced, the coenosteum appears tabulate, but it takes on a vesicular aspect where not so restricted. Bolton's material agrees well with the Brassfield specimens morphologically, and the two assemblages are considered conspecific.

Shimizu, Ozaki, and Obata (1934) described two species from the Silurian of northwestern Korea which appear to belong to *Propora conferta.* The first species which they call *Propora* cfr. *magnifica* is not identical to Počta's species of that name, which has spinose septa. The Korean form has closely spaced corallites, about 1.3 mm in diameter, reportedly circular or crenulate in cross-section, and with definite walls. The tabulae are concave, averaging three to four per mm, and appear to be zoned vertically. The coenosteum is vesiculose.

The second Korean species, *Propora yabei,* has closely spaced corallites, about 1.0 mm in diameter, reportedly circular or slightly crenulate in cross-section. The tabulae are mostly flat or slightly convex, and are spaced about 2.5 per mm, with no evidence of vertical zonation. The coenosteum is vesiculose.

Teichert (1937) illustrated and described *"Plasmopora lambei"* from Iglulik Island in the northeast Canadian Arctic, which agrees in nearly all visible characters with the Brassfield specimens. The corallites are approximately 1.0 to 1.8 mm in diameter, with well-defined, thickened walls bearing 12 projections around the periphery. The corallites are closely spaced, often nearly contiguous laterally, and the coenosteum is definitely vesiculose. The only difference between Teichert's specimen and mine lies in the topography of the tabulae. While the tabular spacing is in good agreement with that in my form, in Teichert's specimen there is a greater tendency toward tabular convexity than in my specimens. (Teichert's pl. 4, fig. 13 shows a specimen upside-down, as indicated by the overlapping of the dissepiments.) Convex tabulae do occur in the Brassfield specimens, but not with the frequency shown in Teichert's section. The importance of this difference is difficult to judge. Though it may seem trivial, the simplicity of Early Paleozoic corals makes it advisable to consider carefully even the least obvious characters in studying relationships between these fossil organisms. Nonetheless, it would be consistent with the practice in the rest of the present paper to consider Teichert's form as probably conspecific with mine. Unfortunately, the only available information on the age is that it is Ordovician.

A number of workers have described and figured specimens from the Maritime Provinces of Canada that bear close resemblance to my corals. *Lyellia affinis,* described by Twenhofel (1928) from Anticosti Island is from zone 4 of the Ellis Bay Formation (variously assigned to the Late Ordovician or Early Llandovery) through the top of the Chicotte Formation (Early Wenlock). Twenhofel's descriptions are meager, and are clearly based upon a number of specimens collectively encompassing a realm of variation. The corallites are separated by distances considerably less than their diameters (though the diameter is not given). Septa are well-developed in some specimens, though their nature is not disclosed.

A recent paper by Bolton (1972) supports the idea that this species ranged from Late Ordovician through Early Silurian time in the Anticosti Island region. Bolton illustrated fossils from the stratigraphic sequence of this region, and *P. conferta,* identified by Bolton as *Propora* sp., appears in several units.

The earliest occurrence is in the upper member of the Vauréal Formation (Late Ordovician, Richmondian). This specimen (Bolton, 1972, pl. 3, figs. 5, 8) has closely spaced corallites, about 1.5 mm in diameter, with a vesiculose coenosteum. Tabulae are spaced about two per mm, and are occasionally incomplete. They are usually flat or very slightly concave or convex. The absence of clear septal projections in the transverse section is the only element of doubt in the identification of this specimen with my material. It should be noted, however, that such projections in my specimens are often not preserved, and Bolton's figure is unclear.

The next specimen is from member 5 of the Ellis Bay Formation, which Berry and Boucot (1970) consider of Early Llandovery age while Bolton (1972) assigned it to the Late Ordovician. In any event, it appears to be near the Ordovician-Silurian boundary. This specimen is essentially identical to that from the Vauréal, but here the tabulae are more widely spaced, about one per mm. Septal projections are unclear in the transverse section, but some serration is present in several of the corallites in this figure.

Bolton next illustrated two specimens from member 6 of the Ellis Bay Formation. The specimens (pl. 6, figs. 5, 11; pl. 6, figs. 9, 10) have corallites of similar size, roughly 1.0 to 1.2 mm in diameter. Here again, the wall structure is not clear enough to discern longitudinal ridges. The tabulae in both specimens are dominantly flat, or shallowly concave, and are spaced about two per mm, in figure 10 (with no clear vertical zonation), while those in fig. 11 are from 1.5 to 4 per mm, and do show some zonation. The coenosteum is in many places quite narrow in both specimens, often limited to the breadth of a single dissepiment.

Another specimen, from the Becscie Formation (Middle Llandovery), shows (pl. 7, figs. 6, 17) closely spaced corallites, about 1.5 mm in diameter, with some evidence of longitudinal ridges within the corallites. The tabulae are dominantly flat, but some are weakly convex or concave. Zonation of tabular spacing is distinct, and ranges from 1 to 4.5 per mm. Here again the coenosteum is often restricted to the width of a single dissepiment.

Finally, Bolton figured a specimen (pl. 10, figs. 2, 3) from the Jupiter Formation (Late Llandovery). The corallite diameters range from about 1.0 to 1.2 mm, and the coenosteum is often restricted, as in Bolton's other figured specimens, to a width of one dissepiment. The tabulae are nearly all horizontal or shallowly concave, as are those of the Brassfield specimens. Their packing is zoned, ranging from 5 per mm down to 1.2 per mm respectively in the denser and sparser zones. The nature of the corallite walls cannot be discerned from the figure.

Bolton thus presented six specimens, ranging in age from Late Ordovician to Late Llandovery, from within essentially the same geographic area. In most respects, they closely resemble the Brassfield specimens. The range of corallite diameters is within the range of variation of my corals, as is the spacing of the corallites. "Septal" projections in the Anticosti Island specimens occur in a pattern that implies that they probably were present initially, but either have been lost through diagenesis by most of the corallites, or have simply not been recorded in the photographic plates. A third possibility is that these structures are not present in all corallites of a corallum, or occur only at certain levels of a corallite. The spacing of the tabulae is variable, but the range of variation encompasses that of the Brassfield corals. Concavity or convexity of the tabulae does not appear to correlate meaningfully with other taxonomic characters, or with the stratigraphic distribution. It seems advisable to treat the predominant concavity of the tabulae in Brassfield specimens of *P. conferta* as a subspecific character. Nonetheless, the question of validity of this assumption would be an important line of study. Certainly, this species (if it is a single species) is sufficiently widespread geographically and stratigraphically to supply abundant material for study.

Northrop (1939) reported corals from the Gaspé region of the Canadian Maritimes which he identified as *Lyellia affinis*. They occur in a rock sequence including the Anse Cascon (Late Llandovery), La Vieille (Late Llandovery through Wenlock) and Gascons, Bouleaux, West Point and Indian Point Formations (Ludlow). The description closely approximates the characters of my specimens, though it is clear that Northrop was describing several specimens, rather than detailing the characters of a single coral. The corallites are closely-spaced, within one-half a diameter from the nearest neighbor, and are separated therefrom by a vesiculose coenosteum. Corallite diameter is 1.0 to 1.7 mm. The corallite rims are exsert. There is no information given concerning septal structures or tabulae. Based on the available information, it seems likely that Northrop was dealing with *P. conferta* .

Shrock and Twenhofel (1939) described a coral from the Pike Arm Formation of northern Newfoundland, associated with a fauna that they considered correlative with that of the Jupiter Formation of Anticosti Island (Late Llandovery). Their description agrees well with the characters of the Brassfield form of *P. conferta*: The corallite diameter is 1.0 to 1.75 mm, with closely packed corallites rarely over half a diameter apart, often nearly contiguous. The corallite walls are crenulated. The tabulae are flat, concave or convex, spaced two to four per mm. The longitudinal section (pl. 28, fig. 11), however, shows one character that is different from typical Brassfield specimens: Incomplete tabulae are somewhat more abundant in this specimen than in mine. Judging by the variation apparent within this species, it seems reasonable to regard Shrock and Twenhofel's specimen as conspecific with mine.

Jones and Hill's (1940) material identified as *Propora conferta* cannot confidently be considered conspecific with my specimens. The three specimens they figured are shown in three transverse and one longitudinal section. The corallite walls are crenulated. Only in plate 11, figure 5 are the corallites particularly closely spaced. These have diameters of about 1 mm, and are spaced within one-half diameter of the nearest neighbor. The coenosteum (fig. 5b) is vesiculose. Tabulae are mostly concave upward, spaced about five per mm. The description refers to short breaks in the corallite walls occasionally visible in transverse sections. Such breaks were not noted in my material, and so the identificaton of these specimens from the Ludlow of Australia must be regarded as questionable.

Sugiyama (1940) described three specimens from the Kawauti Series of northeast Honshu Island, Japan. These are from the *Clathrodictyon* and *Halysites* Limestones, assigned to the Salopian (Wenlock and Early Ludlow) equivalents of England (see Hamada, 1958, p. 95, table 2). These specimens are similar to mine, though Sugiyama identified them as *Propora affinis* (Billings). Their features were described as follows: The corallites are separated by one-half a corallite diameter or less (pl. 24, fig. 1), and are 1.5 to 1.8 mm in diameter. Their margins display varying degrees of crenulation. In one unfigured specimen from the *Clathrodictyon* Limestone, these crenulations result in 12 "minute nodulous protrusions" (presumably referring to the exsert corallite margins typical of *Propora*). In the second specimen (also from the *Clathrodictyon* Limestone), the walls are not so regularly crenulated, and occasionally are even smooth (possibly the result of recrystallization). The tabulae are mostly flat, rarely slightly concave, and are spaced five to seven per mm. The coenosteum is vesiculose. It seems reasonable to consider Sugiyama's specimens conspecific with mine.

Iskyul (1957) described *Propora conferta* from the Siberian Platform (23.5 kilometers upstream from the mouth of the Podkamennaya Tunguska River). The age is reported simply as "Silurian". The corallites are 1.0 to 1.2 mm in diameter, spaced 0.5 to 0.8 mm apart. Their walls are crenulated into 12 projections. The tabulae are mostly complete, and are horizontal, or slightly convex or concave. Their spacing is about four to five per mm. The coenosteum is vesiculose. In all important respects the Siberian form agrees with mine, and they are considered conspecific.

Norford's .(1962a) figure of a coral identified as *Lyellia affinis*, from the Middle Llandovery Fisher Branch Dolomite of Manitoba superficially resembles my material, though the specimen is unfortunately not sectioned. The corallites are about 1 mm in diameter, with exsert rims that are nearly in contact with neighboring corallites. The coenosteum appears to be vesiculose, though it is seen only superficially. Where they are visible, the tabulae appear to be flat, and spaced about three per mm. Crenulation of the corallite walls is unclear in Norford's figure. It seems likely that Norford's specimen does belong to *P. conferta*.

Sokolov (1962) illustrated a specimen which he identified as *Propora conferta* from the Llandovery of the Baltic region (Estonia?). The corallites are about 1.6 mm in diameter, and are spaced less than half their diameter apart. Their walls are thickened in about 12 places around the circumference, forming "septal" projections in transverse section. The coenosteum is vesiculose. The tabulae are flat, or sometimes shallowly concave or convex. The spacing of the tabulae varies rhythmically vertically, from 6 to 2.5 per mm. Rarely the tabulae appear incomplete. In all essential features this specimen agrees with the Brassfield material.

"Heliolitid undet." Bolton and Copeland (1972) from the Silurian of Lake Timiskaming, Canada, is apparently too poorly preserved for determination of the nature of the coenosteum. Therefore, while its features are similar to those of *Propora conferta*, it is not possible to identify it confidently from the figures.

A species quite distinct from *P. conferta*, yet with interesting similarities, is *Plasmopora lambii* Schuchert (1900, p. 154). This species probably belongs to *Plasmoporella* Kiär, 1899. A paralectotype (USNM 28140b) is shown on Plate 35, figures 3, 4. (Flower and Duncan, 1975, p. 186, chose USNM 28140d as the lectotype.) The specimens of the type suite surficially resemble *Propora conferta*: the corallites are about 1.5 mm in diameter, spaced less than a diameter from the nearest neighbor, and are separated by a vesiculose coenosteum. The coralla are massive. In thin-section, however, the distinctive nature of *Plasmopora lambii* Schuchert becomes clear. The corallites have no true walls, but are in direct contact with the dissepiments of the coenosteum. These intrude along the line of the corallite boundaries, and in transverse section (fig. 4) sometimes give the corallite a crenulated appearance. The tabulae are nearly all incomplete, so that the corallite tabularium is vesiculose, and distinguished from the coenosteum solely by the greater width of its vesicles. The type material is from Baffin Land, Arctic Canada, and is reportedly of "Trenton" age (Late Middle and Early Late Ordovician). Flower and Duncan (1975, p. 186) give the locality as "Ordovician beds of Silliman's Fossil Mount. . ., Frobisher Bay, Baffin Island, Canada".

The available data show *Propora conferta* as a form-group of corals with wide geographic and long stratigraphic range, varying in several characters, but apparently forming a morphologic continuum, and thus referable to a single paleontologic species.

Distribution. — Estonia, Late Ordovician, Llandovery(?); Anticosti Island, Late Ordovician (Richmondian) through Late Llandovery, and possibly Early Wenlock; Illinois, Early Llandovery; Manitoulin Island, Early Llandovery; Manitoba, Middle Llandovery; Korea, Silurian; northeast Canadian Arctic, Ordovician; Gaspé region of maritime Canada, Late Llandovery into the Ludlow; northern Newfoundland, Late Llandovery; Japan, Wenlock and Early Ludlow; Siberian Platform, Silurian; Ohio, mid-Llandovery (Brassfield Formation).

Brassfield occurrence. — Localities 7a (near Fairborn), 11 (near West Union), 13 (Todd Fork Creek near Wilmington), 14 (Piqua), 15 (near West Milton), 37 (Turtle Creek near Sharpsville) and 55 (near Ludlow Falls, in clay layer at top of the Brassfield), all in southwestern Ohio.

Propora exigua (Billings)   Pl. 34, figs. 1, 3

1865b. *Heliolites exiguus* Billings, p. 428.
1866. *Heliolites exiguus* Billings, Billings, p. 31, fig. 14.
? 1880. *Heliolites exiguus* Billings, Nicholson & Etheridge, pp. 250-251.
1899. *Propora conferta*, var. *minima* Lindström, p. 95, pl. 9, figs. 24-26.
1899. *Lyellia exigua* (Billings), Lambe, p. 86, pl. 5, figs. 3, 3a.
1928. *Lyellia exigua* (Billings), Twenhofel, pp. 134-135.
Cf. 1939. *Lyellia exigua* (Billings), Shrock & Twenhofel, p. 254, pl. 28, figs. 13, 14.
1939. *Lyellia superba* (Billings), Shrock & Twenhofel, p. 254, pl. 28, figs. 8, 9 (*non*? Billings, 1866; *nec* Lambe, 1899).
Cf. 1940. *Propora conferta* Milne-Edwards & Haime, Jones & Hill (*partim*), p. 209, pl. 11, fig. 4 (*non* figs. 3, 5a, 5b).

Type material. — The holotype is listed in Bolton (1960, p. 28) as GSC No. 2239. It was illustrated by Billings (1866, fig. 14), and possibly by Lambe (1899, pl. 5, figs. 3, 3a).

Type locality and horizon. — According to Bolton (1960), the Ellis Bay Formation (variously considered Late Ordovician or Early Llandovery), at Ellis Bay, Anticosti Island.

Number of specimens examined. — One, found in the field.

Diagnosis. — *Propora* with corallites generally less than 1 mm in diameter, separated from their nearest neighbors by a distance equal to or greater than their diameters; walls crenulate; coenosteal spines apparently absent; tabulae generally complete.

Description. — As preserved, the corallum is 28 mm across and 7 mm thick. It appears originally to have had a low, wavy, lens-like form, and to have grown directly upon the sediment surface, rather than cemented to a solid object, as it was found lying upside-down in the rock (Pl. 34, fig. 1).

The corallites are 0.6 to 0.8 mm in diameter, and are almost all separated from the nearest neighbor by a distance at least equal to their diameters. The maximum separation is about 0.8 to 1.0 mm. The corallites are each bounded by a definite wall, about 1/15 mm thick, which is thickened into 12 longitudinal "septal" ridges. It is assumed that these ridges are continuous, rather than spinal swellings, because they appear in virtually every corallite seen in cross-section. It is, however, possible that they represent extensive stereomal deposits over rows of spines, but this is speculative.

Of 60 tabulae counted in a longitudinal section, 54 were complete. These included flat, as well as shallowly-concave and convex tabulae, with zoned spacing. In the sparser regions, these are spaced about three per mm, while in the denser regions they are seven per mm. Most of the incomplete tabulae occur in the denser regions.

The space between the corallites (the coenosteum) is vesiculose, and from one to five vesicles lie horizontally along the space between tubes. They vary in form from high (0.4 mm) and narrow (0.7 mm) to low (0.2 mm) and wide (1.4 mm). Very small and very large vesicles tend to occur in patches, but on the whole, the coenosteum displays the same zonation as the corallite tabulae, with coarse vesicles on about the same levels as the widely spaced tabulae. The uppermost zone of the corallum contains a concentration of unusually small vesicles, and may reflect a deterioration of conditions that led to the death of the colony.

There appear to be no spines in the coenosteum.

Discussion. — Billings (1865b) described the species as follows [one line = 1/12 inch, or about 2.1 mm]:

> Cells about half a line in diameter and somewhat more than their own width distant from each other, with thin elevated margins, apparently not crenulated. Septa not visible in the only specimen collected. Tabulae numerous, four to six in one line. Coenenchyma minutely vesicular.
>
> As the specimen is somewhat worn, it is possible that the margins of the cells when perfect may be crenulated. The coenenchyma appears to be vesicular, but more specimens are required to decide this point.
>
> The species, on account of the small size of the cells and their greater proportional distance from each other, seems to be distinct from all the others.

Lindström (1899) described, as a new variety of *Propora conferta* Milne-Edwards and Haime, *P. conferta minima*, from his horizon *a* (the *Arachnophyllum* beds, of Middle or Late Llandovery age, according to Manten, 1971, p. 33, 42) near Visby, Gotland. This form is discoid and flat, with very narrow corallite apertures, about 0.5 mm across, separated from one another by at least one diameter. Lindström stressed that the calices are "perfectly smooth, without the least vestige of septa or septal flutings or cannelures." The tabulae are mostly gently concave, about four per mm. The vesiculose coenosteum is free of spines. This form agrees in all respects with my concept of *P. exigua*, except for the absence of wall-fluting. Considering the ease with which this feature can be obscured in the process of fossilization (especially in such small corallites) it seems safe to consider this form conspecific with my own.

Lambe (1899) noted that the type specimen was the only one representing this species in the Canadian Geological Survey collection. He described it as 45 mm long, 30 mm broad, and 13 mm high. The corallites were 0.75 mm wide, "separated from each other by distances generally equal to or less than [*sic*] their width, but varying from 1 to 5 mm wide". The tabulae were horizontal, spaced two to five per mm. The corallites had 12 faint septal ridges on the inner surfaces of their walls, and their rims on the surface of the corallum were exsert. The coenosteum was vesiculose, the vesicles 0.16 to 0.5 mm wide. Lambe's plate 5, figure 3a (the holotype?) does indeed show some of the corallites within half a diameter of their nearest neighbor, but others seem more widely spaced. In view of

Billings' observations, and of the similarity of my specimen to that
described by him and by Lambe, it seems best to regard the spacing
of the corallites as important, but variable. The other Brassfield
species of *Propora*, *P. conferta* and *P. eminula*, do not share the
combination of narrow corallites and relatively wide corallite spacing
seen in *P. exigua*.

Nicholson and Etheridge's reference to *H. exiguus* (1880) only
cited Billings' report (1866). They added nothing to the concept
of the species, but did note that the coenosteum is vesiculose, and
that the corallites are spaced more than a diameter apart.

Twenhofel (1928) reported the present species from Anticosti
Island, in the Ellis Bay Formation (variously regarded as Late
Ordovician or Early Llandovery) and the Jupiter Formation (Late
Llandovery). His specimens were both hemispheric and encrusting,
with corallites 0.5 mm in diameter, separated by 1 to 2 mm from
one another. The corallites contained faint "septa". The calices were
exsert, and Twenhofel noted that there was a hint of the 12 nodules
frequently seen around the corallite apertures of *Propora*. The tabulae were horizontal, spaced about two per mm. The vesicles of the
coenosteum were from 0.25 to 1 mm wide, and 0.25 to 0.35 mm high.
The description fits well with my concept of *P. exigua*, and even
though the corallites appear more widely spaced than is common for
the species, it seems that Twenhofel's identification can be accepted.

Shrock and Twenhofel (1939) described and figured what they
considered to be this species from the Silurian Pike Arm Formation
of northern Newfoundland. This is a problematic case. In almost all
respects this form agrees well with *P. exigua* (Billings): The corallites are 0.75 mm in diameter, with 12 "septal" ridges on their inner
surfaces. The horizontal or gently concave tabulae are spaced three
or four per mm, and the coenosteum is vesiculose. The corallites,
however, are spaced as in *P. conferta*, rarely more than half a
diameter apart. The transverse section (Shrock & Twenhofel, 1939,
pl. 28, fig. 14) shows this to be the rule, not the exception, in this
specimen. It is possible that the corallite diameter is all important,
and that *P. exigua* can have corallites as closely spaced as those of
P. conferta. Yet, it is important to consider the possibility that *P.*
exigua (Billings) and *P. conferta* Milne-Edwards & Haime are

conspecific, representing end points of a morphologic continuum. The decision on this matter seems best deferred until more material has been studied, but for now, the two will be regarded as separate species, with the relationship of *Lyellia exigua* (Billings) Shrock and Twenhofel left in question.

Shrock and Twenhofel (1939) also reported *Lyellia superba* from the same region and formation. The corallites of this species are reportedly 0.5 to 0.6 mm in diameter, and spaced from one to three diameters apart (A transverse section, figure 8, shows corallites about one diameter apart.). Septal ridges are "poorly-developed", and the tabulae, which are usually flat, average four per mm. The coenosteum is vesiculose. This form agrees sufficiently with my concept of *P. exigua* to indicate that it belongs to that species.

Jones and Hill's (1940) plate 11, figure 4, purporting to represent *Propora conferta* Milne-Edwards & Haime from the Late Silurian of New South Wales, Australia, shows a specimen with corallites about 1 mm in diameter. These corallites are widely spaced (most one diameter or less from the nearest neighbor). No longitudinal section is shown, but according to the description, the coenosteum is vesiculose. The authors reported discontinuous corallite walls, a feature not seen in my specimen, or reported in other representatives of *P. exigua*. In addition, some of their specimens have coenosteal spines, and extensions from the outer walls of the corallites into the coenosteum which are lacking in known specimens of *P. exigua*. It seems that this specimen, though similar to my form, has too many distinctive features to be considered conspecific with it.

Distribution. — Anticosti Island, Late Ordovician or Early Llandovery, and Late Llandovery; Newfoundland, Silurian; Gotland, Middle or Late Llandovery; Indiana, mid-Llandovery (Brassfield Formation).

Brassfield occurrence. — Locality 26, at Elkhorn Falls, south of Richmond, Indiana.

Propora eminula (Foerste)

Cf. 1851. *Lyellia americana* Milne-Edwards & Haime, p. 226, pl. 14, figs. 3, 3a.
1906. *Lyellia eminula* Foerste, pp. 306-307, pl. 3, fig. 6; pl. 4, fig. 3.

Type material. — Foerste designated as the holotype the specimen figured in his plate 3, figure 6, and plate 4, figure 3. The repository is unknown.

Type locality and horizon. — North of Irvine, central Kentucky, in the same area as my localities 53 and 53a, from the Waco bed, Noland Member of the Brassfield Formation.

Number of specimens examined. — None found in the field. Analysis based upon Foerste's description and figures.

Description. — As Foerste described it, the upper surface of the corallum is flat or somewhat convex, expanding from a narrower base to reach a height of at least 25 mm and a width of 40 mm in the holotype. The corallites are 1.2 to 1.5 mm in diameter, spaced from 0.70 to 2.0 mm apart. Their rims are exsert. There are 12 septa, consisting of stout spines, and the coenosteum is vesiculose.

Discussion. — If Foerste's description is accurate, this species differs from the other two Brassfield species of *Propora*, in having broader corallites than *P. exigua* (Billings) and more widely spaced corallites than *P. conferta* Milne-Edwards & Haime. It differs from both in its spinose septa. Unfortunately, Foerste figured no thin-sections of his material, so that details of the internal structure cannot be determined.

This species is similar, and perhaps identical to *Lyellia americana* Milne-Edwards & Haime (1851), from the Silurian rocks of Drummond Island, Lake Huron. This latter species has broad corallites (somewhat over 2 mm in diameter) with wide spaces between them (separated by one or two times their diameter), and a vesiculose coenosteum. It is not specifically stated that the septa are spinose, but their figure 3a shows rather pointed septa ends.

Similar forms have been described by Rominger (1876, pp. 15-16) from the Silurian of northern Michigan, and Twenhofel (1928) from the Chicotte Formation (Late Llandovery or Early Wenlock) of Anticosti Island. In the absence of more information on the internal structure of Foerste's material, it seems pointless to speculate further as to the relationships of this species.

Distribution. — Kentucky, mid-Llandovery (Brassfield Formation); Similar and possibly identical forms in the Silurian of northern Michigan, and the Late Llandovery or Early Wenlock of Anticosti Island have been reported.

Brassfield occurrence. — North of Irvine, central Kentucky (localities 53 and 53a), in the Waco bed, Noland Member.

Family **HELIOLITIDAE** Lindström, 1876

(*Pro* Heliolithidae Lindström, 1873, *ex Heliolithes* Lindström, 1873)

Heliolitids with massive coralla; distinct corallites in matrix of tubular, tabulate coenosteum; walls of coenosteal tubes complete; tabulae of both corallites and coenosteal tubes complete; corallites with complete, solid walls, bearing 12 longitudinal rows of spines, or 12 solid longitudinal sheets as septa; pseudocolumella may be present due to axial juncture or flexing of spines; corallites without distinct halo of 12 coenosteal tubes around their perimeters.

Genus **HELIOLITES** Dana, 1848

Non 1770. *Heliolithes* Guettard, p. 454 (non-Linnaean nomenclature).
1848. *Heliolites* Dana, pp. 541-542.
1849. *Palaeopora* M'Coy, p. 129 (objective synonym).
1850. *Geoporites* d'Orbigny, p. 49 (objective synonym).
1895. *Pachycanalicula* Wentzel, p. 503.
1933. *Helioplasma* Kettnerova, p. 181.

Type species. — (by original designation): *Astraea porosa* Goldfuss, 1826, p. 64.

Type locality and horizon. — Devonian, Eifel District of Germany.

Diagnosis. — Corallum of cylindrical corallites separated laterally by tubular coenosteum; corallites with complete tabulae; corallites may possess 12 longitudinal rows of spines; coenosteal tubes with complete walls and tabulae similar to those of corallites, but lacking spines; mural pores absent.

Description. — These are corals whose corallites are separated from one another laterally by a tubular coenosteum. The corallites are traversed by generally complete tabulae, and their walls commonly bear 12 longitudinal rows of spines. The spines may be long, short, or absent, but they never form a complex axial structure as in *Stelliporella* Wentzel (e.g., Hill and Stumm, p. F461, fig. 348, 8a).

The coenosteal tubes are considerably narrower than the corallites, and their walls are solid (in contrast to those in *Palaeoporites* Kiär, *q.v.*). They contain complete tabulae but no spines.

Discussion. — Guettard (1770, p. 454) introduced the concept of this genus, and while he has often been credited with its authorship (by such writers as Hall, 1852; Rominger, 1876; and even by Dana, 1848, where this last author introduced the accepted name, *Heliolites,* for the first time in the literature), Lang, Smith, and Thomas (1940, p. 186) pointed out that Guettard did not use Linnaean nomenclature. They thus consider his term, *Heliolithes,* invalid and accept Dana's (1848) name, *Heliolites,* as the correct form and originally assigned name for the genus under Linnaean strictures. Most recent authors, including Hill and Stumm (1956) and Sokolov (1962) concur in assigning authorship to Dana.

Some confusion has arisen over the date of Dana's initial publication of this name. Both Hill and Stumm (1956, p. 460) and Sokolov (1962, p. 278 in the Russian edition) gave 1846 as the date. Lang, Smith, and Thomas (1940, p. 66) who took considerable care to ascertain precise publication dates, gave 1848 as the correct date. Both Lang, Smith, and Thomas, and Sokolov (1962, p. 282 in the Russian edition), in their bibliographies, stated that "Zoophytes", Dana's publication, appeared in several sections: pp. 1-120 and 709-720 in 1846; pp. 352-364 appeared on pages 178-189 of the 1846 volume of the American Journal of Science. The entire work appeared in March, 1848 (and the Atlas in November 1849). As the description of *Heliolites* appears on pages 541-542, the publication date of this generic name should be 1848, the year in which the entire text was first published.

M'Coy (1849, p. 129) erected the generic name *Palaeopora* for "all the so-called *Porites* of the palaeozoic rocks", which were described by Goldfuss (1826) as *Astraea.* Lang, Smith, and Thomas (1940, p. 94) subsequently designated Goldfuss' *A. porosa* as the type species, thereby purposely making M'Coy's genus an objective junior synonym of *Heliolites* Dana.

D'Orbigny (1850, p. 49) introduced the genus *Geoporites* which he characterized as ". . . *Porites* of which the cellules are only corrugated, and simply emplaced in the midst of a porous mass." This could be accepted as a vague description of *Heliolites* Dana. D'Orbigny included nine species under this genus, among them *Astraea porosa* Goldfuss. Lang, Smith, and Thomas (1940, p. 63) subsequently designated this species as the type of *Geoporites,* remarking that in consequence *Geoporites* was an absolute (objective) synonym of *Heliolites* Dana. This is the same procedure used by these authors to synonymize *Palaeopora* M'Coy with *Heliolites* Dana.

Wentzel (1895) erected the genus *Pachycanalicula* based upon *Heliolites barrandei* Hoernes [*sic*] (see discussion of *Heliolites spongodes*). This genus is characterized by its thick-walled coenosteal tubes that tend to be rounded in cross-section. The corallites each contain 12 "septa", whose inner edges have the form of spines pointing obliquely upward with somewhat thickened ends. *Pachycanalicula* is considered synonymous with *Heliolites* as herein understood. It is possible, however, that future investigations will convincingly show that it is a distinct genus.

Kettnerova (1933, p. 181) described the genus *Helioplasma* based upon material from the Lower Devonian of Bohemia. This genus was supposedly characterized by a coenosteum consisting of both tabulate tubes and vesicular dissepiments. Judging from heliolitid material I have studied, this condition is not greatly abnormal for *Heliolites* Dana, which has (as shown in Kettnerova's text-fig. 2) a dominantly tubular coenosteum with localized vesicular dissepiments. Consequently, *Helioplasma* is considered a junior synonym of *Heliolites*.

Heliolites spongiosus Foerste   PI. 11, figs. 5, 6; PI. 37, figs. 3, 4

    1906. *Heliolites spongiosa* Foerste, pp. 303-304, pl. 3, fig. 3; pl. 4, fig. 6; pl. 5, fig. 5.
Non 1949. *Heliolites spongiosus* Foerste, Amsden, pp. 84-85, pl. 14, figs. 5-8.
  ? 1964. *Heliolites spongiosus* Foerste, Stumm, p. 59, pl. 57, figs. 1-3.

Type material. — Holotype: USNM 84892, the specimen figured by Foerste, and shown here on Plates 11, 37.

Type locality and horizon. — Waco bed, Noland Member of the Brassfield Formation, along the road north of Estill Springs, near Irvine, Kentucky, in the southeast corner of the Cincinnati Arch, central Kentucky (localities 53 and 53a of the present study).

Number of specimens studied. — One (the holotype; none found in the field).

Diagnosis. — *Heliolites* with corallites about 1.5 mm in diameter; coenosteal tubes varying considerably in diameter, giving spongy aspect to this area; corallite tabulae gently convex upward.

Description. — The holotype is the basis for this description, as no specimens were found during this study.

The specimen measures 6.25 x 5.00 cm in diameter, and is about 1.5 cm thick. The corallum appears to have been originally lens-shaped, and probably lay directly upon the sediment. No evidence of a holotheca is preserved.

The corallites are about 1.5 mm in diameter, and are each separated from its nearest neighbor by about 1.1 to 1.5 mm. Septal structures, if originally present, were not preserved. The tabulae are gently convex upward, spaced about three to four per mm, and are complete.

The coenosteum is separated from the corallites by definite walls, and characteristically consists of tubes varying considerably in diameter (0.10 to 0.37 mm), with wide and narrow tubes mixed together. As a result, the coenosteum has a spongy aspect in transverse view. The tube walls are apparently continuous. The tubes contain tabulae spaced four to seven per mm, nearly all gently convex upward or flat, with a few incomplete.

Discussion. — Amsden (1949) ascribed to *H. spongiosus* Foerste some specimens from the Brownsport Formation (Early Ludlow) of western Tennessee. These resemble *H. spongiosus* in having corallites of similar size (slightly less than 2 mm across) with similar spacing. The tabulae of the corallites are also spaced about as in the Brassfield specimen, and are generally flat. Septal structures, if originally present, were not preserved. The tabulae of the coenosteal tubes are also formed and spaced similar to those in the holotype. The coenosteal tubes, however, are more uniform in diameter, and do not present the spongy texture characteristic of the holotype. This feature is of such importance in the concept of this species, that its absence even in a form otherwise so similar cannot be disregarded. While Foerste's holotype and Amsden's specimens may represent end members of a morphologic continuum, it seems best to regard them as distinct species, pending further study.

Stumm (1964) reported *Heliolites spongiosus* from the Louisville Limestone (Late Wenlock and Early Ludlow) in the vicinity of Louisville, Kentucky. He described corallites averaging 2 mm in diameter without septal ridges. The corallite tabulae are described as "complete, horizontal, very closely set, averaging 3 mm apart." [It is likely that this measure of spacing is incorrect, as 3 mm is unusually widespread in corallites 2 mm across.] The coenosteal tubes are polygonal, four to ten extending between neighboring corallites, and average about 0.3 mm in diameter. Their tabular spacing is as in the corallites. Unfortunately, Stumm's figures are all of Brownsport specimens, and he gave insufficient information on the variations in the diameters of the coenosteal tubes to allow certainty of the identification of his material. His figure 3, a transverse section through a corallum, shows a regular aureole of approximately 20 coenosteal tubes around each corallite, a pattern not encountered in the holotype.

Distribution. — Kentucky, mid-Llandovery (Brassfield Formation).

Brassfield occurrence. — Vicinity of localities 53 and 53a (near Irvine, Kentucky), in the Waco bed, Noland Member.

| **Heliolites spongodes** Lindström | Pl. 10, figs. 6, 7; Pl. 35, figs. 5, 6; Pl. 40, figs. 1, 4 |

? 1876. *Heliolites interstinctus* Rominger, p. 12, pl. 1, fig. 1 (*non Madrepora interstincta* Linnaeus, 1767).
? 1880. *Heliolites micropora* Eichwald, Nicholson & Etheridge, p. 245 (*non* Eichwald, 1855).
? 1887. *Heliolites barrandei* Penecke, p. 271, pl. 20, figs. 1-3.
? 1894. *Heliolites barrandei* Penecke, Penecke, p. 591.
  1899. *Heliolites barrandei* Penecke, Lindström, pp. 58-59, pl. 3, figs. 8-12, 17-27 (*non?* Penecke, 1887, *fide* Schouppé, 1954).
  1899. *Heliolites barrandei* var. *spongodes* Lindström, p. 60, pl. 3, figs. 13-16; pl. 4, fig. 1.
? 1906. *Heliolites* sp. Foerste, p. 304, pl. 3, fig. 4.
? 1906. *Heliolites subtubulata-distans* Foerste, pp. 304-306, pl. 3, fig. 5B.
? 1906. *Heliolites subtubulata-nucella* Foerste, pp. 304-306, pl. 3, fig. 5A (*non H. nucella* Amsden, 1949).
? 1954. *Heliolites spongodes* Lindström, Schouppé, pp. 166-168.
  1956. *Heliolites interstinctus* var. *abnormis* Yu, pp. 619-620, pl. 2, figs. 8, 9.
  1964. *Heliolites subtubulatus* (M'Coy) Stumm, p. 59, pl. 57, figs. 6, 7 (*non Palaeopora subtubulata* M'Coy, 1854).
? 1964. *Heliolites romingeri* Stumm, p. 59, pl. 57, figs. 8-12.
  1973. *Heliolites spongodes* Lindström, Galle, p. 31-33, pl. 6, figs. 1-4; text-fig. 7.

Type material. — Lectotype (Schouppé, 1954, p. 166): specimen shown by Lindström (1899), plate 3, figs. 13-16.

Type locality and horizon. — Silurian of Gotland: strata *b* and *c* in the Visby area, according to Lindström. Based on Manten's work in Gotland (1971, pp. 33-35, 43, map) these are Late Llandovery to Early Wenlock in age.

Number of specimens examined. — Two, collected in the field.

Diagnosis. — *Heliolites* with corallites approximately 1 mm in diameter; septa consisting of twelve longitudinal rows of spines that flex upward in the axis; corallites generally spaced more than one diameter from nearest neighbor; coenosteal tubes of more or less constant diameter in any one corallum; corallite walls thickened.

Description. — The corallum appears usually to have been hemispherical in life. Specimens have been found ranging upward from 2 cm in diameter (Pl. 10, figs. 6, 7).

Corallite diameter is fairly constant at 1 mm. The corallite walls are about three times as thick as those of the coenosteal tubes. Around the corallite perimeter are 12 longitudinal rows of spines which flex upward in the corallite axis. Here they appear at times to fuse with one another, forming a sort of axial structure. At other times, they remain separate, appearing in transverse section as dots in the axis (Pl. 40, fig. 4).

A distance generally greater than its diameter separates each corallite from its nearest neighbor. This distance is usually about 1 to 2 mm.

Nearly all corallite tabulae appear complete. Most are slightly concave upward, and spaced about three per mm. It could not be determined if the spacing is rhythmic.

In transverse view, the coenosteal tubes range in diameter from about 0.1 to 0.4 mm, with the wider tending to occur midway between corallites, and the narrower ones closer to the corallites. This zonation of the diameters results in a smooth, homogeneous aspect of the coenosteum, in contrast to the spongy appearance in *H. spongiosus* Foerste, where broad and narrow tubes are intimately mingled.

The coenosteal tubes appear to arise interstitially with the divergence of two extant tubes, though it is unclear if budding actually takes place. The corallites, on the other hand, were not observed in contact with one another, and may arise through direct differentiation of the coenosteal tissue.

Discussion. — Rominger (1876) described, under the name *Heliolites interstinctus* (Linnaeus), material from the Niagara Group (probably Wenlockian) at Louisville, Kentucky. The corals are *Heliolites* with corallite diameters of 1 to 1.5 mm, containing longitudinal rows of upward-pointing spines. The corallites are separated from one-another by at least one diameter. Rominger noted that "specimens from the Silurian strata of Bohemia perfectly correspond with the American form". His illustration and description indicate a form similar and possibly identical to the Brassfield material, but insufficient data is provided in this reference for certainty.

Lindström (1899) described specimens which he identified as *Heliolites barrandei* Penecke from the Silurian of Gotland. Manten (1971, p. 33) pointed out that Lindström's designations for Gotland's stratigraphic divisions were based upon incorrect assumptions (using lithology to determine horizon and relative age), and are of limited use. By comparing the localities cited by Lindström (1899, p. 59), it appears that his material occurs in the Wenlock, and possibly the Upper Llandovery and Ludlow of Gotland, as shown on Manten's geologic map, and correlated by him on p. 43.

The name *Heliolites barrandei* was first validly used by Penecke (1887), who noted its origin as a manuscript name of R. Hörnes. I have been unable to obtain a copy of Penecke's paper of 1887, and so, have no grounds for judging the virtue of Lindström's identification. The name appeared several times more before 1899, including a reference by Penecke (1894, p. 591) to the species (insufficiently described to allow validation) in the Lower Devonian of the Graz, Austria area, and Wentzel's (1895) proposal of this species as the type of a new heliolitid genus, *Pachycanalicula* (apparently a synonym of *Heliolites*).

Lindström (1899) is the earliest work I have seen that includes a clear description of supposed *H. barrandei* specimens. He described the corals as large, flat or somewhat domed discs with a wrinkled holotheca. The corallite diameters in what he regarded as the older specimens (though these age assignments are unreliable) are about 0.5 mm, while those of the "younger" specimens reach 1.0 mm. The septa are spinose, judging from the figures (though his description implies they are proximally solid, protruding little from the wall, and only distally break up into spines), and the spines curve upward distally, so that their intertwined ends may appear as a columellar structure in transverse section.

The tabulae in the corallites appear generally to be about 0.3 mm apart, comparable to the spacing in my specimens, and flat.

The coenosteal tubes are polygonal and appear to be fairly uniform in diameter. It is not clear from Lindström's figures if the tubes close to the corallites tend to be larger than those midway between neighboring corallites.

Lindström presented (pl. 3, fig. 27) a series of four serial transverse sections, showing the development of a new corallite which gradually takes shape directly from several coenosteal tubes between a number of older corallites. It may be significant that the differentiating coenosteum is in part contiguous with a neighboring corallite.

H. barrandei Penecke of Lindström appears to be identical to the Brassfield *H. spongodes* insofar as the two forms are understood.

Lindström (1899, p. 60) described the new variety *H. barrandei spongodes*, a form that appears to occur in the Högklint, and possibly the Visby beds of Gotland (Early Wenlock and Late Llandovery, according to Manten, 1971, p. 43). Because Lindström used his previously-mentioned system of stratigraphic designation to cite the horizons of occurrence of this coral, there is a problem in determining just what the horizons are. Manten's geologic map, however, indicates that Lindström's stated localities are probably from the above horizons.

Lindström distinguished this "variety" from other members of the species by its growth form of "sponge-like, irregularly lobate masses expanding from a narrow basis, having short fingerlike processes from the rounded edges, and with calicles on all sides." The corallites are quite narrow, "scarcely" 0.5 mm across. The septa consist of sheets of variable length, which split up distally into spines that curve upward in the corallite axis. The coenosteum consists of narrow, polygonal tubes. Lindström cited another growth form characterized by branches, 4 to 5 mm in diameter. The figures accompanying this description closely resemble Brassfield *H. spongodes*, save for their narrower corallites (0.5 mm *versus* approximately 1 mm in the Brassfield form). In light of other material described in the literature (see below), it seems safe to consider the Gotland corals conspecific with mine despite this difference. This is the earliest occurrence in the literature of a name that can be confidently identified with my specimens.

Lindström (1899, p. 60) synonymized his variety with *Heliolites micropora* Nicholson and Etheridge (1880, *non* Eichwald, 1855). Nicholson and Etheridge described this species not for the purpose of reporting it from Girvan (with which area their paleontologic work dealt), but included it in a list of the species of *Heliolites* Dana of which they were aware. Their description is brief and sketchy, limited to an account of corallite diameters and spacing, as well as growth form. Concerning the septa, only their number (12), and the fact that they are short and delicate are mentioned. The description is insufficient for synonymizing Nicholson and Etheridge's material with Lindström's. Lindström supplied Nicholson and Etheridge with the only specimen of *H. spongodes* they were able to study (Nicholson & Etheridge, 1880, p. 245). Lindström (1899, p. 60) stated that the material he supplied them was identical with *H. barrandei spongodes*, but that he had previously confused it with *H. micropora* Eichwald, sending Nicholson the specimen under that name. Having subsequently studied Eichwald's types at the Museum of the University of St. Petersburg, Lindström became persuaded that *H. micropora* Eichwald differed from his Gotland form, which he then designated *Heliolites barrandei spongodes*.

From the Waco bed of the Noland Member, Brassfield Formation, Foerste (1906) reported *Heliolites* sp.. The locality, north of Irvine, Kentucky, is the equivalent of my localities 53-53a. The specimen appears hemispherical, with corallites about 1 mm in diameter, their centers 1 to 2 mm apart. Examination of the specimen (USNM 87172) substantiates Foerste's report of septal spines in the corallites, extending toward the axis. Dots in the axes of the corallites as seen in transverse section suggest axial flexing of the spines, presumably upward, as in my form. Arrangement of the tabulae in the corallites is difficult to determine, but those of the coenosteal tubes appear to be spaced about three per mm. The coenosteal tubes are about 0.3 mm in diameter. This form may be conspecific with mine, but its features are not clear enough for me to be certain of this.

In the same work, Foerste described what he considered two "varieties" of *Heliolites subtubulatus* (M'Coy) from the Waco bed, Noland Member of the Brassfield, in central Kentucky. The first, *Heliolites subtubulatus distans*, from half a mile east of Waco, Kentucky, is represented by a flat corallum with concentric holotheca. The corallites are about 0.75 mm in diameter, separated from one-another by 1.5 to 2.0 mm. This combination of narrow corallites and their wide spacing, Foerste considered to set the form apart from more typical members of *H. subtubulatus* (M'Coy). The coenosteal tubes are 0.2 to 0.25 mm in diameter, in the narrow part of the range found in the Brassfield specimens. The corallite tabulae are spaced three to four per mm, while those in the coenosteal tubes are 4.5 to 5.5 per mm. In these measurements, the form is in basic agreement with my specimens. As Foerste was unable to get a good look at the septal structure, I cannot be certain of the relationship between his material and my own, despite the fact that they agree in most other respects.

Foerste's second "variety," *Heliolites subtubulata nucella,* is from the same locality and horizon as his *Heliolites* sp. (see above). This form seems identical to *H. s. distans,* except that it grew more vertically than horizontally. Again, there is insufficient information to be certain how closely it is related to my specimens.

Schouppé (1954) chose as lectotype of *Heliolites spongodes* (elevating Lindström's "variety" to species level), the specimen figured by Lindström (1899) in his plate 3, figures 13-16. He reported this species in the horizon which he designated as ef₁ (Upper Silurian; see his p. 170) of the Paleozoic sequence at Graz, in the Carnic Alps. He considered the material, described by Lindström (1899) as *H. barrandei,* to be a species distinct from *H. barrandei* Penecke (1887), and he assigned the former to *Heliolites spongodes.*

Schouppé's material consists of flat coralla with bulbous surface protrusions. The corallite diameters are mostly 0.5 mm, in isolated instances reaching 0.7 mm (thus, narrower than mine). The walls of the corallites and coenosteal tubes are thin. From one to six tubes lie between adjacent corallites. The coenosteal tubes are regularly polygonal, 0.1 to 0.15 mm in diameter, rather narrow compared with mine. The corallites each contain 12 short septa, consisting of longitudinal rows of spines pointing upward axially. The corallite tabulae are fairly flat with about three per mm. The tabulae of the coenosteal tubes are more closely spaced. Schouppé furnished no figures with his description. This fact, as well as the unusually narrow coenosteal tubes, makes it best to reserve judgement on the relationship between his material and mine.

Yu (1956) reported a heliolitid from the Middle Silurian of northwestern Kansu Province, China, which he referred to as *Heliolites interstinctus abnormis*, a new "variety". His specimen was dome-shaped, with a concentrically wrinkled base. The corallites were 0.8 to 1.1 mm in diameter, most commonly 0.9 mm, and these were separated from one-another by about 1 to 2 mm. Their walls were about 0.05 mm thick, somewhat thicker than the walls of the coenosteal tubes (as in my form). Septal spines in the corallites reached to the axis where they often united in pairs. (It is unclear from the description whether it is adjacent or opposite spines that unite.) The corallite tabulae were generally complete, sometimes slightly concave or convex, spaced 2.6 to 3.0 per mm. The coenosteal tubes were polygonal, between 0.2 and 0.4 mm in maximum diameter. Apparently, this form agrees in all important respects with mine, and the two are regarded as conspecific.

Stumm (1964) described a coral under the name *Heliolites subtubulatus* (M'Coy) from the Louisville Limestone (Late Wenlock to Early Ludlow) of northern Kentucky and southern Indiana. His description is cursory, but the thin-section of Stumm's plate 57, fig. 7 (UMMP 3908) shows that this specimen is probably conspecific with the Brassfield material.

In the same work, Stumm introduced *Heliolites romingeri*, a species including the specimen identified as *H. interstinctus* (Linnaeus) by Rominger (1876, discussed above), from the Louisville Limestone (Late Wenlock to Early Ludlow) at Louisville, Kentucky. This form closely resembles mine in most respects and may be identical. The corallites are about 1 mm in diameter, and are spaced as in my specimens. The coenosteal tubes average 0.3 mm in diameter. The corallite tabulae are complete, horizontal, and average 1 mm apart, while those of the coenosteal tubes are twice as closely spaced. Stumm described 12 long septa typically joined axially. His figure 12, however, suggests the situation may be more complex, and the possibility of upward-curving spines cannot be precluded.

Galle (1973) reported *H. spongodes* from Bohemia, in the upper part of the Liten Formation (Wenlock), and the Kopanina Formation (Ludlow). The coralla are discoid and holothecate. Corallite diameters are 1.0 to 1.1 mm, with the corallite walls slightly thicker than those of the coenosteal tubes, but not greatly thick. The corallites are spaced somewhat closer than mine, usually within a diameter of one another. Long spines in longitudinal rows curve upward and intertwine in the axis, sometimes forming a plaited axial structure. The corallite tabulae are complete, sometimes a bit wavy in vertical section, usually 2 to 2.4 per mm. The coenosteal tubes are 0.2 to 0.4 mm in diameter, with tabulae spaced about 3 to 3.4 per mm. Only the closer spacing of the corallites in Galle's material differs from mine, a difference which does not seem as important as the similarities between these forms. They are therefore considered conspecific.

Distribution. — Gotland, Late Llandovery(?), Wenlock, Ludlow(?); Falls of the Ohio region (northern Kentucky, southern Indiana), Late Wenlock to Early Ludlow; Bohemia, Wenlock, Ludlow; Kansu Province, China, Middle Silurian.

Brassfield occurrence. — Localities 1 (near Panola) and 46 (near College Hill), in central Kentucky, from the Noland Member.

Family PALAEOPORITIDAE Kiär, 1899

(*Nom. transl.* Sokolov, 1962 [*ex.* Palaeoporitinae Kiär, 1899])

Longitudinal skeletal elements (wall, "septa") trabecular, commonly thickened and porous; coenosteum tubular; tabulae not obvious, due to sparseness of spacing, as well as camouflage by irregular ("spongy") structure of coenosteum and corallites; corallite walls not well-defined.

Genus PALAEOPORITES Kiär, 1899

Type species. — (by monotypy): *Palaeoporites estonicus* Kiär (1899), pp. 18-21, pl. 3, figs. 1-4.

Type locality and horizon. — Porkuni Stage (F_2), Lower Llandovery of Estonia.

Description. — As there are only one or two species assigned to this genus (Sokolov, 1962, p. 278 in Russian edition), the following description is based largely on Kiär's account (1899) of the type species:

The corallum is spherical or knobby. The corallites stand out more on slightly weathered surfaces of the corallum than in thinsection, due to the irregular structure of the interior. The skeletal material is made of trabeculae, bundles of carbonate mineral shaped like rods. The corallites have 12 "septa", and lie in a matrix of tubular coenosteum. A weakly developed columella consisting of the distal ends of 12 to 15 trabeculae, appears as an aggregate of bumps in the axial floor of the calices.

The outstanding feature of this genus is the irregularity of the internal structure. In a sense, *Palaeoporites* among the heliolitids is similar to *Calostylis* among the rugosans. The coenosteal tubes, the corallite walls and the septa are frequently pierced by openings, which may be pores, or possibly even irregular gaps. The corallite walls are not well-developed, in that they are not significantly thicker than the walls of the coenosteal tubes (compare this with *Heliolites spongodes*). This fact, and the frequent perforation of the corallite walls results in the corallites being visually lost in a sea of coenosteum, discernible only on close inspection, where they are marked by the radial "septa". The "septa", too, are often discontinuous, making the corallites even more obscure.

The "septa" consist of two or more rows of trabeculae "per septum", narrowing often to a single row toward the axis. The "septal" trabeculae vary in length. They curve distally from their base at the corallite wall, so that their distal ends may appear in the axis, to help form the knobby columella. Synapticulae may connect neighboring "septa" laterally.

Tabulae occur in the corallites and in the coenosteal tubes, usually obscured by the chaotic pattern of the longitudinal skeletal elements. Kiär referred to the tabulae as "thickened but irregular".

Another important feature of *Palaeoporites* (and one seen clearly in the Brassfield material) is the fact that the thickening of the coenosteal tubes as seen in transverse section is quite inhomogeneous. Zones of considerable thickening are found next to thinner zones, contributing further to the irregular appearance of the internal structure.

Palaeoporites sp.   Pl. 35, figs. 1, 2; Pl. 40, fig. 3

? 1890. *Heliolites subtubulatus* Foerste, pp. 232-233 (*non Palaeopora subtubulata* M'Coy, 1854).

Type material. — Specimen figured is UCGM 41932.

Type locality and horizon. — Specimen figured is from the Brassfield Formation at locality 11, near West Union, Ohio, from cross-bedded talus block.

Number of specimens examined. — Two, found in the field.

Description. — The corallum is laminar, two inches or more in diameter, with apparent growth discontinuities (sharp decreases in corallum width followed by recovery of original diameter) at certain levels. The thickest corallum is approximately 5 mm thick.

As is typical of this genus, the corallites are obscure, and not immediately differentiable visually from the coenosteum (Pl. 35, fig. 2). They are more apparent on the surfaces of weathered coralla. The corallites are approximately 1.0 mm in diameter, and appear (based on observations on relatively small clear portions of thin-sections) to be about one diameter from the nearest neighbor. Their walls do not seem to be much thicker than those of the coenosteal elements.

In transverse section, the corallites display 12 "septal" elements which vary somewhat in length. Some of these seem to join axially with their neighbors, but preservation is too poor to be certain of this. The "septal" elements probably are continuous lamellae, rather than longitudinal rows of spines, as there are invariably 12 of them in the section, but this too is not certain. The proximal end of each element is thickened, and connects directly with the coenosteal tissue, so that the former have the appearance of being one with the coenosteal wall structures, while effecting a radiating configuration, their sole distinguishing mark.

The coenosteum consists of tubes with discontinuous walls. The perimeters of the corallites may also be discontinuous (Pl. 40, fig. 3). There are approximately four to five coenosteal tubes in a length of 1 mm across a transverse section. The walls of these tubes are sometimes thickened to almost equal the tube diameter (though this may be an artifact of recrystallization, as the walls are elsewhere only about half so thick).

Little is discernible in vertical section (Pl. 35, fig. 1). The corallites are not readily differentiable from the coenosteum, and no tabulae were observed. Again, the possibility that this is due to recrystallization of the narrow tubes cannot be ruled out.

Discussion. — As this is the only species of heliolitid with a *Heliolites*-like tubular coenosteum that I have found in the Brass-field north of the Ohio River, it is worth comparison with material identified as *Heliolites subtubulatus* (M'Coy) by Foerste (1890) from the Clinton beds (probably Brassfield) at Ludlow Falls, Ohio (my locality 55). This is a laminar form with corallites 0.6 to 0.75 mm in diameter, apparently spaced about a diameter or more apart. The corallites are circular in plan view and contain 12 septa. The coenosteal tubes are about 0.2 mm in diameter. Judging from the wording of his description (there are no figures) Foerste did not section his material. In light of this fact and in the absence of more detailed information on his specimens, I can only questionably identify them with my specimens of *Palaeoporites* sp.

Distribution. — Brassfield Formation (mid-Llandovery) of southwestern Ohio.

Brassfield occurrence. — Localities 11 (West Union) and 15 (West Milton) in southwestern Ohio.

REFERENCES CITED

Agassiz, J. L. R.
1833. *Recherches sur les poissons fossiles.* 5 vols. in 3, text; 5 vols. in 3, Atlas. Neuchâtel.

Alexander, F. E. S.
1947. *On* Phaulactis versatilis, *sp. n., from the English Upper Silurian.* Ann. Mag. Nat. Hist., ser. 11, vol. 14, pp. 175-182, 2 pls.

Amsden, T. W.
1949. *Stratigraphy and paleontology of the Brownsport Formation (Silurian) of western Tennessee.* Peabody Mus. Nat. Hist. (Yale Univ.), Bull. 5, 138 pp., 34 pls.

Baly, W.
1864. *Descriptions of new genera and species of* Phytophaga. Ann. Mag. Nat. Hist., ser. 3, vol. 14.

Barbour, E. H.
1911. *A new Carboniferous coral,* Craterophyllum verticillatum. Nebraska Geol. Surv., Pub., vol. 4, pt. 3, pp. 38-49, pls. 1-4.

Bargatzky, F. A.
1881. *Die Stromatoporen des Rheinischen Devons.* Naturhist. ver. Rheinlande. Verh., vol. 38, 76 pp.

Bassler, R. S.
1915. *Bibliographic index of American Ordovician and Silurian fossils.* U.S. Natl. Mus., Bull., vol. 92, No. 1, pp. viii + 1-718; No. 2, pp. iv + 719-1521, pls. 1-4.
1944. *Faunal lists and descriptions of Paleozoic corals.* Geol. Soc. America, Mem. 44, 315 pp., 20 pls.

Bate, C.
1857. *A synopsis of the British Edriophthalmous Crustacea, pt. 1, Amphipoda.* Ann. Mag. Nat. Hist., ser. 2, vol. 19, pp. 135-152.

Bayer, F. M.
1956. *Octocorallia.* pp. F166-F231 in R. C. Moore (editor), *Treatise on invertebrate paleontology, Part F, Coelenterata.* Geol. Soc. America & Univ. Kansas Press, 494 pp.

Bergroth, E. E.
1906. *Aphylinae und Hyocephalinae, zwei neue Hemipteren-Subfamilien.* Zool. Anz., vol. 29, pp. 644-649, 4 text-figs.

Berry, W. B. N.
1967. *A new subspecies of* Climacograptus scalaris *(Linné?) (Hisinger) from Kentucky.* Jour. Paleont., vol. 41, No. 2, pp. 507-511, 51 text-figs.

Berry, W. B. N., and Boucot, A. J. (editors)
1970. *Correlation of the North American Silurian rocks.* Geol. Soc. America, Spec. Paper 102, 289 pp.
1972. *Correlation of the Southeast Asian and Near Eastern Silurian rocks.* Geol. Soc. America, Spec. Paper 137, 65 pp.

Bigsby, J. J.
1824. *Notes on the geography and geology of Lake Huron.* Geol. Soc. London, Trans., ser. 2, vol. 6, pp. 175-209, pls. 25-31.

Billings, E.
1858. *Report for the year 1857 of E. Billings, Esq., Palaeontologist, addressed to Sir W. E. Logan, F. R. S., Director of the Geological Survey of Canada.* Geol. Surv. Canada, Rept. Progress for 1857, pp. 147-192.
1862. *Paleozoic fossils, vol. I, containing descriptions and figures of new or little known species of organic remains from the Silurian rocks.* Geol. Surv. Canada, 426 pp. (advance sheets of 1865 publication, cited below).
1865a. *Paleozoic fossils. Vol. I, containing descriptions and figures of new or little known species of organic remains from the Silurian rocks.* Geol. Surv. Canada, 426 pp.
1865b. *Notice of some new genera and species of Palaeozoic fossils.* Canadian Naturalist and Geologist, n.s., vol. 2, pp. 425-432.
1866. *Catalogues of the Silurian fossils of the Island of Anticosti, with descriptions of some new genera and species.* Geol. Surv. Canada, 99 pp.

Bolton, T. E.
1957. *Silurian stratigraphy and palaeontology of the Niagara Escarpment in Ontario.* Geol. Surv. Canada, Mem. 289, 145 pp., 13 pls.
1960. *Catalogue of type invertebrate fossils of the Geological Survey of Canada, vol. 1*: Ottawa, Canada Geol. Survey, 215 pp.
1966. *Illustrations of Canadian fossils — Silurian faunas of Ontario.* Geol. Surv. Canada, Paper 66-5, 46 pp., 19 pls.
1968. *Silurian faunal assemblages, Manitoulin Island, Ontario,* in B. Liberty and F. D. Sheldon (editors), *Geology of Manitoulin Island.* Michigan Basin Geol. Soc., 101 pp.
1972. *Geological map and notes on the Ordovician and Silurian litho- and biostratigraphy, Anticosti Island, Quebec.* Geol. Surv. Canada, Paper 71-19, pp. vi + 45, 12 pls.

Bolton, T. E., and Copeland, M. J.
1972. *Paleozoic formations and Silurian biostratigraphy, Lake Timiskaming region, Ontario and Quebec.* Geol. Surv. Canada, Paper 72-15, 49 pp., 13 pls.

Briden, J. C., Morris, W. A., and Piper, J. D. A.
1973. *Paleomagnetic studies in the British Caledonides — VI. Regional and global implications.* Roy. Astron. Soc., Geophys. Jour., vol. 34, pp. 107-134.

Bromell, M. von
1728. *Lithographiae Svecanae . . .* Acta Liter. Sveciae, vol. 2, pp. 363-470 [370], 408-415, plate opposite p. 410.

Budge, D. R.
1972. *Paleontology and stratigraphic significance of Late Ordovician Silurian corals from the eastern Great Basin.* Unpublished Ph.D. dissertation, Univ. of California, Berkeley, 527 pp., 22 pls.

Buehler, E. J.
1955. *The morphology and taxonomy of the Halysitidae.* Peabody Mus. Nat. Hist. (Yale Univ.), Bull. 8, 79 pp., 12 pls.

Bulvanker, E. Z.
1952. *Korally rugoza Silura Podolii ("Rugose corals of the Silurian of Podolia")*. Vsesoyuz. Nauch.-Issled. Geol. Inst., Tr., 33 pp.

Butler, A. J.
1937. *A new species of Omphyma (Silurian, Worcestershire, England) and some remarks on the Pycnactis-Phaulactis group of Silurian corals.* Ann. Mag. Nat. Hist., ser. 10, vol. 19, pp. 87-96.

Caramanica, F. P.
1975. *Stabilization of the spelling of the generic name Paleofavosites Twenhofel.* J. Paleontol., vol. 49, no. 6, p. 1126.

Chapman, E. J.
1893. *On the corals and coralliform types of Paleozoic strata.* Roy. Soc. Canada, Proc. Trans., vol. 10, sect. 4, pp. 39-48 (for 1892).

Chen, M.-C.
1959. *Nekotorye Siluriiskie i Devonskie stromatoporoide i korally iz raiona Lushanya vostochnoi chasti provintsii Guichzhou ("Some Silurian and Devanian stromatoporoids and corals from the region of Lushan in the eastern part of Kweichow Province")*. Acta Palaeont. Sinica, vol. 7, pp. 285-317.

Cherepnina, S. K.
1962. *O novom rode tetrakorallov iz Ordovikskikh otlozheniy Gornogo Altaya ("On a new genus of tetracoral from the Ordovician strata of Gornogo Altay")*, pp. 140-142, 1 pl., in M. V. Kasyanov and F. G. Gurari (editors), *Materialy po paleontologii i stratigrafii zapadnoi Sibiri*. Sibirsk. Nauch.-Issled. Inst. Geol. Geofiz. i Min. Syrya, Tr., vyp. 23, Ser. Neft. Geol., 179 pp.
1965. *Novyy rod semeystva Lykophyllidae iz Siluriyskikh otlozheniy Gornogo Altaya ("New genus of the Family Lykophyllidae from the Silurian strata of Gornogo Altay")*, pp. 31-32, in B. S. Sokolov and A. B. Ivanovskiy (editors), *Rugozy Paleozoya SSSR*. Izdatel'stvo "Nauka", Moscow, 90 pp., 26 pls.

Cherkesov, V. Yu.
1936. *Lower Silurian corals of the Leningrad region (Russia)*. Inst. Mines Leningrad, Ann. (Zapiski) tom 9, vyp. 2, pp. 41-46 (in Russian, with English summary).

Chernyshev, B. B.
1937a. *The Upper Silurian and Devonian Tabulata of Novaya Zemlya and Taimyr*. Arctic Inst., Trans., vol. 91, pp. 67-134, pls. 1-13.
1937b. *Silurian and Devonian Tabulata of the Mongolia and Tuva Republics*. Akad. Nauk SSSR, Mongol. Kom., Tr., No. 30, 34 pp. (in Russian, with English summary).
1938a. *Tabulata ostrova Vaygacha ("Tabulata of the Island of Vaygach")*. Arkticheskogo Instituta, Tr., tom 101, Leningrad, pp. 109-145, 7 pls.
1938b. *O nekotorykh verkhne Siluriyskikh Tabulata s Reki Letney ("On some Upper Silurian Tabulata from the River Letney")*. Arkticheskogo Instituta, Tr., tom 101, Leningrad, pp. 147-153.
1941a. *Siluriyskie i nizhnedevonskie korally basseyna Reki Tarei (Yugozapadnyy Taimyr) ("Silurian and Lower Devonian corals from the basin of the River Tarei, southwestern Taimyr")*, pp. 9-64, pls. 1-14, in D. V. Nalivkin (editor), *Paleontologiya Sovyetskoy Arktiki*. Arkticheskogo Nauch.-Issled. Inst., Glavnogo Upravleniya Severnogo Morskogo Puti Pri SNK SSSR, tom 158, 157 pp., 29 pls.
1941b. *O nekotorykh Verkhnesiluriyskikh korallakh vostochnogo Verkhoyansk ("On some Upper Silurian corals from eastern Verkhoyan")*, pp. 65-74, pls. 1-3, in D. V. Nalivkin (editor), *Paleontologiya Sovyetskoy Arktiki*. Arkticheskogo Nauch.-Issled. Inst., Glavnogo Upravleniya Severnogo Morskogo Puti Pri SNK SSSR, tom 158, 157 pp., 29 pls.

Chi, Y.-S.
1933. *Lower Carboniferous syringoporas of China.* Palaeont. Sinica, ser. B, vol. 12, fasc. 4, pp. 1-48 + 5, pls. 1-7.
1935. *Note on two aseptate corals from the upper part of the Nanshan Series in Kansu.* Geol. Soc. China, Bull., vol. 14, pp. 47-49.

Colter, V. S.
1956. *On Heliolites caespitosa Salter.* Geol. Mag., vol. 93, pp. 229-232.

Cowper-Reed, F. R.
1912. *Ordovician and Silurian fossils from the central Himalayas.* Geol. Surv. India, Mem., Palaeontologia Indica, ser. 15, vol. 7, No. 2, Calcutta, pp. 1-168, 20 pls.

Cox, I.
1937. *Arctic and some other species of* Streptelasma. Geol. Mag., vol. 74, pp. 1-19, pls. 1, 2.

Dall, W. H.
1876. *Note on "Die Gasteropoden Fauna Baikalsees".* Boston Soc. Nat. Hist., Proc., vol. 19, pp. 43-47.

Dana, J. D.
1846-1849*. *Zoophytes,* in *United States Exploring Expedition during the years 1838, 1839, 1840. 1841, 1842, under the command of Charles Wilkes, U.S.N..* pp. x + 1-740, Atlas, 61 pls.

Davis, W. J.
1887. *Kentucky fossil corals — a monograph of the fossil corals of the Silurian and Devonian rocks of Kentucky, Part 2.* Kentucky Geol. Surv., pp. xiii + 4, pls. 1-139.

Deffeyes, K. S., Lucia, F. J., and Weyl, P. K.
1965. *Dolomitization of Recent and Plio-Pleistocene sediments by marine evaporite waters on Bonaire, Netherlands Antilles,* pp. 71-88, in L. C. Pray and R. C. Murray (editors), *Dolomitization and limestone diagenesis, a symposium.* Soc. Econ. Paleont. Mineral., Spec. Pub. No. 13, 180 pp.

Defrance, M. J. L.
1816-1830. *Dictionnaire des Sciences naturelles.* Paris, 60 vols.

Desparmet, R.
1969. *Nouvelles données sur le paléozoïque ancien d'Afghanistan central.* Acad. Sci., Paris, Compt. Rend., sér. D, vol. 268, pp. 2389-2391.

Duncan, H.
1956. *Ordovician and Silurian coral faunas of western United States.* U.S. Geol. Surv., Bull. 1021-F, pp. 209-236, pls. 21-27.

Duncan, P. M.
1872. *Third report on the British fossil corals.* British Assoc. Advanc. Sci., Rept. 41st Meeting (1871), pp. 116-137.

Dunham, R. J.
1962. *Classification of carbonate rocks according to depositional texture,* pp. 108-121, in W. E. Ham (editor), *Classification of carbonate rocks — a symposium.* Amer. Assoc. Petrol. Geol., Mem. 1, 279 pp.

Dybowski, W. N.
1873. *Monographie der Zoantharia sclerodermata rugosa aus der Silur-formation Estlands, Nord-Livlands und der Insel Gotland. . . .* Dorpat, Archiv Naturk. Liv-, Ehst- und Kurlands, Ser. 1, Bd. 5, Lf. 3, pp. 257-414, pls. 1, 2.
1874. *Monographie der Zoantharia sclerodermata rugosa aus der Silur-formation Estlands, Nord-Livlands und der Insel Gotland (Fortsetzung):* Dorpat, Archiv Naturk. Liv-, Ehst- und Kurlands, Ser. 1, Bd. 5, Leif. 4, pp. 415-531, pls. 3-5.

---

*See Lang, *et al.* 1940, p. 173 for account of publication dates for the various portions of this work.

Dzyubo, P. S.

1960. Karagemia, *novyy rod geliolitid iz Ordovika Altaya* ("Karagemia, *a new genus of heliolitid from the Ordovician of Altay*"), pp. 86-88, Pl. 3, in L. L. Khalfin (editor), *Materialy po paleontologii i stratigrafii zapadnoy Sibiri.* Sibirsk. Nauch.-Issled. Inst. Geol., Geofiz. i Min. Syrya, Leningrad, Tr., vyp. 8, Ser. Neft. Geol., 284 pp.

1962. *Novyy rod tabulyat iz Ordovika Altaya* ("*A new tabulate genus from the Ordovician of Gornogo Altay*"), pp. 154-157, in M. V. Kasyanov and F. G. Gurari (eds.), *Materialy po paleontologii i stratigrafii zapadnoy Sibiri.* Sibirsk. Nauch.-Issled. Inst. Geol., Geofiz. i Min. Syrya, Leningrad, Tr., vyp. 23, Ser. Neft. Geol., 179 pp.

1971. *O nakhodkakh landoveriyskoy fauny v zapadnoy chasti Kul'dzhuktau (yuzhnyy sklon Bel'tau)* ("*On the discovery of a Llandoverian fauna in the western part of the Kul'dzhuktau, southern slope of the Bel'tau*"). Uzbek Geol. Zhur., No. 1, p. 90.

Ehrenberg, C. G.

1834. *Beiträge zur physiologischen Kenntniss der Corallenthiere im allgemeinen, und besonders des rothen Meeres, nebst einem Versuche zur physiologischen Systematik.* K. Akad. Wissen., Berlin, Abh., 1832, Th. 1, pp. 225-380.

Eichwald, C. E. von

1829. *Zoologia Specialis quam expositis animalibus tum vivis, tum fossilibus potissimum Rossiae in universum, et Poloniae in specie . . .* Pt. 1. Vilna, pp. xix + 314, pls. 1-5.

1855, 1860. *Lethaea Rossica ou paléontologie de la Russie, I.* Stuttgart, pp. xix + 17-26 + 1-681 (Atlas, 1855; Text, 1860).

Etheridge, Jr., R.

1904. *A monograph of the Silurian and Devonian corals of New South Wales, Part 1. The genus* Halysites. Geol. Surv. New South Wales, Mem., Palaeont., No. 13, pp. ix + 39, pls. 1-9.

1907. *A monograph of the Silurian and Devonian corals of New South Wales. Part 2. The genus* Tryplasma. Geol. Surv. New South Wales, Mem., Palaeontol., No. 13, pp. ix + 41-102, pls. 10-28.

Fischer von Waldheim, G.

1813. *Zoognosia tabulis synopticis illustrata . . . Editio tertia, Part 1.* Moscow, pp. xiv + 465, pls. 1-8.

1828. *Notice sur les polypiers tubipores fossiles.* Programme pour la seance publique de la Société Impériale des Naturalistes, pp. 9-23, 1 pl. (Moscow).

Fitzinger, L. J. F. J.

1826. *Neue classification de reptilien nach ihren naturlichen Verwandtschatten . . .* Vienna, viii + 66 pp.

Flower, R. H.

1961. Montoya *and related colonial corals.* State Bur. Mines and Mineral Res., New Mexico Inst. of Mining and Technol., Mem. 7, pt. 1, 97 pp., 52 pls.

Flower, R. H., and Duncan, H. M.

1975. *Some problems in coral phylogeny and classification,* pp. 175-192, 3 pls. in J. Pojeta, Jr. and J. K. Popé (eds.), *Studies in paleontology and stratigraphy*: Bull. Amer. Paleont., vol. 67, No. 287, 456 pp.

Floyd, J. C., Kahle, C. F., and Hoare, R. D.

1972. *A new species of* Favosites *from the Tymochtee Formation (Silurian), northwestern Ohio.* Jour. Paleont., vol. 46, No. 4, pp. 533-535.

Flügel, H.
1956. *Revision der ostalpinen Heliolithina.* Graz, Landesmus. "Joanneum", Mus. Bergbau, Geol. u. Tech., Mitt., H. 17, pp. 53-101, 4 pls.
1962. *Korallen aus dem Silur von Ozbak-Kuh (NE Iran).* Austria Geol. Bundesanst., Wien, Jahrb., vol. 105, pp. 287-330, pls. 20-23.

Flügel, H., and Saleh, H.
1970. *Die palaeozoischen Korallen-faunen Ost-Irans; I — Rugose Korallen der Niur Formation (Silur.).* Austria Geol. Bundesanst., Wien, Jahrb., vol. 113, pp. 267-302.

Foerste, A. F.
1888. *Notes on a geological section at Todd's Fork, Ohio.* Amer. Geol., vol. 2, pp. 412-419.
1889 (1890). *Notes on Clinton Group fossils, with special reference to collections from Indiana, Tennessee and Georgia.* Boston Soc. Nat. Hist., Proc., vol. 24, pp. 263-355.
1893. *Fossils of the Clinton Group in Ohio and Indiana.* Geol. Surv. Ohio, Rept., vol. 7, pp. 516-601.
1896. *An account of the Middle Silurian rocks of Ohio and Indiana.* Cincinnati Soc. Nat. Hist., Jour., vol. 18, Nos. 3, 4, pp. 161-200, pl. 7.
1904. *The Ordovician-Silurian contact in the Ripley Island area of southern Indiana, with notes on the age of the Cincinnati geanticline.* Amer. Jour. Sci., ser. 4, vol. 18, pp. 321-342, pl. 17.
1906. *The Silurian, Devonian, and Irvine formations of east-central Kentucky.* Kentucky Geol. Surv., Bull. 7, 369 pp., 8 pls.
1909a. *Fossils from the Silurian formations of Tennessee, Indiana, and Illinois.* Denison Univ., Sci. Lab., Bull., vol. 14, pp. 61-107, pls. 1-4.
1909b. *Silurian fossils from the Kokomo, West Union and Alger horizons of Indiana, Ohio, and Kentucky.* Cincinnati Soc. Nat. Hist., Jour., vol. 21, No. 1, 41 pp.
1917. *Notes on Silurian fossils from Ohio and other central states.* Ohio Jour. Sci., vol. 17, pp. 187-204, 233-267, pls. 8-12.
1931. *The Silurian fauna,* pp. 167-213, in *The Palaeontology of Kentucky.* Kentucky Geol. Surv., ser. 6, vol. 36, 469 pp.
1935. *Correlations of Silurian formations in southwestern Ohio, southeastern Indiana, Kentucky, and western Tennessee.* Denison Univ., Sci. Lab., Jour., vol. 30, pp. 119-205.

Fontaine, H.
1964. *Madréporaires paléozoïques du Viêt-Nam, du Laos, du Cambodge, et du Yunnan; nouvelles déterminations et notes bibliographiques.* Viêt-Nam, Arch. Géol., No. 6, pp. 75-90.

Frech, F.
1890. *Die Korallenfauna der Trias.* Palaeontographica. Bd. 37, pp. 1-116, pls. 1-21.

Galle, A.
1968. *Two new heliolitid species from the Silurian of Bohemia (Anthozoa).* Czech. Ústřed. Ústav. Geol., Vestnik, vol. 43, pp. 53-55.
1973. *Family Heliolitidae from the Bohemian Paleozoic.* Geologických Věd: Paleontologie, Sborník, řada P, sv. 15, pp. 7-48, 12 pls.

Gerth, H.
1921. *Die Anthozöen der Dyas von Timor.* Paläont. von Timor, Lf. 9, Abt. 16, pp. 67-147, pls. 145-150.

Gignoux, M.
1950. *Stratigraphic geology* (4th edition), (transl. to English by Gwendolyn G. Woodford, 1955): San Francisco, 682 pp.

Goldfuss, G. A.
1826-1833. *Petrefacta Germaniae.* Düsseldorf, vol. 1, lf. 1, pp. 1-76, pls. 1-25 (1826); lf. 2, pp. 77-164, pls. 26-50 (1829); lf. 3, pp. 165-240, pls. 51-71 (1831); lf. 4, pp. 241-252 (1833).

Grabau, A. W.
1899. *Moniloporidae, a new family of Palaeozoic corals.* Boston Soc. Nat. Hist., Proc., vol. 28, pp. 409-424, pls. 1-4.
1901a. *Guide to the geology and paleontology of Niagara Falls and vicinity.* New York State Mus., Bull., vol. 9, No. 46, 284 pp.
1901b. *Guide to the geology and paleontology of Niagara Falls and vicinity.* Buffalo Soc. Nat. Sci., Bull., vol. 7, 284 pp.
1926. *Silurian faunas of Eastern Yunnan.* Palaeont. Sinica, ser. B, vol. 3, fasc. 2, 100 pp, 4 pls.

Gray, J., and Boucot, A. J.
1972. *Palynological evidence bearing on the Ordovician-Silurian paraconformity in Ohio.* Geol. Soc. America, Bull, vol. 83, pp. 1299-1314.

Greene, G. K.
1898-1904 (1900). *Contribution to Indiana paleontology.* vol. 1, pt. 4, pp. 26-33, pls. 10-12.

Gregory, J. W.
1938. *Second collection of fossil corals from the Kenya coast lands made by Miss McKinnon Wood.* Glasgow Univ., Hunterian Mus., Geol. Dept., Mon. 5, pp. 90-97.

Guettard, J. E.
1770. *Mémoires sur différentes parties de l'histoire naturelle des sciences et arts.* Paris; t. 2, pp. lxxii + 1-530; t. 3, pp. iv + 1-544, pls. 1-71.

de Haan, W.
1833-1850. *Crustacea elaborante, vol. 1* (1842), in Siebold, P. F. von, Fauna Japonica. Lugduni Batavorum, xvi, xxxi, vii-xvii + 243 pp., 64 pls., 5 vols. in 4.

von Hagenow, K. F.
1851. *Die Bryozoen der Maastrichter Kreidelbildung.* Cassel, xv + 111 pp., 12 pls.

Hall, J.
1843. *Geology of New York,* Part 4 (Fourth Geological District): Albany, 525 pp.
1847. *Natural History of New York, Pt. 6. Palaeontology of New York.* Vol. 1. Albany, 338 pp., 33 pls.
1851. *New genera of fossil corals from the report by James Hall, on the Palaeontology of New York.* Amer. Jour. Sci., ser. 2, vol. 11, pp. 398-401.
1852. *Natural History of New York, Pt. 6. Palaeontology of New York.* Vol. 2. Albany, pp. viii + 363, 85 pls.
1877. Plate 3-34, with preliminary text sheet (unpaginated) preceding Pl. 1, *in* New York State Museum of Natural History, 28th Annual Report (for 1875), 97 pp., 34 pls.
1882. *Fossil corals of the Niagara and upper Helderberg groups:* N.Y. State. Mus. Natur. Hist., 35th Ann. Rep., pp. 409-464, 482, pls. 23-30.

Hall, J., and Simpson, G. B.
1887. *Natural History of New York. Pt. 6. Palaeontology of New York,* vol. 6, *Corals and bryozoa:* Albany, 298 pp., 66 pls.

Hall, J., and Whitfield, R. P.
1877. *Paleontology in U.S. Geol. Explor. 40th Parallel (King). Rep.,* vol. 4, pt. 2, pp. 199-302.

Hamada, T.
1956. Halysites kitakamiensis *Sugiyama from the Gotlandian Formation in the Kuraoka District, Kyûshû, Japan.* Japanese Jour. Geol. Geog., vol. 27, Nos. 2-4, pp. 133-141, pl. 9.
1957. *On the classification of the Halysitidae, I.* Univ. Tokyo, Jour. Faculty Sci., sec. 2, vol. 10, pp. 383-430.
1958. *Japanese Halysitidae.* Univ. Tokyo, Jour. Faculty Sci., sec. 2, vol. 11, pp. 91-114, pls. 6-10.

1959. *Corallum growth in the Halysitidae.* Univ. Tokyo, Jour. Faculty Sci., sec. 2, vol. 11, pp. 273-289, pls. 12-15.

Hicken, A., Irving, E., Law, L. K., and Hastie, J.
1972. *Catalogue of paleomagnetic directions and poles (1st issue).* Earth Physics Branch, Publ., Dept. of Energy, Mines, and Res., Ottawa, Canada, 135 pp.

Hickson, S. J.
1911. *On* Ceratopora, *the type of a new family of Alcyonaria.* Roy. Soc. London, Proc., vol. 84B, pp. 195-200, pl. 6.

Hill, D.
1935. *British terminology for rugose corals.* Geol. Mag., vol. 72, pp. 481-519, figs. 1-21.
1940. *The Silurian Rugosa of the Yass-Bowning district, New South Wales.* Linn. Soc. New South Wales, Proc., vol. 65, pts. 3-4, pp. 388-420, pls. 11-13.
1942. *Some Tasmanian Palaeozoic corals.* Roy. Soc. Tasmania, Pap. and Proc., 1941, pp. 3-11, pl. 2.
1954. *Coral faunas from the Silurian of New South Wales and the Devonian of Western Australia.* Australian Bur. Min. Res., Geol. Geophys., Bull. 23, 46 p.
1955. *Ordovician corals from Ida Bay, Queenstown and Zeehan, Tasmania.* Roy. Soc. Tasmania, Papers and Proc., vol. 89, pp. 237-253.
1956. *Rugosa,* pp. F233-F324, in R. C. Moore (editor), *Treatise on invertebrate paleontology, Part F, Coelenterata.* Geol. Soc. America and Univ. Kansas Press, 494 pp.
1957. *Ordovician corals from New South Wales.* Roy. Soc. New South Wales, Jour. and Proc., vol. 91, pp. 97-106.
1959a. *Distribution and sequence of Silurian coral faunas.* Roy. Soc. New South Wales, Jour. and Proc., vol. 92, pp. 151-173.
1959b. *Some Ordovician corals from New Mexico, Arizona, and Texas.* State Bur. of Mines and Mineral Res., New Mexico Inst. of Mining and Technol., Bull. 64, vi + 25 pp., 2 pls.

Hill, D., and Butler, A. J.
1936. *Cymatelasma, a new genus of Silurian rugose corals.* Geol. Mag., vol. 73, pp. 516-527, pl. 16.

Hill, D., and Edwards, A. B.
1941. *Note on a collection of fossils from Queenstown, Tasmania.* Roy. Soc. Victoria, Proc., n.s., vol. 53, pp. 222-230, 1 pl.

Hill, D., and Jell, J. S.
1970. *The tabulate coral families Syringolitidae Hinde, Roemeriidae Počta, Neoroemeridae Radugin, and Chonostegitidae Lecompte, and Australian species of* Roemeripora *Kraicz.* Roy. Soc. Victoria, Proc., vol. 86, pp. 171-189.

Hill, D., and Smyth, L. B.
1938. *On the identity of* Monilopora *Nicholson and Etheridge, 1879, with* Cladochonus *McCoy, 1847.* Roy. Irish Acad., Proc., vol. 45B, pp. 125-138, pls. 22, 23.

Hill, D., and Stumm, E. C.
1956. *Tabulata,* pp. F444-477, text-figs. 340-357, in R. C. Moore (ed.), *Treatise on invertebrate paleontology, Part F, Coelenterata.* Geol. Soc. America and Univ. Kansas Press, 494 pp.

Hinde, G. H.
1879. *On a new genus of favosite coral from the Niagara Formation (Upper Silurian), Manitoulin Island, Lake Huron.* Geol. Mag., n.s., dec. 2, vol. 6, pp. 244-246.

Hisinger, W.
1831. *Bidrag till Sveriges Geognosie. Physik och Geognosie under Resor uti Sverige och Norrige.* Stockholm, 174 pp.

1837-1841. *Lethea Svecica seu Petrificata Sveciae, iconibus et characteribus illustrata.* Stockholm, pp. i-iv + 1-112, + supp., pp. 113-124, pls. A-C, 1-36 (1837); supp. *secund.*, pp. 1-11, pls. 37-39 (1840); supp. secund. *cont.*, pp. 1-6, pls. 40-42 (1841).

Holmes, M. E.
1887. *The morphology of the carinae upon the septa of rugose corals.* Bradley Whidden, Boston, 63 pp. (31 numbered), pls. 1-16.

Holtedahl, O.
1914. *On the fossil faunas from Per Schei's Series B in South Western Ellesmereland.* Rept. 2d Norwegian Arctic Expedition in the "Fram", 1898-1902, vol. 4, No. 32, Kristiania, 48 pp., 8 pls.

Huebner, J.
1819. *Verzeichniss bekannter Schmetterlinge* . . . . Augsburg, 431 pp., + 72 pp.

Hume, G. S.
1925. *The Palaeozoic outlier of Lake Timiskaming, Ontario and Quebec.* Geol. Surv. Canada, Mem. 145, Geol. Ser. 125, 129 pp., pls. 1-16.

Hurley, P.M., Cormier, R. F., Hower, J., Fairbairn, H. W., and Pinson, W. H., Jr.
1960. *Reliability of glauconite for age measurement by K-Ar and Rb-Sr methods.* Amer. Assoc. Petrol. Geol., Bull., vol. 44, pp. 1793-1808.

International Commission on Zoological Nomenclature
1976. *Opinion 1059. Suppression of* Calamopora Goldfuss, *1829 (Anthozoa, Tabulata).* Bull. Zool. Nomencl., vol. 33, pp. 24-26.

Iskyul, N. V.
1957. *Korally s Podkamennoi Tunguski ("Corals from the Podkamennaya Tunguska").* Akad. Nauk SSSR, Geol. Muz., Tr., vyp. 1, pp. 84-102.

Ivanovskiy, A. B.
1959a. *O nekotorykh kolonial'nykh korallakh Rugosa s r. Sukhaya Tunguska ("On some colonial rugose corals from the River Sukhaya Tunguska"),* pp. 135-143, 2 pls., in L. L. Khalfin (editor), *Materialy po paleontologii i stratigrafii zapadnoy Sibiri.* Sibirsk. Nauch.-Issled. Inst. Geol., Geofiz. i Min. Syrya, Tr., vyp. 2, 139 pp., Leningrad.

1959b. *K voprosu o sistematicheskom polozhenii Ordoviskikh i Siluriyskikh zaphrentoidnykh korallov ("On the question of the systematic position of the Ordovician and Silurian zaphrentoid corals").* Akad. Nauk SSSR, Doklady, n.s., vol. 125, 1959, pp. 895-897.

1960. *Novy vidy roda* Dinophyllum Lindström *iz Silura Sibirskoy Platformy ("New species of the genus* Dinophyllum Lindström *from the Silurian of the Siberian Platform"),* pp. 92-94, pls. 9, 10, in L. L. Khalfin (ed.), *Materialy po paleontologii i stratigrafii zapadnoy Sibiri.* Sibirsk. Nauch.-Issled. Inst. Geol., Geofiz, i Min. Syrya, Tr., vyp. 8, Ser. Neft. Geol., 284 pp. Leningrad.

1961. *Nekotorye Streptelasmatida Srednego i Verkhnego Ordovika s r. Podkamennaya Tunguska ("Some Streptelasmatida of the Middle and Upper Ordovician from the River Podkamennaya Tunguska"). Materialy po paleontologii i stratigrafii zapadnoy Sibiri.* Sibirsk. Nauch-Issled. Inst. Geol., Geofiz. i Min. Syrya, Tr., vyp. 15, pp. 197-207, 3 pls. Leningrad.

1962. *Dva novykh roda Siluriyskikh rugoz ("Two new genera of Silurian rugosans"),* pp. 126-133, 2 pls., in M. Y. Kasyanov and F. G. Gurari (eds.), *Materialy po paleontologii i stratigrafii zapadnoy Sibiri.* Sibirsk. Nauch.-Issled. Inst. Geol., Geofiz. i Min. Syrya, Tr., vyp. 23, Ser. Neft. Geol., 179 pp. Leningrad.

1963. *Rugozy Ordovika i Silura Sibirskoy Platformy ("Rugosans from the Ordovician and Silurian of the Siberian Platform").* Akad. Nauk SSSR, Sibirsk. Otdel., Inst. Geol. i Geofiz., Moscow, 158 pp., 33 pls.

1969. *Korally semeystvo Tryplasmatidae i Cyathophylloididae (rugozy) ("Rugose corals of the families Tryplasmatidae and Çyathophylloididae").* Akad. Nauk SSSR, Sibirsk. Otdel., Inst. Geol., Geofiz., Moscow, 108 pp.

James, U. P.
1878. *Descriptions of Cincinnatian and other Paleozoic fossils.* The Paleontologist, vol. 1, No. 2.

Jones, O. A.
1936. *The controlling effect of environment upon the corallum in Favosites; with a revision of some massive species on this basis.* Ann. Mag. Nat. Hist., ser. 10, vol. 17, pp. 1-24.

1937. *The Australian massive species of the coral genus Favosites.* Australian Mus. Rec., vol. 20, pp. 79-102.

1944. *Tabulata and Heliolitida from the Wellington district, New South Wales.* Roy. Soc. New South Wales, Jour. and Proc., vol. 77, pp. 33-39, 1 pl.

Jones, O. A., and Hill, D.
1940. *The Heliolitidae of Australia, with a discussion of the morphology and systematic position of the family.* Roy. Soc. Queensland, Proc., vol. 51, pp. 183-215, 5 pls.

Jones, O. T.
1925. *On the geology of the Llandovery district, Part 1.* Geol. Soc. London, Quart. Jour., vol. 81, pp. 344-388.

Kal'o, D. L.
1956a. *O streptelazmidnykh rugozakh pribaltiyskogo Ordovika ("On streptelasmatid rugosans of the Ordovician of the Baltic region").* Eesti NSV Teaduste Akad., Geol. Inst., Uurimused, tom 1, pp. 68-73, illus.

1956b. *Rody Primitophyllum gen. n. i Leolasma gen. n. ("The genera Primitophyllum gen. n. and Leolasma gen. n."),* pp. 35-37, pls. 9, 10, in L. D. Kiparisova, B. P. Markovski, and G. P. Radchenko (eds.), *Materialy po paleontologii: Novye semeystva i rody.* Vsesoyuz. Nauch.-Issled. Geol. Inst., Mater., n.s., vyp. 12, Paleont., 267 pp.

1957. *Codonophyllacea Ordovika i Llandoveri Pribaltiki ("Codonophyllacea of the Ordovician and Silurian of the Baltic region").* Eesti NSV Teaduste Akad., Juures Asuva, Loodusuurujate Seltsi, Aastaraamat 1957, tom 50, köide, pp. 153-168, pls. 16, 17.

1958a. *K sistematike roda Streptelasma Hall, opisanie nekotorykh novykh tetrakorallov ("On the systematics of the genus Streptelasma Hall, descriptions of some new tetracorals").* Eesti NSV Teaduste Akad., Geol. Inst., Uurimused, tom 2, pp. 19-26, illus.

1958b. *Nekotorye novye i maloizvestnye rugozy Pribaltiki ("Some new and poorly-known rugosans from the Baltic region").* Eesti NSV Teaduste Akad., Geol. Inst., Uurimused, tom 3, pp. 101-123, 5 pls.

1961. *Dopolneniya k izucheniyu streptelazmid Ordovika Estònii ("Addition to the study of streptelasmatids from the Ordovician of Estonia").* Eesti NSV Teaduste Akad., Geol. Inst., Uurimused, tom 6, pp. 51-67.

Kal'o, D. L., and Klaamann, E.
1965. *The fauna of the Portrane Limestone, 3, The corals.* British Mus., Nat. Hist., Bull., Geology, vol. 10, pp. 413-434.

Kal'o, D. L., and Reiman, V. M.
1958. *Dva novykh vida roda* Calostylis *iz Nizhnego Silura Estonia* ("*Two new species of the genus* Calostylis *from the Lower Silurian of Estonia"*). Eesti NSV Teaduste Akad., Geol. Inst., Uurimused, tom 2, pp. 27-31, 1 pl.

Kettnerova, M.
1933. Helioplasma kolihai *n. g., n. sp. (family Heliolitidae) from the Koněprusy limestones (Etage f, Lower Devonian, Bohemia)*. Stat. Geol. Ústav. Ceskoslov. Repub. Věstnik, vol. 9, Nos. 3-4, pp. 180-183.

Kiär, J.
1899. *Die Korallenfaunen der Etage 5 des norwegischen Silursystems.* Palaeontographica, Bd. 46, pp. 1-60, pls. (1) + 1-7.

King, W.
1850. *A monograph of the Permian fossils of England.* Palaeontol. Soc. London Monogr., 258 pp.

Kiryushina, M. T., Gatiev, I. D., and Chernyshev, B. B.
1939. *Geology and mineral deposits of the Chukchee district, Part 5, The district of the Chukchee Peninsula.* Arctic Inst., Trans., vol. 131, 188 pp. (in Russian).

Kissling, D. L.
1970. *Recent and Silurian circumrotatory corals (abstr.).* Geol. Soc. America, Abstr. with Progr., vol. 2, No. 3, p. 225.
1973. *Circumrotatory growth form in Recent and Silurian corals,* pp. 43-58, in R. S. Boardman, A. H. Cheetham and W. A. Oliver, Jr. (eds.), *Animal colonies: Development and function through time.* Dowden, Hutchinson and Ross, Inc., Stroudsburg, Pennsylvania, 603 pp.

Kjerulf, T.
1865. *Veiviser ved geologiske Excursioner i Christiania omegn.* Christiania, pp. i-iv + 43, illus.

Klaamann, E. R.
1959. *O faune tabulyat yuuruskogo i tamsaluskogo gorizontov ("On the tabulate fauna of the Juuru and Tamsalu stages").* Eesti NSV Teaduste Akad., Toimetised, tom 8, Teh. ja Püüs.-Mat. Tead. ser. No. 4, pp. 256-270.
1961a. *Tabulyaty i geliolitidy Venloka Estonii ("Tabulates and heliolitids from the Wenlock of Estonia").* Eesti NSV Teaduste Akad., Geol. Inst., Uurimused, tom 6, pp. 69-112, 13 pls.
1961b. *Estonian earliest favositids.* Eesti NSV Teaduste Akad., Toimetised, tom 10, Teh. ja Füüs.-Mat. Tead. ser. No. 2, pp. 120-129, 3 pls.
1964. *Pozdneordovikskie i Rannesiluriyskie Favositida Estonii ("Late Ordovician and Early Silurian Favositida of Estonia").* Eesti NSV Teaduste Akad., Geol. Inst., Tallin, Estonia, 118 pp.
1966. *Inkommunikatnye tabulyaty Estonii ("Incommunicate tabulates of Estonia").* Eesti NSV Teaduste Akad., Geol. Inst., Tallin, Estonia, 96 pp., 22 pls.
1971. *Tabulyaty verkhnego Korallovogo izvestnyaka Norvegii ("Tabulates of the upper Coralline limestone of Norway").* Eesti NSV Teaduste Akad., Toimetised, Keemia Geol., vol. 20, p. 356-363.

Kříž, J., and Pojeta, J., Jr.
1974. *Barrande's colonies concept and a comparison of his stratigraphy with the modern stratigraphy of the Middle Bohemian Lower Paleozoic rocks (Barrandian) of Czechoslovakia.* Jour. Paleontol., vol. 48, No. 3, pp. 489-494.

Lafuste, J., and Desparmet, R.
1969 (1970). *Tabulés siluriens de Sar-e-Pori, Afghanistan.* Mus. Natl. Hist. Nat. (Paris), Bull., vol. 41 (1969), pp. 1299-1305.

Lamarck, J. B. P. A., de
1801. *Systême des animaux sans vertèbres.* Paris, pp. viii + 432.
1816. *Histoire naturelle des animaux sans vertèbres.* Paris, vol. 2, pp. 1-568.

Lambe, L. M.
1899. *A revision of the genera and species of Canadian Paleozoic corals: The Madreporaria Perforata and the Alcyonaria.* Geol. Surv. Canada, Contr. Canadian Paleontology, vol. 4, pt. 1, pp. 1-96, pls. 1-5.
1901. *A revision of the genera and species of Canadian Paleozoic corals: The Madreporaria Aporosa and the Madreporaria Rugosa.* Geol. Surv. Canada, Contr. Canadian Paleontol., vol. 4, pt. 2, pp. 97-197, pls. 6-18.
1906. *Notes on the fossil corals collected by Mr. A. P. Low at Beechy Island, Southampton Island and Cape Chidley in 1904,* appendix 4, pp. 322-328, *in A. P. Low, Report on the Dominion Government Expedition to Hudson Bay and the Arctic Islands on board the D. G. S. Neptune, 1903-1904.* Ottawa.

Lang, W. D.
1926. *Naos pagoda (Salter), the type of a new genus of Silurian corals.* Geol. Soc. London, Quart. Jour., vol. 82, pp. 428-435, pl. 30.

Lang, W. D., and Smith, S.
1927. *A critical revision of the rugose corals described by W. Lonsdale in Murchison's "Silurian System".* Geol. Soc. London, Quart. Jour., vol. 83, pp. 448-490, pls. 34-37.
1939. *Some new generic names for Palaeozoic corals.* Ann. Mag. Nat. Hist., ser. 11, vol. 3, pp. 152-156, pl. 4.

Lang, W. D., Smith, S., and Thomas, H. D.
1940. *Index of Palaeozoic coral genera.* British Mus. (Nat. Hist.), London, 231 pp.

Laub, R. S.
1972. *The auloporid genus* Cladochonus *M'Coy, 1847: new data from the New York Devonian.* Jour. Paleont., vol. 46, No. 3, pp. 364-370, 1 pl., 2 text-figs.
1975. *The ancestry, geographical extent, and fate of the Brassfield coral fauna (Middle Llandovery, North America),* p. 273-286 *in* J. Pojeta, Jr. and J. K. Pope (eds.), Studies in paleontology and stratigraphy. Bulls. Amer. Paleontol., vol. 67, No. 287, 456 pp.
1978. *Axial torsion in corals.* Jour. Paleont., vol. 52, No. 3, p. 737-740.

Lavrusevich, A. I.
1965. *Predstavitel' maloizvestnogo roda* Ceriaster *(Rugosa) iz tsentral'nogo Tadzhikistana ("Representatives of the poorly-known genus* Ceriaster *[Rugosa] from central Tadzhikstan"),* pp. 27-30, in B. S. Sokolov and A. B. Ivanovskiy, *Rugozy Paleozoya SSSR.* Izdatel'stvo "Nauka", Moscow, 90 pp., 26 pls.
1971a. *New Late Ordovician Rugosa of the Zeravshan-Gissar Mountain region.* Paleontol. Jour., vol. 5, pp. 421-423.
1971b. *Rugozy Rannego Silura Zeravshano-Gissarskoy gornoy oblasti ("Early Silurian rugosans of the Zeravshan-Gissar Mountain Region").* Trudy Upravleniya Geologii Sovyeta Ministrov Tadzhikskoy SSR — Paleont. i Strat., No. 3, pp. 38-136.

Lecompte, M.
1936. *Revision des tabulés Dévoniens décrits par Goldfuss.* Mus. Hist. Nat. Belgique, Mém. 75, 112 pp., 14 pls.

Leleshus, V. L.
1972. *Siluriyskie tabulyaty Tadzhikistana ("Silurian tabulates of Tadzhikstan").* Acad. Sci. Tadzhik. SSR, Inst. Geol., 85 pp., 26 pls.

Lindström, G.
1868. *Om tvenne nya ofversiluriska Koraller från Gotland.* Öfver. Kl. Vetenskaps-Akad., Förh., Årg. 25, No. 8, pp. 419-428, pl. 6.

Linnae
1

Linney
1

Locke,
1

Lonsda
1

Manter
1
M'Coy,
1

1

Melvill
1

1870. *A description of the Anthozoa Perforata of Gotland.* K. Svenska Vetens.-Akad., Handl., Bd. 9, No. 6, pp. 1-6, 1 pl.

1871. *On some operculate corals, Silurian and Recent.* Geol. Mag., ser. 1, vol. 8, pp. 122-126.

1873. *Några anteckningar om Anthozoa tabulata.* Öfvers. K. Vetens.-Akad., Förhandl., Årg. 30, No. 4, pp. 3-20.

1876. *On the affinities of the Anthozoa Tabulata.* Ann. Mag. Nat. Hist., ser. 4, vol. 18, pp. 1-17.

1882a. *Anteckningar om Silurlagren på Carlsöarne.* Öfvers. K. Vetens.-Akad., Förhandl., Årg. 39, No. 3, pp. 5-30.

1882b. *Silurische Korallen aus Nord-Russland und Sibirien.* Bihang K. Svenska Vetens.-Akad., Handl., Bd. 6, No. 18, pp. 1-23, pl. 1.

1885. *List of the fossils of the Upper Silurian Formation of Gotland.* Stockholm, 20 pp.

1896. *Beschreibung einiger Øbersilurischer Korallen aus der Insel Gotland.* Bihang. K. Svenska Vetens.-Akad., Handl., Bd. 21, Afd. 4, No. 7, pp. 1-51, pls. 1-8.

1899. *Remarks on the Heliolitidae.* K. Svenska Vetens.-Akad., Handl., Bd. 32, No. 1, pp. 1-140, 12 pls.

Linnaeus, C.

1758. *Systema naturae.* Tomus I, Editio Decima Reformata: Stockholm, pp. iv + 823 (10th ed.).

1767. *Systema naturae.* Tomus I, pars II, Editio Duodecima Reformata: Stockholm, pp. 533-1327 (12th ed.).

Linney, W. M.

1882a. *Report on the geology of Garrard County (Kentucky).* Frankfort, 30 pp.

1882b. *Report on the geology of Lincoln County (Kentucky).* Frankfort, 36 pp.

Locke, J.

1838. *Prof. Locke's geological report,* pp. 201-274, in W. W. Mather, *Geological Survey of the State of Ohio,* Second Annual Report. 286 pp.

Lonsdale, W.

1839. *Corals,* pp. 675-694, pls. 15, 15 *bis,* 16, 16 *bis,* in R. I. Murchison, *The Silurian System,* Vol. 2, London, pp. 579-768, 37 pls.

1845. *Description of some Palaeozoic corals of Russia.* Appendix A, pp. 591-634, 1 pl., in R. I. Murchison, E. de Verneuil, and A. von Keyserling, *The geology of Russia in Europe and the Ural Mountains.* Vol. 1, London, 700 pp.

Manten, A. A.

1971. *Silurian reefs of Gotland.* Amsterdam, London, New York, 539 pp.

M'Coy, F.

1844. *A synopsis of the characters of the Carboniferous limestone fossils of Ireland.* University Press, Dublin, pp. viii + 5-207, pls. 1-29.

1847. *On the fossil botany and zoology of the rocks associated with the coal of Australia.* Ann. Mag. Nat. Hist., ser. 1, vol. 20, pp. 145-157, 226-236, 298-312, pls. 9-17.

1849. *On some new genera and species of Palaeozoic corals and Foraminifera.* Ann. Mag. Nat. Hist., ser. 2, vol. 3, pp. 1-20 (January); pp. 119-136 (February).

1854. *A systematic description of the British Palaeozoic fossils in the Geological Museum of the University of Cambridge.* Part 2, in A. Sedgwick, *A synopsis of the classification of the British Palaeozoic rocks* . . . London and Cambridge, 661 pp., 20 pls.

Melville, R. V.

1978. Paleofavosites/Palaeofavosites. J. Paleont., vol. 52, No. 1, p. 121.

Michelin, J. L. H.
1841-1848. *Iconographie zoophytologique, description par localités et terrains des polypiers fossiles de France et pays environnants.* Paris, pp. xii + 348, Atlas, 79 pls. (pp. 1-40, 1841; pp. 42-72, 1842; pp. 73-104, 1843; pp. 105-144, 1844; pp. 145-184, 1845; pp. 185-248, 1846; pp. 249-328, 1847; pp. 329-348, 1848).

Miller, S. A.
1877. *The American Palaeozoic fossils.* Publ. by the author, Cincinnati, 334 pp.

Milne-Edwards, H. and Haime, J.
1849. *Mémoire sur les polypiers appartenant aux groupes naturels des Zoanthaires perforés et des Zoanthaires tabulés.* Acad. Sci., Paris, Compt. Rend., vol. 29, pp. 257-263.
1850-1854. *A monograph of the British fossil corals.* Palaeontogr. Soc., London, pp. lxxxv + 322, 72 pls.
1851. *Monographie des polypiers fossiles des terrains Palaeozoïques.* Mus. Hist. Nat., Paris, Arch., t. 5, 502 pp., 20 pls.

Minato, M.
1961. *Ontogenetic study of some Silurian corals of Gotland.* Stockholm Contr. Geol., vol. 8, pp. 37-100, pls. 1-22, 31 text-figs.

Murray, R. C., and Lucia, F. J.
1967. *Cause and control of dolomite distribution by rock selectivity.* Geol. Soc. America, Bull., vol. 78, pp. 21-36.

Murray, R. C., and Pray, L. C.
1965. *Dolomitization and limestone diagenesis — An introduction,* pp. 1-2, in L. C. Pray and R. C. Murray (editors), *Dolomitization and limestone diagenesis — A symposium.* Soc. Econ. Paleont. Mineral., Spec. Pub. 13, 180 pp.

Musper, K. A. F. R.
1938. *Over het voorkomen van Halysites wallichi Reed op Nieuw-Guinea.* De Ingenieur in Nederl.-Indie, Jahrg. 5, No. 10, pt. 4, pp. 156-157, 1 pl.

Naumenko, A. I.
1970. *Kompleksy Rannesiluriyskikh tabulyatomorfnykh korallov zapadnogo Sayana i ikh ekologicheskiye osobennosti ("Early Silurian tabulate coral assemblages of the western Sayan region and their ecologic characteristics"),* pp. 60-74, 9 text-figs., in *Zakonomernosti rasprostraneniya paleozoyskikh korallov SSSR* ("Sequence and distribution of Paleozoic corals of the USSR"). Vsesoyuz. Simp. Izuch. Iskop. Korallov SSSR, Tr. 2, vyp. 3, Izdatel'stvo "Nauka", Moscow, 103 pp.

Nelson, S. J.
1963. *Ordovician paleontology of northern Hudson Bay lowland.* Geol. Soc. America, Mem. 90, 152 pp., 37 pls.

Neuman, B.
1968. *Two new species of Upper Ordovician rugose corals from Sweden.* Geol. Fören. i Stockholm Förhandl., vol. 90, pp. 229-240, 6 text-figs.
1969. *Upper Ordovician streptelasmatid corals from Scandinavia.* Univ. Uppsala, Geol. Inst., Bull., n.s., vol. 1, pp. 1-73.

Nicholson, H. A.
1874. *On Columnopora, a new genus of tabulate corals.* Geol. Mag., n.s. 2, vol. 1, pp. 253-254.
1875. *Descriptions of the corals of the Silurian and Devonian Systems.* Geol. Surv. Ohio, vol. 2, *Geology and Palaeontology,* Pt. 2. *Palaeontology,* pp. 181-242, pls. 21-23.
1879. *On the structure and affinities of the "Tabulate corals" of the Palaeozoic period.* Edinburgh and London, xii + 342 pp., 15 pls.
1887. *On Hemiphyllum siluriense, Tomes.* Geol. Mag., dec. 3, vol. 4, pp. 173-174.

Nicholson, H. A., and Etheridge, R., Jr.
1877. *Notes on the genus* Alveolites *Lamarck, and on some allied forms of Palaeozoic corals.* Linn. Soc. London Jour., (Zool.), vol. 13, pp. 353-370, pls. 19, 20.
1878-1880. *A monograph of the Silurian fossils of the Girvan District in Ayrshire, with special reference to those contained in the "Gray Collection".* Fasc. 1, 3, Edinburgh and London. Fasc. 1, pp. i-ix + 1-135, pls. 1-9, 1878; Fasc. 3, pp. i-vi + 237-341, pls. 16-24, 1880.
1879. *On the microscopic structure of three species of the genus* Cladochonus *M'Coy.* Geol. Mag., n.s. 2, vol. 6, pp. 289-296.

Nicholson, H. A., and Hinde, G. H.
1874. *Notes on the fossils of the Clinton, Niagara, and Guelph formations of Ontario, with descriptions of new species.* Canadian Jour., n.s., vol. 14, p. 137-152, 137-144 (*bis*).

Nicholson, H. A., and Lydekker, R.
1889. *A manual of palaeontology for the use of students.* 3d edition: Edinburgh and London; Vol. 1, pp. 1-885; Vol. 2, pp. 886-1624.

Norford, B. S.
1962a. *Illustrations of Canadian fossils — Cambrian, Ordovician and Silurian of the western Cordillera.* Geol. Surv. Canada, Paper 62-14, 24 pp., 10 pls.
1962b. *The Silurian fauna of the Sandpile Group of northern British Columbia.* Geol. Surv. Canada, Bull. 78, 51 pp., 16 pls.

Northrop, S. A.
1939. *Paleontology and stratigraphy of the Silurian rocks of the Port Daniel-Black Cape region, Gaspé.* Geol. Soc. America, Spec. Paper 21, 302 pp., 28 pls.

Nutting, C. C.
1900, 1904. *American Hydroids. pt. 1, The Plumularidae.* U.S. Natl. Mus., Spec. Bull. No. 4, pp. 1-285, pls. 1-34 (1900); *pt. 2, The Sertularidae.* U.S. Natl. Mus. Spec. Bull. No. 4, pp. 1-325, pls. 1-41 (1904).

O'Donnell, E.
1967. *The lithostratigraphy of the Brassfield Formation (Lower Silurian) in the Cincinnati Arch area.* Unpublished Ph.D. dissertation, Univ. of Cincinnati, 143 pp.

Oekentorp, K.
1971. Palaeofavosites *Twenhofel, 1914 (Anthozoa, Tabulata): proposed validation under the Plenary Powers, Z.N. (S.) 1961.* Bull. Zool. Nomencl., vol. 28, pts. 5/6, pp. 158-160.
1974. *Comment on* Palaeofavosites *Twenhofel, 1914 (Anthozoa, Tabulata): proposed validation under the Plenary Powers.* Bull. Zool. Nomencl., vol. 31, pt. 3, pp. 112-113.
1976. *Revision und Typisierung des Genus* Paleofavosites *Twenhofel, 1914 (Coelenterata, Tabulata).* Palaont. Z., vol. 50, nos. 3/4, p. 151-192 (152?), 3 figs. 3 pls.
1977. Palaeofavosites *Twenhofel in place of* Calamopora *Goldfuss and the spelling of* Palaeofavosites. *Fossil Cnidaria, vol. 6, no. 1, p. 13.*

Okulitch, V. J.
1943. *The Stony Mountain Formation of Manitoba.* Roy. Soc. Canada, Trans., sect. 4 (Geology), ser. 3, vol. 37, pp. 59-74, 2 pls.

Oliver, W. A., Jr.
1960. *Rugose corals from reef limestones in the Lower Devonian of New York.* Jour. Paleontol., vol. 34, No. 1, pp. 59-100, pls. 13-19.
1962a. *Silurian rugose corals from the Lake Témiscouata area, Quebec.* U.S. Geol. Surv., Prof. Paper 430, pp. 11-19, pls. 5-8.
1962b. *A new* Kodonophyllum *and associated rugose corals from the Lake Matapedia area, Quebec.* U.S. Geol. Surv., Prof. Paper 430, pp. 21-31, pls. 9-14.

1963. *Redescription of three species of corals from the Lockport Dolo-mite in New York.* U.S. Geol. Surv., Prof. Paper 414-G, 9 pp., 5 pls.

d'Orbigny, A.
1850. *Prodrome de Paléontologie stratigraphique universelle des animaux mollusques et rayonnés.* Vol. 1: Paris, pp. ix + 394.

Orlov, Yu. A.
1930. *O nekotorykh novykh verkhne-Siluriyskikh favozitakh Fergany ("On some new Upper Silurian favositids from Ferghana").* Geol. Prosp. Serv. USSR, Bull., vol. 49, pp. 121-127, 2 pls.

Orton, E.
1871. *Report on geology of Montgomery County.* Geol. Surv. Ohio, Rept. 1869, pt. 3, pp. 143-171.

Owen, D. D.
1844. *Description of some organic remains figured in this work, sup-posed to be new,* pp. 69-86, pls. 11-18, in *Report of a geological exploration of part of Iowa, Wisconsin, and Illinois . . . in the autumn of the year 1839.* U.S. 28th Congress, 1st session, Senate Executive Document 407, 191 pp.

1857. *Second and Third reports of the Geological Survey in Kentucky made during the years 1856 and 1857.* Frankfort, 391 + 589 pp.

Penecke, K. A.
1887. *Ueber die Fauna und das Alter einige palaozoischer Korallriffe der Ostalpen.* Deutsch. Geol. Gesell., Zeitschr., Bd. 39, pp. 267-276.

1894. *Das Grazer Devon.* Jahrb. K.-K. Geol. Reichsanst., Jahrg. 1893, vol. 43, pp. 567-616, pls. 7-12.

Philcox, M. E.
1970. *Coral bioherms in the Hopkinton Formation (Silurian), Iowa.* Geol. Soc. America, Bull., vol. 81, pp. 969-973.

Phillips, J.
1841. *Figures and descriptions of the Palaeozoic fossils of Cornwall, Devon, and West Somerset.* Geol. Surv. Great Britain and Ireland, pp. xii + 231, pls. 1-60, London.

Počta, P.
1902. *Anthozoaires et alcyonaires,* Vol. 8, tome 2, of J. Barrande, *Systême silurien du centre de la Bohême, Pt. 1, Recherches Paléontologiques.* Pp. viii + 347, pls. 20-117, Prague.

Poltavtseva, N. V.
1965. *Novy nakhodki tabulyat v otlozheniyakh nizhnego Venloka Kazakh-stana (zapadnoe Pribalkhash'e) ("Newly-discovered tabulates in the Lower Wenlock beds of Kazakhstan, western Pribalkhash"),* pp. 40-50, in B. S. Sokolov and V. N. Dubatolov (eds.), *Tabulyato-morfnye korally Ordovika i Silura SSSR.* Vsesoyuz. Simp. Izuch. Iskop. Korallov SSSR, Tr. 1, vyp. 1, Izdatel'stvo "Nauka", Mos-cow, 138 pp., 34 pls.

Porfiriev, V.
1937. *On some corals of the group Tabulata from the eastern slope of the Urals.* [Russia], Central Geological and Prospecting Inst. Paleont., Stratigraphy, Mater., Leningrad, mag. 3, pp. 22-34, pls. 1-5.

Prantl, F.
1940. *Příspěvek k poznáni českých silurských korálu (Rugosa).* Ceská Akad., Třída 2, Rozpravy, r. 49, č. 14, 11 pp.

1957. *O rodu Helminthidium Lindström z českého siluru (Rugosa).* Czechoslovakia, Ústřed. Ústav. Geol., Sborník sv. 23 (1956), Geol., pp. 475-495.

Preobrazhenskiy, B. V.

1964. *Novye vidy roda* Rhaphidophyllum *v Verkhnem Ordovike bassena r. Kolyma ("New species of the genus* Rhaphidophyllum *in the Upper Ordovician of the basin of the Kolyma River")*. Materialy po Geol. i Poleznym Iskopaemym Severo-Vostoka SSSR, No. 17, pp. 68-73, 2 pls.

1968. *Pozdneordovikskiye desmidoporidy Omulevskikh gor, basseyn r. Kolymy ("Late Ordovician desmidoporids from the Omulevsk Mountains, Kolyma River basin")*. Paleont. Zhur., No. 4, pp. 89-93 (*also*, Paleontol. Jour., vol. 2, No. 4 (for 1968, publ. in 1969), pp. 518-522.

Rafinesque, C. S., and Clifford, J. D.

1820. *Prodrome d'une monographie des Turbinolies fossiles du Kentucky (dans l'Amérique Septentrionale)*. Ann. Gén. Sci. Phys., Bruxelles, vol. 5, pp. 231-235.

Regnéll, G.

1941. *On the Siluro-Devonian fauna of Chöl-tagh, eastern T'ien-shan; Part 1, Anthozoa. Reports from the scientific expedition to the north-western provinces of China under the leadership of Dr. Sven Hedin*, The Sino-Swedish Expedition, Pub. 17, Vol. 5, Invertebrate Paleontology 3, 64 pp., 12 pls.

1961. *Supplementary remarks on the Siluro-Devonian of Chöl-Tagh, eastern T'ien Shan*. Uppsala Univ., Geol. Inst., Bull., vol. 40, pp. 413-427.

Reiman, V. M.

1956. *Semeystvo Cyphophyllidae Wedekind, 1927 ("The Family Cyphophyllidae Wedekind, 1927")*, pp. 37-39, in L. D. Kiparisov, B. P. Markovskiy, and G. P. Radchenko (eds.), *Materialy po paleontologii: Novye semeystva i rody*. Vsesoyuz. Nauch.-Issled. Geol. Inst., Mater., n.s. vyp. 12, Paleont., 267 pp., illus.

1958. *Novye rugozy iz Verkhneordovikskikh i Llandoveriyskikh otlozhenii Pribaltiki ("New rugosans from the Upper Ordovician and Llandoverian strata of the Baltic region")*. Eesti NSV Teaduste Akad., Geol. Inst., Uurimused, tom 2, pp. 33-47.

Rexroad, C. B.

1967. *Stratigraphy and conodont paleontology of the Brassfield (Silurian) in the Cincinnati Arch area*. Geol. Surv. Indiana, Bull. 36, 64 pp., 4 pls.

Rexroad, C. B., Branson, E. R., Smith, M. O., and Summerson, C. S.

1965. *The Silurian formations of east central Kentucky and adjacent Ohio*. Geol. Surv. Kentucky, ser. 10, 34 pp.

Roemer, C. F.

1860. *Die silurische Fauna des westlichen Tennessee*. Breslau, 99 pp., 5 pls.

1861. *Die fossile Fauna der silurischen Diluvial-Geschiebe von Sadewitz bei Oels in Nieder-Schlesien*. Breslau, pp. xvi + 81, 8 pls.

1883. *Lethaea palaeozoica*, pp. 113-544, in *Lethaea geognostica*, Thiel 1, Bd. 1, Lf. 2: Stuttgart.

Rofe, J.

1869. *Note on the cause and nature of the enlargement of some crinoidal columns*. Geol. Mag., dec. 1, vol. 6, pp. 351-353.

Rominger, C.

1876. *Lower Peninsula (1873-1876)*, Part 2. *Palaeontology. Fossil corals*. Geol. Surv. Michigan, vol. 3, 161 pp., pls. 1-55.

Ross, J. P.

1961. *Liscombea, a new Silurian tabulate coral genus from New South Wales, Australia*. Jour. Paleontol., vol. 35, No. 5, pp. 1017-1019, pl. 122.

Rózkowska, M.
1946. *Koralowce Rugosa z gotlandu Podola, cześć 1 ("Rugose corals from the Gotlandian of Podolia, part 1").* Polsk. Tow. Geol., Rocznik, tom 16, pp. 139-157.

Rukhin, L. B.
1936. *Opisanie nekotorykh favozitid iz Nizhne-Devonskikh otlozheniy Zabaykal'ya ("Descriptions of some favositids from the Lower Devonian strata of Transbaikal").* Leningrad State Univ., Ann. No. 10, Ser. Geol., issue 3, pp. 96-110, 2 pls.
1938. *The Lower Paleozoic corals and stromatoporoids of the upper part of the Kolyma River.* Russia, State Trust Dalstroy, Contr. Knowledge Kolyma-Indighirka Land, ser. 2, Geol. Geomorph., fasc. 10, 119 pp., Moscow.

Ryder, T. A.
1926. *Pycnactis, Mesactis, Phaulactis, gen. nov., and Dinophyllum Lind.* Ann. Mag. Nat. Hist., ser. 9, vol. 18, pp. 385-401, pls. 9-12.

Sahni, M. R., and Gupta, V. J.
1968. *Silurian corals from the Kashmir Himalaya.* Paleont. Soc. India, Jour., vols. 5-9, (1960-1964), pp. 71-72.

Saleh, H.
1969. *A new coral fauna from the Niur Formation (Silurian) of east Iran.* Australia Geol. Bundesanst., Wien, Verh., No. 1, pp. 33-34.

Salter, J. W.
1873. *Catalogue of the Cambrian and Silurian fossils contained in the Geological Museum of the University of Cambridge.* Cambridge, pp. xlviii + 204.

Savage, T. E.
1913. *Stratigraphy and paleontology of the Alexandrian Series in Illinois and Missouri, Part 1.* Geol. Surv. Illinois, Bull. 23 (extract), 124 pp.

Scheffen, W.
1933. *Die Zoantharia rugosa des Silurs auf Ringerike im Oslogebiet.* Norske Videns.-Akad., Oslo, Skrift. utgitt, 1. Math.-Naturv. Kl. Skrift., vol. 2, No. 5, (1932), 63 pp., 11 pls.

Schlotheim, E. F. von
1820. *Die Petrefactenkunde auf ihrem jetzigen Standpunkte durch die Beschreibung seiner Sammlung . . . erläutert.* Gotha, pp. lxii + 437.

Schlüter, C.
1880. *Ueber Zoantharia rugosa aus dem rheinischen Mittel- und Ober-Devon.* Sitz-Ber. Gesellsch. naturforsch. Freunde Berlin, No. 3, pp. 49-53.
1885a. *Ueber einige neue Anthozoen aus dem Devon.* Naturhist. Vereines Preuss., Verhandl., Rheinlande, Westfalens, . . . Jahrg. 42, Sitz-ber. [niederrhein, Gesellsch. Natur- und Heilkunde Bonn,] pp. 144-151.
1885b. *Dünnschliffe von Zoantharia rugosa, Zoantharia tabulata und Stromatoporiden aus den paläontologischen Museum der Universität Bonn, Aussteller Professor Dr. C. Schlüter in Bonn.* Exposition géologique, Congrès géol. internat., 3d sess., Catalogue, Berlin, pp. 52-56.
1889. *Anthozoen des rheinischen Mittel-Devon.* Geol. Specialkarte Preuss. Thüring. Staat, Abhandl., vol. 8, No. 4, pp. x + 259-465, pls. 1-16.

Schouppé, A.
1954. *Die Korallenfauna aus dem ef des Paläozoikums von Graz.* Naturw. Vereines für Steiermark, Mitt., Bd. 84, pp. 159-171.

Schuchert, C.
1900. *On the Lower Silurian (Trenton) fauna of Baffin Land.* U.S. Natl. Mus., Proc., vol. 22, pp. 143-177, pls. 12-14.

Schweigger, A. F.
1819. *Beobachtungen auf Naturhistorischen Reisen.* Berlin, pp. xii + 127, pls. 1-8, 12 tables.

Scrutton, C. T.
1971. *Palaeozoic coral faunas from Venezuela, 1. Silurian and Permo-Carboniferous corals from the Mérida Andes.* British Mus., Nat. Hist., Bull., Geol., vol. 20, pp. 186-227, 5 pls.

Scudder, S. H.
1875. *Notes on Orthoptera from northern Peru, collected by Professor James Orton.* Boston Soc. Nat. Hist., Proc., vol. 17 (for 1874-1875) pp. 257-282.

Sharkova, T. T.
1964. *Nekotorye novye vidy Siluriyskikh i Devonskikh tabulyat yugo-vostochnogo Kazakhstana ("Some new species of Silurian and Devonian tabulates from southeastern Kazakhstan").* Paleont. Zhur., No. 1, pp. 20-25.

Sheehan, P. M.
1973. *The relation of Late Ordovician glaciation to the Ordovician-Silurian changeover in North American brachiopod faunas.* Lethaia, vol. 6, pp. 147-154.

Shimizu, S., Ozaki, K.-E., and Obata, T.
1934. *Gotlandian deposits of Northwest Korea.* Shanghai Sci. Inst., Jour., sect. 2, vol. 1, pp. 59-88.

Shinn, E. A., Ginsburg, R. N., and Lloyd, R. M.
1965. *Recent supratidal dolomite from Andros Island, Bahamas,* pp. 112-123, in L. C. Pray and R. C. Murray (editors), *Dolomitization and limestone diagenesis — a symposium.* Soc. Econ. Paleont. Mineral., Spec. Pub. 13, 180 pp.

Shrock, R. R., and Twenhofel, W. H.
1939. *Silurian fossils from northern Newfoundland.* Jour. Paleont., vol. 13, no. 3, pp. 241-266.

Simpson, G. B.
1900. *Preliminary descriptions of new genera of Paleozoic rugose corals.* New York State Mus., Bull., vol. 8, No. 39, pp. 199-222.

Sinclair, G. W.
1955. *Some Ordovician halysitoid corals.* Roy. Soc. Canada, Trans., vol. 49, ser. 3, sect. 4, pp. 95-103.

Smith, S.
1930a. *The Calostylidae Roemer: a family of rugose corals with perforate septa.* Ann. Mag. Nat. Hist., ser. 10, vol. 5, pp. 257-278, pls. 10-12.
1930b. *Some Valentian corals from Shropshire and Montgomeryshire, with a note on a new stromatoporoid.* Geol. Soc. London, Quart. Jour., vol. 86, pp. 291-330, pls. 26-29.
1945. *Upper Devonian corals of the Mackenzie River region.* Geol. Soc. America, Spec. Paper 59, 126 pp., 35 pls.

Smyth, L. B.
1915. *On the faunal zones of the Rush-Skerries Carboniferous section, County Dublin.* Roy. Dublin Soc., Sci. Proc., vol. 14, pp. 535-562, pls. 35-37.

Sokolov, B. S.
1949a. *Tabulata i Heliolitida Silura SSSR ("Tabulata and Heliolitida of the Silurian of the USSR").* Atlas rukovoddyashchikh form isko-paemikh faun SSSR, vol. 2, pp. 75-98, pls. 6-10, Moscow.
1949b. *Filogeneticheskie otnosheniya Syringoporidae i Favositidae ("Phylogenetic relationships between Syringoporidae and Favositidae").* Akad. Nauk SSSR, Dokl., n.s., tom 64, No. 1, pp. 133-135.
1950. *Sistematika i istoriya razvitiya paleozoiskikh korallov Anthozoa Tabulata ("Systematics and evolution of the Paleozoic corals Anthozoa Tabulata").* Voprosy Paleont., Leningrad Univ., tom 1, pp. 134-242, 8 pls.

1962. *Subclass Tabulata*, pp. 293-404, *and Subclass Heliolitoidea*, pp. 405-431, in B. S. Sokolov (editor), *Fundamentals of paleontology (Osnovy paleontologii), Vol. 2, Porifera, Archaeocyatha, Coelenterata, Vermes.* Izdatel'stvo Akad. Nauk SSSR, Moscow (Transl. 1971 by Israel Program for Scientific Transl., Jerusalem), 900 pp.

Soshkina, E. D.
1937. *Korally Verkhnego Siluria i Nizhnego Devona vostochnogo i zapadnogo sklonov Urala ("Corals of the Upper Silurian and Lower Devonian of the eastern and western slopes of the Urals").* Akad. Nauk SSSR i Vsesoyuz. Inst. Mineral. Syrya NKPT, Tr., tom 6, vyp. 4, 116 pp., 21 pls.
1955. *Korally,* pp. 118-128 of E. A. Ivanova, E. D. Soshkina, G. G. Astrova, and V. A. Ivanova, *Fauna Ordovika i Gotlandiya nizhnego techeniya r. Podkamennoy Tunguski, yeyo ekologiya i stratigraficheskoe znacheniye ("The Ordovician and Gotlandian fauna of the lower course of the Podkamennaya Tunguska River, its ecology and stratigraphic significance"),* pp. 93-196, in T G. Sarycheva (editor), *Materialy po faune i flore Paleozoya Sibiri.* Izdatel'stvo Akad. Nauk SSSR, Moscow.

Soshkina, E. D., and Kabakovich, N. V.
1962. *Suborder Streptelasmatina,* pp. 480-491 in B. S. Sokolov (editor), *Fundamentals of paleontology (Osnovy paleontologii), vol. 2, Porifera, Archaeocyatha, Coelenterata, Vermes.* Izdatel'stvo Akad. Nauk SSSR, Moscow (Transl. 1971 by Israel Progr. for Scientific Trans., Jerusalem), 900 pp.

Spjeldnaes, N.
1961. *A new silicified coral from the Upper Ordovician of the Oslo region.* Norsk Geol. Tidsskr., Bd. 41, H. 1, pp. 79-84.
1964. *Two compound corals from the* Tretaspis *beds of the Oslo-Asker district.* Norsk Geol. Tidsskr., Bd. 44, H. 1, pp. 1-10.

Stainbrook, M. A.
1946. *Corals of the Independence shale of Iowa.* Jour. Paleont., vol. 20, No. 5, pp. 401-427, pls. 57-61.

Stasínska, A.
1967. *Tabulata from Norway, Sweden and from the erratic boulders of Poland (Tabulata z Norwegii, Szwecji i z Glazów narzutowych Polski).* Palaeont. Polonica, No. 18, 112 pp.

Stearn, C. W.
1956. *Stratigraphy and paleontology of the Interlake Group and Stonewall Formation of southern Manitoba.* Geol. Soc. Canada, Mem. 281, 162 pp.

Stechow, E.
1922. *Zur Systematic der Hydrozoen, Stromatoporen, Siphonophoren, Anthozoen und Ctenophoren.* Arch. Naturg., vol. 88a, No. 3, pp. 141-155.

Stoll, N. R., Dollfus, R. P., Forest, J., Riley, N. D., Sabrosky, C. W., Wright, C. W., and Melville, R. V.
1964. *International code of zoological nomenclature adopted by the XV International Congress of Zoology.* Internat. Trust Zool. Nomencl., London, xx + 176 pp.

Strand, E.
1934. *New name for* Parallelopora Holtedahl, 1914, *not Bargatzky, 1881.* Folia zoologica hydrobiologica, vol. 6, p. 271.

Strusz, D. L.
1961. *Lower Palaeozoic corals from New South Wales.* Palaeontology, vol. 4, pp. 334-361.

Stuckenberg, A.
1895. *Korallen und Bryozoen der Steinkohlenablagerungen der Ural und des Timan.* Com. Géol. St. Pétersbourg, Mém., n.s., vol. 10, No. 3, pp. (viii) + 1-244, pls. 1-24.

Stumm, E. C.
1952. *Species of the Silurian rugose coral Tryplasma from North America.* Jour. Paleont., vol. 26, No. 5, pp. 841-843, pl. 125.
1962. *Silurian corals from the Moose River Synclinorium.* U.S. Geol. Surv., Prof. Paper 430, pp. 1-9, pls. 1-4.
1963. *Ordovician streptelasmid rugose corals from Michigan.* Univ. of Michigan, Mus. Paleont., Contr., vol. 18, No. 2, pp. 23-31, 2 pls.
1964. *Silurian and Devonian corals of the Falls of the Ohio.* Geol. Soc. America, Mem. 93, 184 pp., 80 pls.

Sugiyama, T.
1940. *Stratigraphical and palaeontological studies of the Gotlandian deposits of the Kitakami mountainland (Japan).* Tôhoku Imp. Univ., Sci. Rept., ser. 2 (Geol.), vol. 21, No. 2, pp. 81-146, 21 pls.

Sultanbekova, Zh. S.
1971. *Rugozy Verkhnego Ordovika Khr. Chingiz (Kazakhstan)* ("Rugosans from the Upper Ordovician of the Chingiz Ridge, Kazakhstan"), pp. 86-87, in *Mezhdunarodnyy paleontologicheskiy simpozium po korallam (Coelenterata)*, Tezisy Dokladov, Akad. Nauk SSSR, Sibirsk. Otdel., Inst. Geol. Geofiz. - Otdel. Obshch. Biol., Kom. Izuch. Iskop. Korallov, Novosibirsk.

Sutherland, P. K.
1965. *Rugose corals of the Henryhouse Formation (Silurian).* Geol. Surv. Oklahoma, Bull. 109, 92 pp., 34 pls.

Sutton, I. D.
1966. *The value of corallite size in the specific determination of the tabulate corals Favosites and Paleofavosites.* Mercian Geol., vol. 1, pp. 255-263, pl. 16.

Swartz, C. K., Alcock, F. J., Butts, C., Chadwick, G. H., Cumings, E. R., Decker, C. E., Ehlers, G. M., Foerste, A. F., Gillette, T., Kindle, E. M., Kirk, E., Northrop, S. A., Prouty, W. F., Savage, T. E., Shrock, R. R., Swartz, F. M., Twenhofel, W. H., and Williams, M. Y.
1942. *Correlation of the Silurian formations of North America.* Geol. Soc. America, Bull., vol. 53, pp. 533-538.

Sytova, V. A.
1968. *Tetrakorally skal'skogo i borshchovskogo gorizontov Podolii* ("Tetracorals from the Skala and Borshchov horizons of Podolia"), pp. 51-71, in *Siluriysko-devonskaya fauna Podolii.* Leningrad Univ., Nauch.-Issled. Inst. Zemnoy Kory, Paleont. Lab., Leningrad.

Teichert, C.
1937. *Ordovician and Silurian faunas of Arctic Canada.* Fifth Thule Exped. (1921-1924), Rept., vol. 1, No. 5, 169 pp., 25 pls.

Tesakov, Yu. I.
1965. *Tsepochechnye favozitidy* ("Cateniform favositids"), pp. 14-20, in Sokolov, B. S. and Dubatolov, V. N. (editors), *Tabulyatomorfnye korally Ordovika i Silura SSSR.* Izdatel'stvo "Nauka", Moscow, 138 pp., 34 pls.

Thomas, H. D.
1963. *Silurian corals from Selangor, Federation of Malaya.* Overseas Geol. Min. Res., vol. 9, pp. 39-46.

Thomas, H. D., and Scrutton, C. T.
1969. *Palaeozoic corals from Perak, Malaya, Malaysia.* Overseas Geol. Min. Res., vol. 10, pp. 164-171.

Thomas, H. D., and Smith, S.
1954. *The coral genus Halysites Fischer von Waldheim.* Ann. Mag. Nat. Hist., ser. 12, vol. 7, pp. 765-774, pls. 20-22.

Thorslund, P.
1948. *De Silurska lagren ovan* Pentamerus *Kalkstenen i Jämtland.* Sveriges Geol. Undersökning, Årsb. 42, No. 3, ser. C, No. 494, 37 pp.

Tolmachev, I. P.
1924, 1931. *Faune du calcaire carbonifere du bassin houiller de Kousnetzk.* Com. Geol. Mat. Geol. Gen. Appl., 25 (Part 1, pp. 4 + 1-320 + i-xii, pls. 1-5, 8-11, 18-20, 1924. Part 2. pp. 321-663, pls. 6-7, 12-17, 21-23, 1931).

Tomes, R. F.
1887. *On two species of Palaeozoic Madreporaria hitherto not recognized as British.* Geol. Mag., dec. 3, vol. 4, pp. 98-100.

Tripp, K.
1933. *Die Favositen Gotlands.* Palaeontographica, Bd. 79, abt. A, lf. 3-6, pp. 75-142.

Troedsson, G. T.
1928. *On the Middle and Upper Ordovician faunas of northern Greenland, Part 2.* Jubilaeums-ekspeditionen Nord om Grønland, 1920-23, Nr. 5, 197 pp., 56 pls., København.

Troost, G.
1840. *Organic remains discovered in the state of Tennessee.* Fifth Geol. Rept. 23d General Assembly, Tennessee, pp. 45-75, Nashville.

Tsin, H.-P.
1956. *New material of Silurian fossils from the vicinity of Kueiyang, central Kueichou.* Acta Palaeont. Sinica, vol. 4, pp. 621-639.

Turmel, R. J., and Swanson, R. G.
1976. *The development of Rodriguez Bank, a Holocene mudbank in the Florida Reef Tract.* J. Sed. Petrol., vol. 46, pp. 497-518.

Twenhofel, W. H.
1914. *The Anticosti Island faunas.* Geol. Surv. Canada, Mus. Bull. 3, Geol. Ser., No. 19, pp. 1-38, 1 pl.
1928. *Geology of Anticosti Island.* Geol. Surv. Canada, Mem. 154, Geol. Ser. No. 135, 481 pps.

Vinassa de Regny, P.
1918. *Coralli mesodevonici della Carnia.* Palaeontogr. Italica, vol. 24, pp. 59-120.

Voynovskiy-Kriger, K. G.
1971. *Yavleniye zavivaniya u rugoz ("Occurrence of coiling in rugose corals"),* pp. 16-28, in *Rugozy i stromatoporoidey paleozoya SSSR.* Vsesoyuz. Simp. Izuch. Iskop. Korallov SSSR, 2d, Tr. No. 2, Izdatel'stvo "Nauka", Moscow, 152 pp.

Waagen, W., and Wentzel, J.
1886. *Salt Range fossils,* Vol. 1, Productus-*Limestone fossils, 6. Coelenterata*: Geol. Surv. India, Palaeontologia Indica, ser. 13, pp. 836-924, pls. 97-116.

Walliser, O. H.
1964. *Conodonten des Silurs.* Hess. Landes. Bodenf., Abh. No. 41, 106 pp., 32 pls.

Wang, H. C.
1944. *The Silurian rugose corals of northern and eastern Yunnan.* Geol. Soc. China, Bull., vol. 24, pp. 21-32, 1 pl.
1947. *New material of rugose corals from Yunnan.* Geol. Soc. China, Bull., vol. 27, pp. 171-188, 2 pls.
1948. *Notes on some rugose corals in the Gray Collection from Girvan, Scotland.* Geol. Mag., vol. 85, pp. 97-106, 1 pl.

Webby, B. D.
1971. *The new Ordovician genus* Hillophyllum *and the early history of rugose corals with acanthine septa.* Lethaia, vol. 4, pp. 153-168.

Webb

Wede

Wells

lf

Weiss

Wenh

White

Willia

Willia

Willia

Wilso

Wu,

Yabe

Yu,

Zhe

1972. *The rugose coral* Palaeophyllum *Billings from the Ordovician of central New South Wales.* Linn. Soc. New South Wales, Proc., vol. 97, pt. 2, pp. 150-157, pls. 8, 9.

Webby, B. D., and Semeniuk, V.
1969. *Ordovician halysitid corals from New South Wales.* Lethaia, vol. 2, pp. 345-360.

Wedekind, P. R.
1927. *Die Zoantharia Rugosa von Gotland (besonders Nordgotland)* ... Sveriges Geol. Undersökning, Avh. och uppsatser i 4:0, ser. Ca, No. 19, 95 pp., 30 pls.

Wells, J. W.
1944. *New Tabulate Corals from the Pennsylvanian of Texas.* Jour. Paleont., vol. 18, No. 3, pp. 259-262, pls. 40, 41.
1956. *Scleractinia,* pp. F328-F444 *in* R. C. Moore (ed), *Treatise on invertebrate paleontology, Part F, Coelenterata.* Geol. Soc. America & Univ. Kansas Press, 494 pp.

Weissermel, H. W.
1894. *Die Korallen der Silurgeschiebe Ost-preussens und östlichen Westpreussens.* Deutsch. Geol. Gesell., Zeitschr., Bd. 46, pp. 580-674, pls. 47-53.

Wentzel, J.
1895. *Zur Kenntniss der Zoantharia Tabulata.* K. Akad. Wissen., Wien, Math.-Naturw. Cl., Denkschr., Bd. 62, pp. 479-516, 5 pls.

Whiteaves, J. F.
1904. *Description of a new genus and species of rugose corals from the Silurian rocks of Manitoba.* Ottawa Naturalist, vol. 18, pp. 113-114.

Williams, A.
1951. *Llandovery brachiopods from Wales with special references to the Llandovery district.* Geol. Soc. London, Quart. Jour., vol. 107, pp. 85-136.

Williams, M. Y.
1919. *The Silurian geology and faunas of the Ontario Peninsula, and Manitoulin and adjacent islands.* Geol. Surv. Canada, Mem. 111, 195 pp.

Wilson, A. E.
1926. *An Upper Ordovician fauna from the Rocky Mountains, British Columbia.* Geol. Surv. Canada, Dept. Mines, Bull. 44, pp. 1-34.

Wu, W.-S.
1958. *Some Silurian corals from the vicinity of Beiyin Obo, Inner Mongolia.* Acta Palaeont. Sinica, vol. 6, pp. 59-70.

Yabe, H.
1915. *Einige Bemerkungen über die* Halysites-*Arten.* Tôhoku Imp. Univ., Sci. Rept., ser. 2 (Geol.), vol. 4, pp. 25-38, 5 pls.

Yu, C.-M.
1956. *Some Silurian corals from the Chuichuan Basin, western Kansu.* Acta Palaeont. Sinica, vol. 4, pp. 599-620.
1960. *Late Ordovician corals of China.* Acta Palaeont. Sinica, vol. 8, pp. 65-132, 15 pls.

Zheltonogova, V. A.
1965. *Znachenie rugoz dlya stratigrafii Silura Gornogo Altaya i Salaira* *("Significance of rugose corals for the stratigraphy of the Silurian of Gornogo Altay [the Altay Mountains] and Salair"),* pp. 33-44, in B. S. Sokolov and A. B. Ivanovskiy (eds.), *Rugozy Paleozoya SSSR.* Izdatel'stvo "Nauka", Moscow, 90 pp., 26 pls.

Zhizhina, M. S.
1968. *Nekotorye favozitidy iz Llandoveriyskikh i Venlokskikh otlozheniy Noril'skogo rayona ("Some favositids from Llandoverian and Wenlockian deposits of the Norilsk region"*). Nauch.-Issled. Inst. Geol. Arktiki, Uch. Zap., Paleontol. Biostratig., No. 23, pp. 81-103, 11 pls.
1969. *Ordovikskiye tabulyaty i geliolitidy yuga Novoy Zemli, Vaygacha i Pay-Khoya ("Ordovician tabulates and heliolitids of southern Novaya Zemlya, Vaygach and Pai Khoi"*). Nauch.-Issled. Inst. Geol. Arktiki, Uch. Zap., Paleontol. Biostratig., No. 26, pp. 13-22.

Zhizhina, M. S., and Smirnova, M. A.
1957. *Novye favozitidy Llandoveri i Venloka vostochnogo Taimyra ("New favositids of the Llandovery and Wenlock of eastern Taimyr"*). Nauch.-Issled. Inst. Geol. Arktiki, Sbornik Statei, Paleontol. Biostratig., vyp. 6, pp. 15-43, pls.

Ziegler, A. M., Hansen, K. S., Johnson, M. E., Kelly, M. A., Scotese, C. R., and Van Der Voo, R.
1977. *Silurian continental distributions, paleogeography, climatology, and biogeography.* Tectonophysics, vol. 40, pp. 13-51.

Ziegler, A. M., Rickards, R. B., and McKerrow, W. S.
1974. *Correlation of the Silurian rocks of the British Isles.* Geol. Soc. America, Spec. Paper 154, 154 pp.

Ziegler, A. M., Scotese, C. R., McKerrow, W. S., Johnson, M. E., and Bambach, R. K.
1977. *Paleozoic biogeography of continents bordering the Iapetus (Pre-Caledonian) and Rheic (Pre-Hercynian) Oceans,* pp. 1-22 *in* Robert M. West (ed), *Paleontology and plate tectonics with special reference to the history of the Atlantic Ocean*: Milwaukee Pub. Mus., Sp. Publ. Biol. Geol. No. 2, vi + 109 pp.

KEYS FOR THE IDENTIFICATION OF THE CORALS
OF THE BRASSFIELD FORMATION
IN THE CINCINNATI ARCH REGION

RUGOSA

1 a. Colonial 2
   b. Solitary 6
2 a. Dissepimentarium present (Septa not acanthine) 3
   b. Dissepimentarium absent (Septa acanthine, corallum dendroid)..**Tryplasma sp.**
3 a. Corallum cerioid 4
   b. Corallum not cerioid 5
4 a. Calice pits on top of mammiform projections, with thamnasterioid septa ....**Arachnophyllum mamillare**
   b. Calice pit regions each enclosed in polygonal area.......................
      Arachnophyllum granulosum
5 a. Corallum fasciculate only ....**Petrozium pelagicum**
   b. Corallum with corallites alternately fasciculate and cerioid ....**Strombodes socialis**
6 a. Dissepimentarium present 7
   b. Dissepimentarium absent 15
7 a. Calice reflexed distally into broad, horizontal platform (Corallite patelloid) ....**Craterophyllum(?) solitarium**
   b. Calice not reflexed into horizontal platform 8

8 a. Septa acanthine (Tabularium not clearly marked from
    dissepimentarium) ....**Cystiphyllum spinulosum**
  b. Septa not acanthine **9**
9 a. Axial region contains septal lamellae and (or) pali **10**
  b. Axial region does not contain septal lamellae or pali **12**
10 a. Axial space in neanic and ephebic stages narrow, about
    one-fifth total corallite diameter ....**Paliphyllum primarium**
  b. Axial space in neanic and ephebic stages broad, one-fourth
    to one-third total diameter **11**
11 a. Major septa straight, roughly equal in length (Axial space
    distinct; corallite cross-section with strong radial sym-
    metry) ....**Paliphyllum regulare**
  b. Major septa sinuous, unequal in length, so borders of axial
    space not distinct ....**Paliphyllum suecicum brassfieldense**
12 Talon present **13**
. Talon absent ....**Schizophaulactis densiseptatus**
13 a. Major septa in axial contact in ephebic stage ....
    Protocyathactis cf. P. cybaeus
  b. Axial region open in ephebic stage, occupied only by
    septal lobes **14**
14 a. No supplementary plates between tabularium and dissepi-
    mentarium ....**Cyathactis typus***
  b. Supplementary plates between tabularium and dissepi-
    mentarium ....**Cyathactis sedentarius***
15 a. Septa acanthine **16**
  b. Septa not acanthine **17**
16 a. Spines on upper surfaces of tabulae (Septal spines diverge
    distally from septal planes) ....**Tryplasma cylindrica**
  b. Upper surfaces of tabulae without spines (Septal spines re-
    main in septal planes) ....**Tryplasma radicula**
17 a. Septa perforate **18**
  b. Septa not perforate **19**
18 a. Corallite scolecoid (Theca rarely preserved above extreme
    proximal end; no fossula) ....**Calostylis lindstroemi**
  b. Corallite trochoid to sub-cylindrical (Theca generally well-
    preserved; weak fossula present in ephebic stage) ....
    Calostylis spongiosa
19 a. Corallite patelloid in ephebic stage (Septocoels in calice
    platform reduced to sutures, or nearly so, by pronounced
    broadening of all septa) **20**
  b. Corallite not patelloid (Septocoels not reduced to sutures
    in calice platform) **22**
20 a. Prominent axial boss present ....**Schlotheimophyllum patellatum**
  b. No axial boss present **21**
21 a. Septa in peripheral calice platform with "feathery" texture
    (Septocoels narrow, smooth-bottomed valleys in peri-
    pheral platform) ....**Schlotheimophyllum benedicti**
  b. Septa in peripheral calice platform with smooth surfaces
    (Septocoels reduced to sutures) ......**Schlotheimophyllum ipomaea**
22 a. Linguoid axial structure present in ephebic stage, appear-
    ing as football-like profile in transverse sections **23**
  b. No linguoid axial structure present **24**
23 a. Corallite trochoid (Major septa bear carinae) ....
    Dalmanophyllum linguliferum
  b. Corallite ceratoid (Major septa lack(?) carinae) ....
    Dalmanophyllum(?) obliquior

Cyathactis typus Soshkina, 1955, and *Cyathactis sedentarius* (Foerste) 1906
may be conspecific.

24 a. Corallite scolecoid (Septa withdraw from axis in ephebic
stage, leaving only septal lobes) ....**Streptelasma scoleciforme**
   b. Corallite not scolecoid **25**
25 a. Cardinal fossula not present (Cardinal septum on convex
side of corallite is the most prominent septum through
much of ontogeny, being longest and sometimes thickest;
septa carinate) ....**Pycnactis tenuiseptatus**
   b. Cardinal fossula present, at least in ephebic stage **26**
26 a. Cardinal fossula reaches entirely to corallite axis (Fossula
usually straight-sided, narrowing axially; septa commonly carinate) ....**Rhegmaphyllum daytonensis**
   b. Cardinal fossula blocked from corallite axis by juncture of
bounding septa, and possibly tabulae, so that its axial
end is commonly rounded (Septa non-carinate; in ephebic
stage septa always affected by counter-clockwise axial
whorl) **27**
27 a. Stereome fills septocoels on cardinal side in neanic stage
after those on counter side have become open ....
Dinophyllum semilunum
   b. Septocoels either uniformly filled or vacant in neanic stage **28**
28 a. Corallite trochoid (Septa below calice floor thick, equal to
or greater than width of septocoels throughout onto-
geny) ....**Dinophyllum stokesi**
   b. Corallite ceratoid to conical (Septa narrower than septo-
coels throughout ontogeny) ....**Dinophyllum hoskinsoni**

TABULATA

1 a. Corallum cateniform **2**
   b. Corallum not cateniform **6**
2 a. Ranks long, often with 10 or more corallites, seldom
closing to form lacunae (Ranks not closely packed to-
gether) ....**Halysites(?) meandrinus?**
   b. Ranks close to form frequent lacunae **3**
3 a. Interstitial tubes between some or all corallites **4**
   b. Interstitial tubes absent **5**
4 a. Spines absent or weakly developed (Corallite inside
diameter about 1.5 × 1.0 mm) ....**Halysites catenularius**
   b. Spines well developed, in 12 longitudinal rows (Corallite
inside diameter about 1.2-1.5 × 0.7-1.0 mm) ....**Halysites nitidus**
5 a. Spines well developed (Inside corallite diameters typically
1.0-1.5 mm) ....**Catenipora gotlandica**
   b. Spines absent or weakly developed (Inside corallite
diameters 1.8-2.0 mm) ....**Catenipora favositomima**
6 a. Corallites distinct, separate, with later contact only
through transverse, cylindrical stolons or through ap-
proximation when corallites bend **7**
   b. Corallites not separate, all in lateral contact throughout **10**
7 a. Corallum has two phases: proximal corallites prostrate-
unattached; distal corallites elongate into phaceloid
phase (Lateral contact between corallites by rare horizontal stolons and by approximation when corallites
bend ....**Syringopora(?) reteformis**
   b. Corallum consists only of prostrate corallites, without
distal phaceloid phase **8**
8 a. Distal portion of corallite (beyond offset) is of greater
diameter than proximal portion ....**Cladochonus** sp. A
   b. Corallite diameter does not expand distally **9**

9 a. Dichotomous offsets rare or non-existent ....**Cladochonus** sp. B
  b. Dichotomous offsets common ....**Cladochonus** (?) sp. C
10 a. Calice apertures crescentic, oblique to corallum surface **11**
  b. Calice apertures non-crescentic, approximately normal to
     corallum surface **12**
11 a. Corallite walls thick, apertures separated from each other
     by distance about equal to their greatest width ....
                                                          Alveolites labrosus
  b. Corallite walls thin, apertures separated from each other
     by distance much less than their greatest width ....
                                                          Alveolites labechii
12 a. Tabulae infundibuliform ....**Syringolites vesiculosus**
  b. Tabulae not infundibuliform **13**
13 a. Corallum ramose ....**Striatopora flexuosa**
  b. Corallum not ramose **14**
14 a. Mural pores in corners of corallites (Corallite aperture
     diameters typically 1.0-1.5 mm) ....**Paleofavosites prolificus**
  b. Mural pores on corallite wall faces **15**
15 a. Tabulae predominantly convex upward (Spines abundant,
     on upper tabular surfaces as well as walls) ....**Favosites favosus**
  b. Tabulae not commonly convex upward **16**
16 a. Corallites of greatly varying diameter, with wider ones
     scattered in matrix of narrower ones ....:........**Favosites discoideus**
  b. Corallites within a given area of relatively uniform
     diameter **17**
17 a. Corallite diameters typically less than 1.0 mm, rarely
     reaching 1.5 mm ....**Favosites hisingeri**
  b. Corallite diameters typically more than 2.0 mm **18**
18 a. Corallite diameters approximately 5 mm (wall-to-wall) ....
                                                       Favosites densitabulatus
  b. Corallite diameters approximately 2.5-3.0 mm (wall-to-wall)...............
                                                          Favosites sp. A

HELIOLITIDA

1 a. Coenosteum vesiculose **2**
  b. Coenosteum tubular **4**
2 a. "Septa" spinose (Corallite diameter 1.2-1.5 mm; corallites
     spaced 0.7-2.0 mm apart) ....**Propora eminula**
  b. "Septa" continuous ridges, not spinose **3**
3 a. Corallite apertures typically separated from nearest
     neighbor by less than their diameter (Corallite diameter
     approximately 1-2 mm) ....**Propora conferta**
  b. Corallite apertures separated from nearest neighbor by
     distance equal to or greater than their diameter (Coral-
     lite diameter typically less than 1 mm) ....**Propora exigua**
4 a. Corallite walls not well-defined, discontinuous ......**Palaeoporites** sp.
  b. Corallite walls well-defined, continuous **5**
5 a. Coenosteal tubes typically of equal diameter (Corallite
     diameter approximately 1 mm; septal spines flex upward
     in corallite axis) ....**Heliolites spongodes**
  b. Coenosteal tubes vary greatly in diameter, giving spongy
     appearance to coenosteum (Corallite diameter approxi-
     mately 1.5 mm) ....**Heliolites spongiosus**

PLATES

In the following plates, magnifications are standardized where possible to facilitate size comparison. Plates 1-11 show whole specimens. Plates 12-38 show sections of corals which were photographed by using the thin-sections as negatives in a photographic enlarger, and projecting the images onto photographic paper. Where necessary, optical filters were used to clarify the photograph. Plates 39-42 show details of thin-sections as photographed through a light microscope under polarized light, with nicols crossed.

Libraries are making increased use of the microfilm and microfiche processes to gain space. An unfortunate result, in addition to loss of original plate quality, is the loss of the proper size scale in plates where size is expressed as magnification (*e.g.*, "$\times 2$"), rather than by a physical scale (*e.g.*, a bar labelled as representing 5 mm in the figure). I used the first of these methods to indicate size of specimens in the figures. Therefore, in the event of micro-reproduction of this work, please note that in the original, the word PLATE on each photographic plate is 8 mm long.

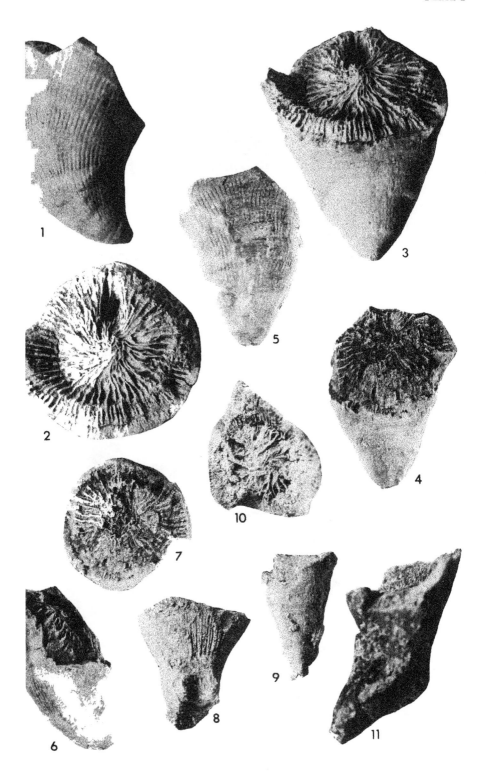

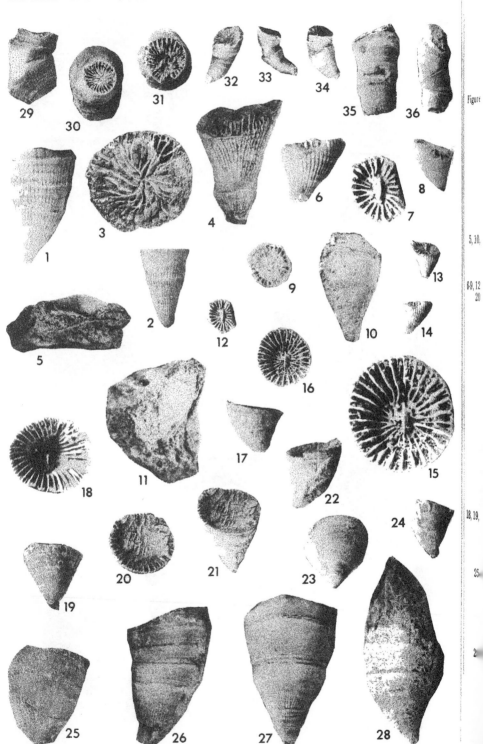

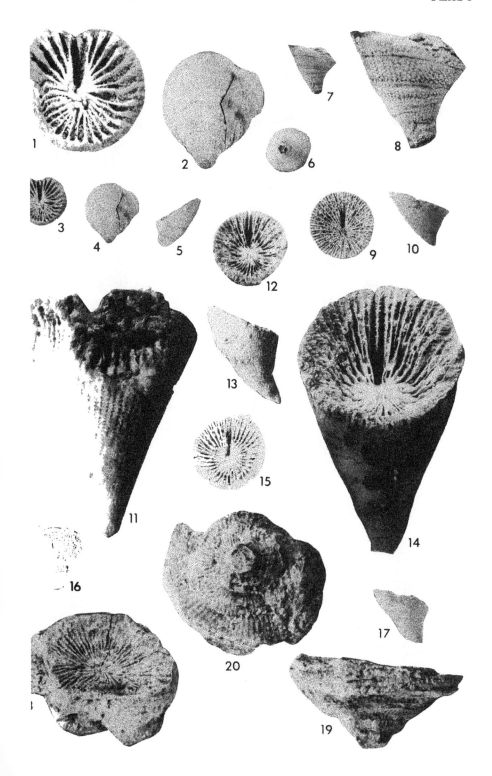

PLATE 4

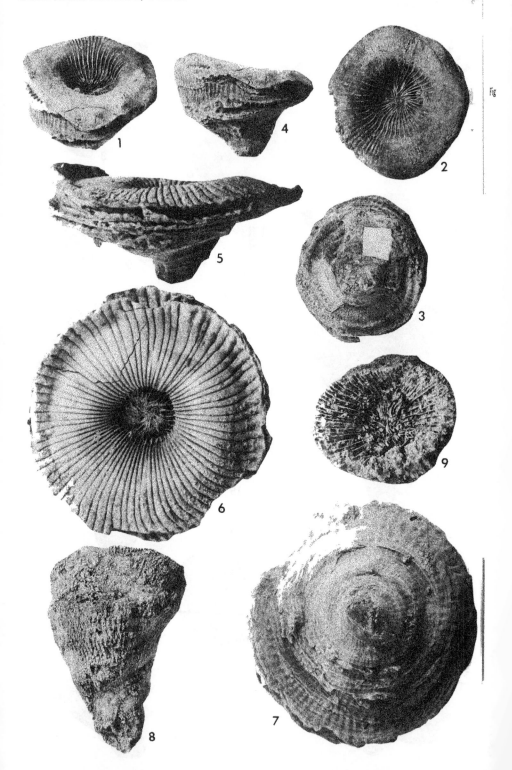

fig

Figure   r

1. **Schlotheimophyllum patellatum** (Schlotheim) ....
UCGM 41971, Specimen from talus block of Brassfield Formation at locality 7a, near Fairborn, Ohio.

2-5. **Protocyathactis cf. P. cybaeus** Ivanovskiy ....
2-4. UCGM 41956, Specimen from locality 46, Noland Member, Brassfield Formation near College Hill, Kentucky: 2, lateral view; 3, lateral view 90° from fig. 2; 4, calice view.
5. Another specimen, UCGM 41985, from same locality and horizon, shown in lateral view.

6, 7. **Cyathactis typus** Soshkina ....
UCGM 41943, Specimen from locality 1, in top layer of the Noland Member, Brassfield Formation near Panola, Kentucky: 6, lateral view; 7, calice view.

8-11. **Cyathactis sedentarius** (Foerste) ....
8, 10. UCGM 87177, Paralectotype of *Cyathophyllum sedentarium* Foerste from the Brassfield Formation north of Irvine, Kentucky (equivalent of locality 53-53a): 8, calice view; 10, lateral view. This is one of several specimens comprising this catalogue number.
9, 11. USNM 87177, Lectotype of *Cyathophyllum sedentarium* Foerste, the original of Foerste's plate 6, fig. 3B, from same locality and horizon as previous specimen: shown in two lateral views, front and back. This is one of several specimens under this catalogue number.

12-15. **Tryplasma cylindrica** (Wedekind) ....
12, 13. UCGM 41937, Specimen from locality 1, at top of Noland Member, Brassfield Formation near Panola, Kentucky. Two views of single specimen (fig. 13 magnified approximately ×4). Note the two orders of acanthine septa, and the irregular orientation of the spines, in contrast with those of *T. radicula* (q.v.).
14, 15. UCGM 41948, Specimen from same locality and horizon as previous specimen: 14, lateral view; 15, enlarged view of calice floor, magnified approximately ×4, showing its covering of spines. Compare with *T. radicula*, Pl. 6, fig. 2, 3, which lacks spines on its tabular surfaces.

PLATE 5

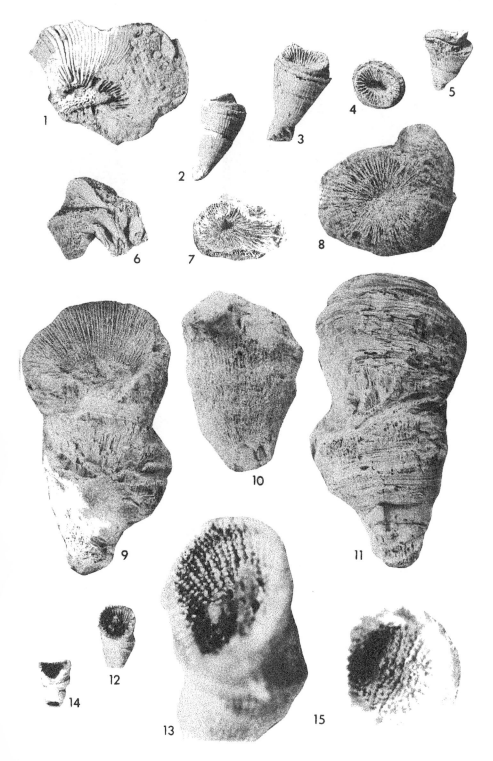

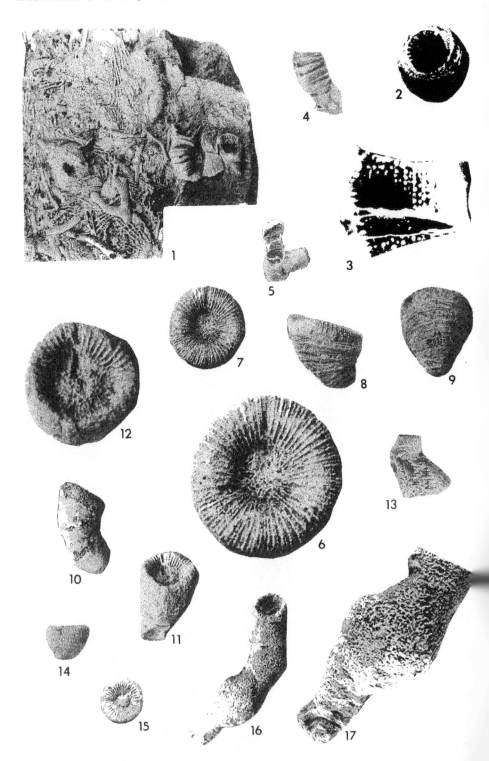

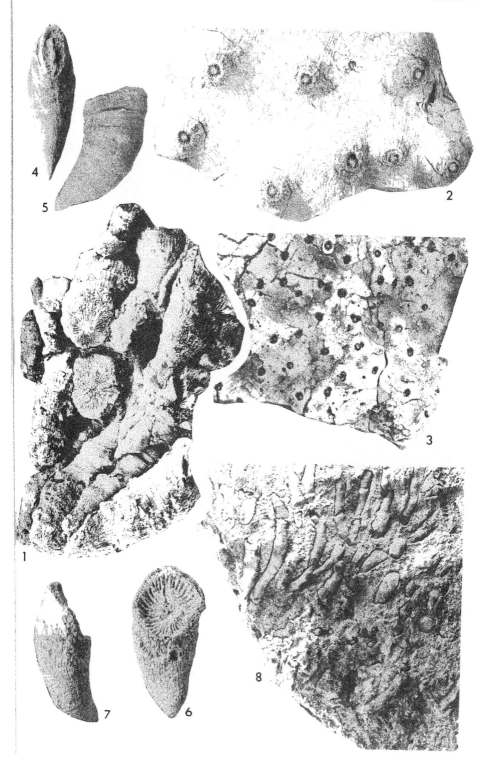

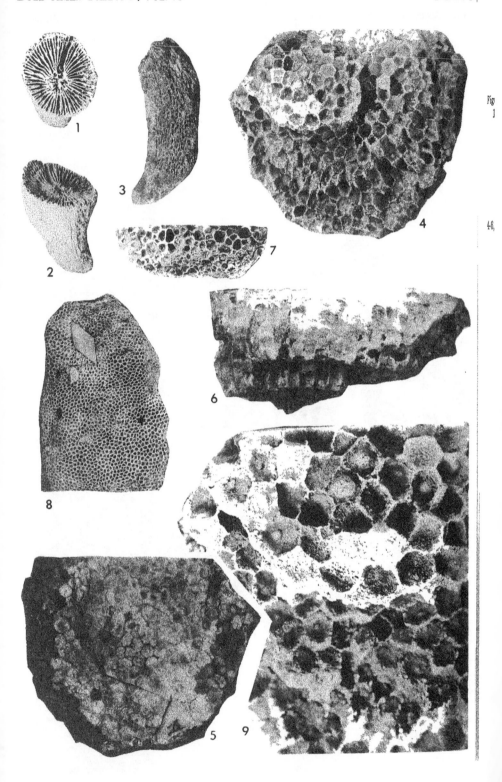

Fig
1

4-6,

EXPLANATION OF PLATE 9

(Figures ×1 unless otherwise indicated)

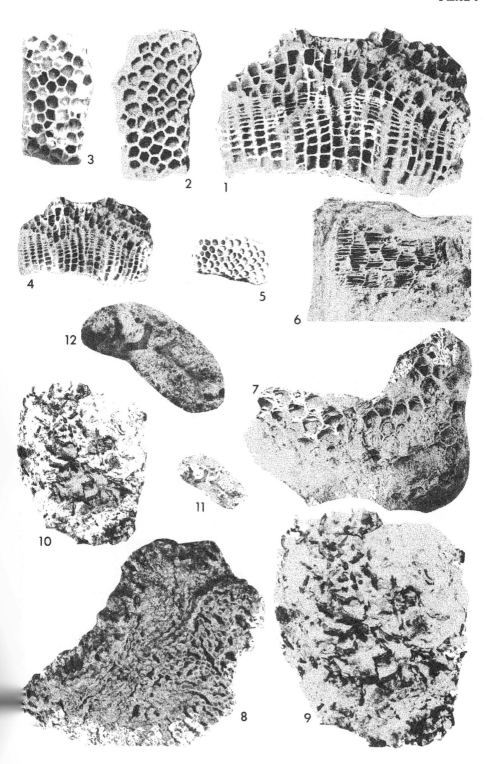

EXPLANATION OF PLATE 11

(Figures ×1 unless otherwise indicated)

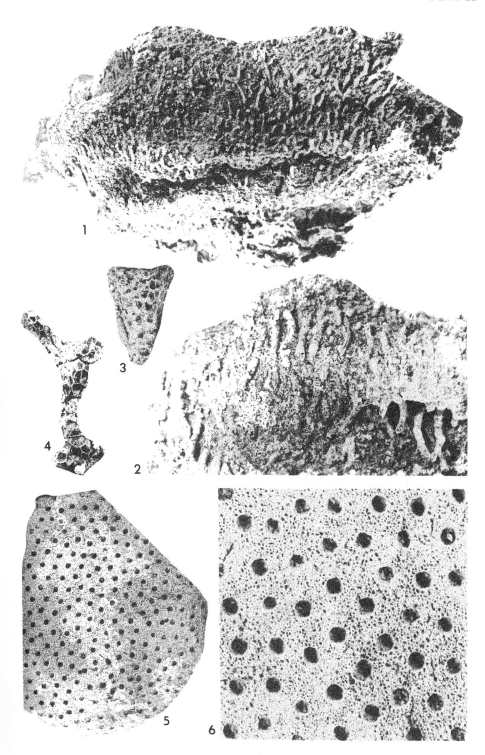

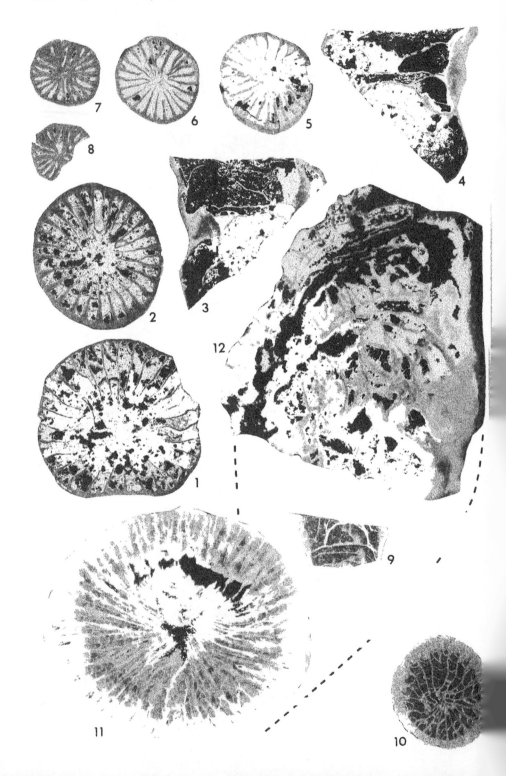

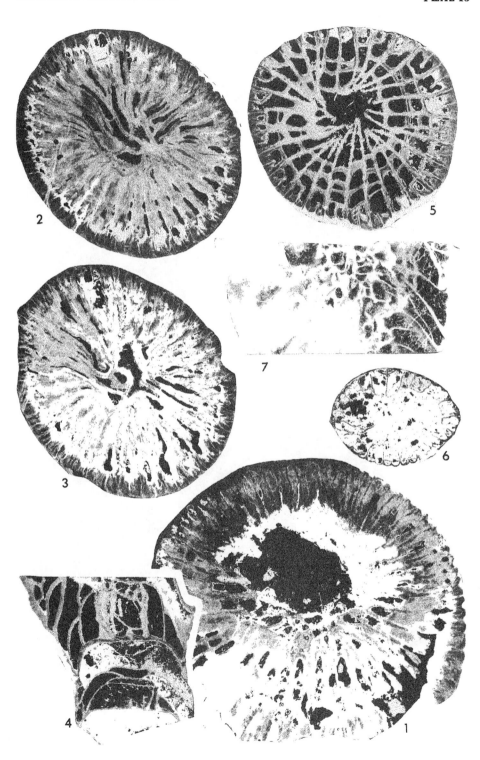

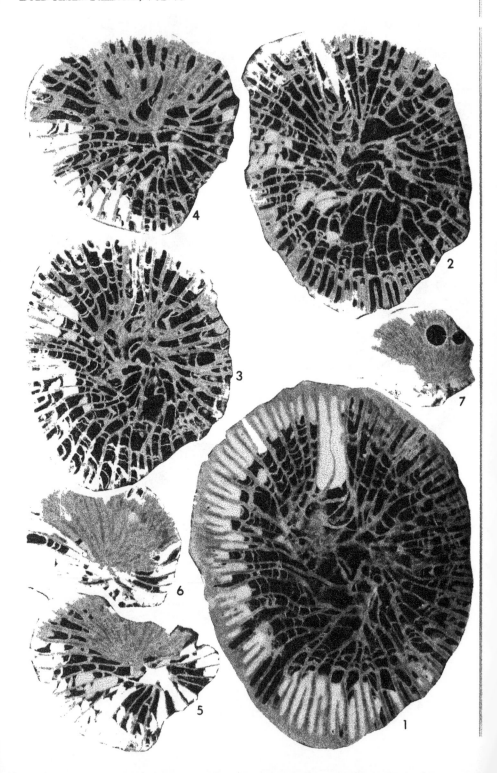

EXPLANATION OF PLATE 15

(All figures ×3.5)

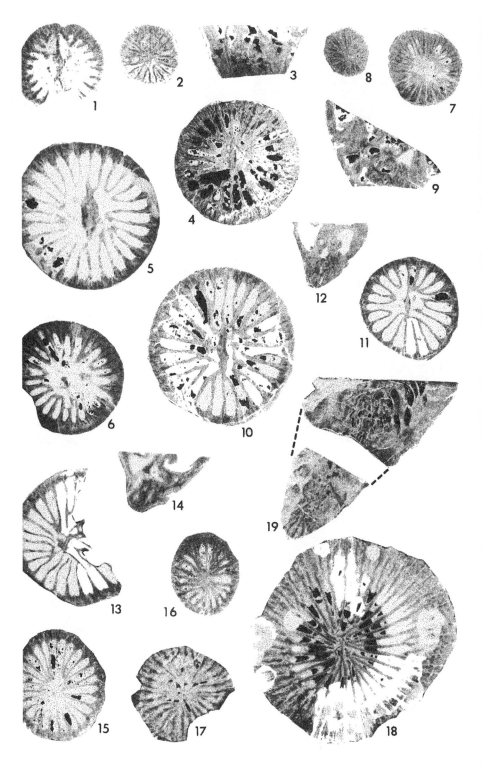

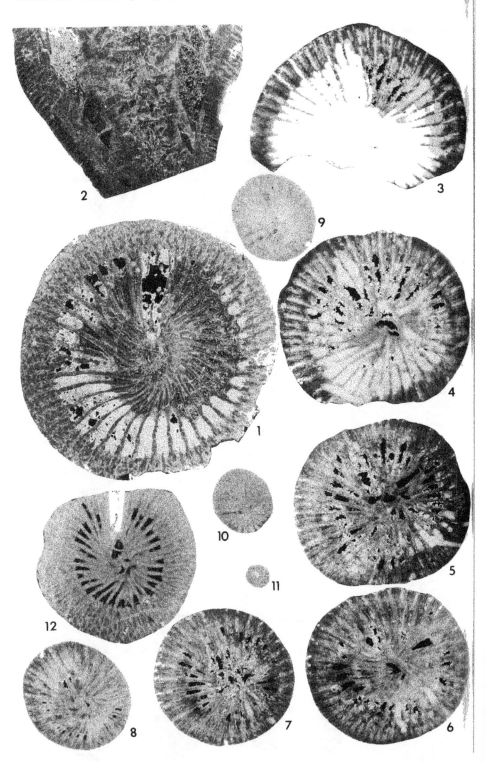

EXPLANATION OF PLATE 17

(All figures ×3.5)

Figure   Page

1-5. **Schlotheimophyllum benedicti** (Greene) **115**

    1, 2. UCGM 41700, Specimen from the Brassfield Formation at locality 15, near West Milton, Ohio: 1, longitudinal section; 2, transverse section. The dark segment of the transverse section (toward the top of the plate) is the rejuvenescent side of the specimen, where the septocoels are filled with calcite spar (corresponding to the left side of the lower surface of fig. 1). The remaining 60% of the calice cross-section, which is of a lighter shade, is filled with sediment, and represents the right side of the lower surface of fig. 1. Whole specimen shown on Pl. 3, fig. 18-20.

    3-5. AMNH 23505, Lectotype of *Ptychophyllum benedicti* Greene, the original of Greene's fig. 5, from the Louisville Limestone near Louisville, Kentucky: 3, 5, transverse sections, respectively lower and higher in the corallite, mounted right-side-up, but with their mutual radial orientations uncertain; 4, longitudinal section. Entire specimen shown on Pl. 4, fig. 1-4.

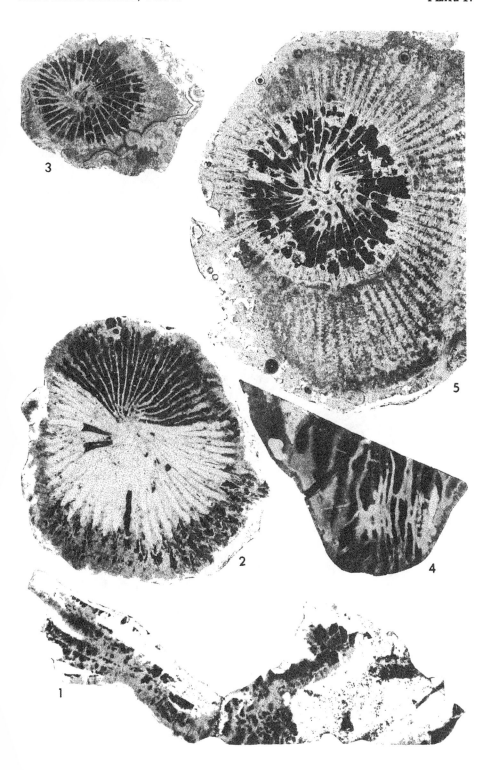

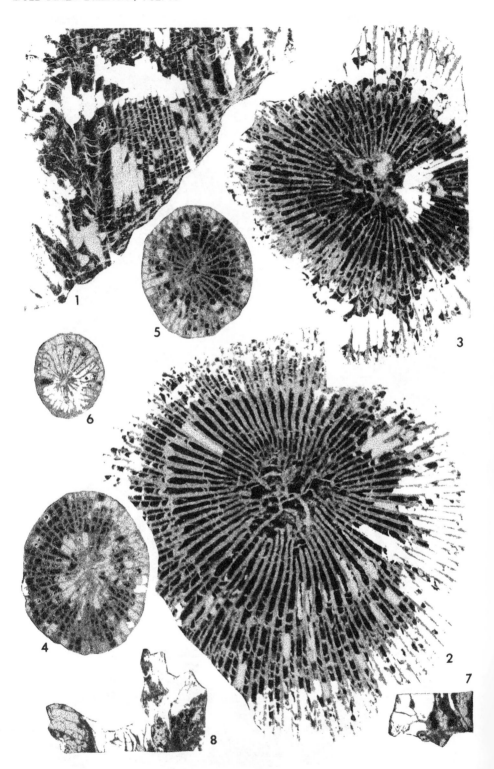

Figure   Page

1-3. **Paliphyllum suecicum** Neuman **brassfieldense,** subsp. nov. 137
UCGM 41970, Holotype of the subspecies from locality 7a, the
Brassfield Formation near Fairborn, Ohio, found loose, but
probably comes from clay-coral bed described under "Brassfield Lithosome": 1, 2, transverse sections, respectively higher
and lower in the corallite, both mounted right-side-up and with
same radial orientation. Note pali in axis of both sections,
and contratingent septa. 3, longitudinal section. Fig. 1 comes
from upper surface of this longitudinal section, fig. 2 comes
from its lower surface. Entire specimen shown on Pl. 4, fig. 8, 9.

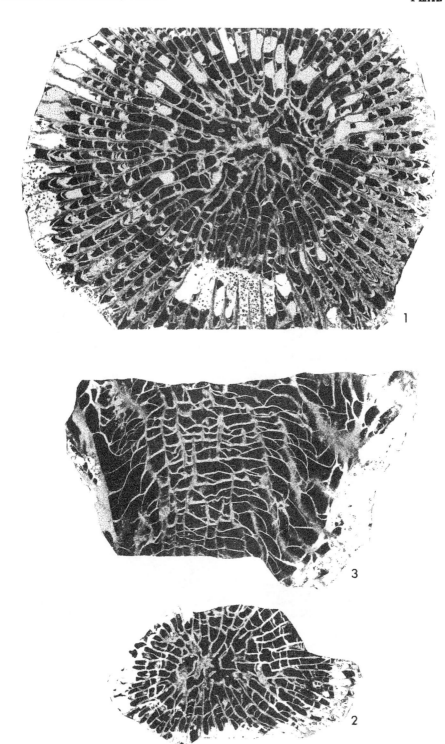

PLATE 20

EXPLANATION OF PLATE 21
(All figures ×3.5)

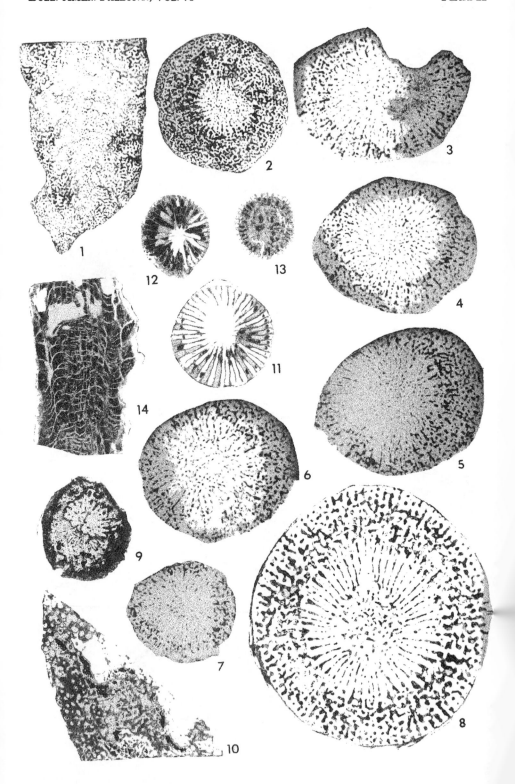

EXPLANATION OF PLATE 22

(All figures ×3.5)

     1, 2. UCGM 41941, Specimen from locality 1, top of Noland Member, Brassfield Formation near Panola, Kentucky: 1, longitudinal section; 2, transverse section.

     3-7. UCGM 41995, Specimen from same locality and horizon as the above specimen: five of seven transverse sections from approximately equal intervals along a corallite which was about 3.8 cm long. The sections are from progressively lower levels in the corallite (fig. 3 being the highest), and the second and third lowest sections (which would be between figs. 7 and 6) are missing. All sections mounted right-side-up, with same radial orientation.

     UCGM 41954, Specimen from same locality and horizon as figs. 1-7 of this plate: 8, 9, transverse sections, respectively higher and lower in the corallite; 10, longitudinal section, with convex side of corallite to the left. Fig. 8 is from the top edge of fig. 10 (here shown sloping down) and fig. 9 is from the bottom edge. Both transverse sections are mounted right-side-up, with same radial orientation, with the convex side of the corallite toward the top of the plate. Note cardinal fossula (fig. 8, above axial region).

     UCGM 41763, Specimen from locality 15, Brassfield Formation near West Milton, Ohio, *in place* in a clay pocket just below the upper contact (top) of the Brassfield: 11-13, three transverse sections through a single corallite, progressively lower in the corallite, with mutual orientations uncertain; 14, longitudinal section from same corallite.

    1, 2. UCGM 41921, Specimen from Brassfield Formation at locality 7, near Fairborn, Ohio, *in place* near top of unit: 1, longitudinal section; 2, transverse section, mounted right-side-up, from upper edge of fig. 1. Both sections are from a single corallite of the specimen shown on Plate 7, fig. 1.

    3, 4. UCGM 41925, Corallum from Brassfield Formation at locality 7a, near Fairborn, Ohio: 3, longitudinal section of several corallites; 4, transverse section of several corallites. Note thecal layer between contiguous corallites.

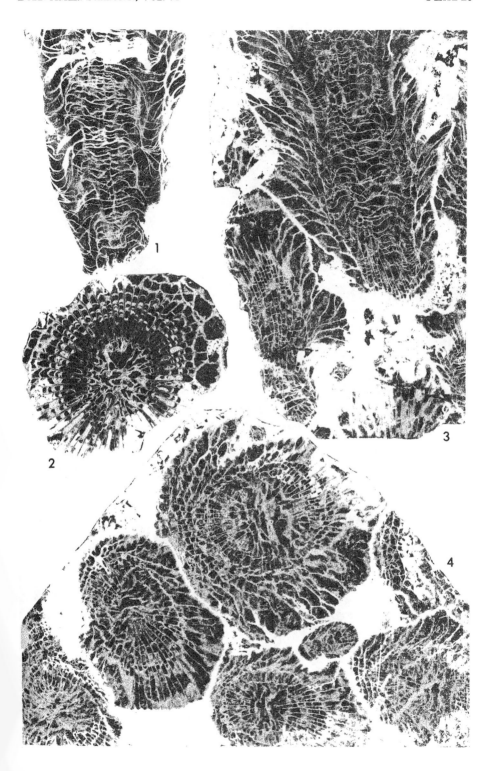

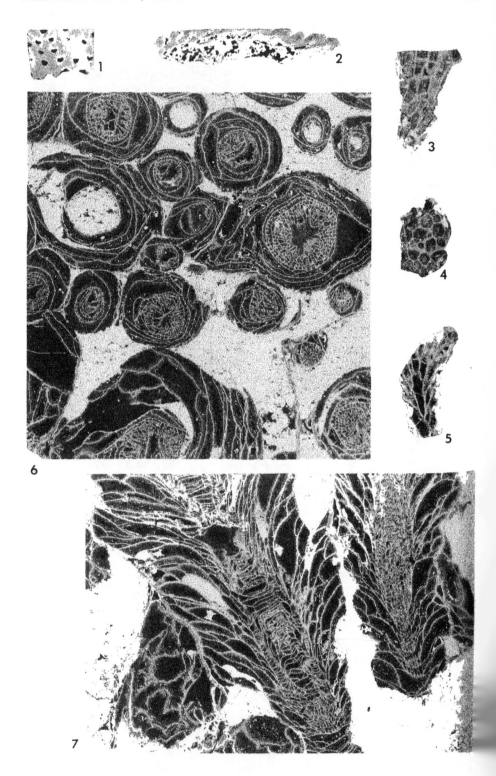

Explanation of Plate 25

(All figures ×3.5)

PLATE 25

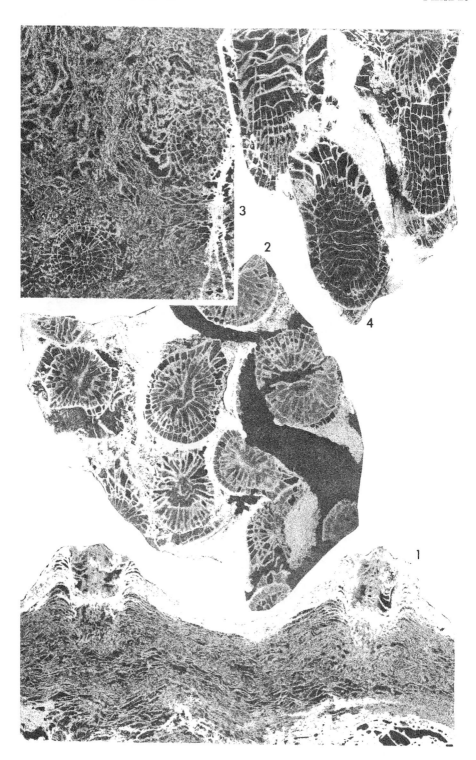

PLATE 26

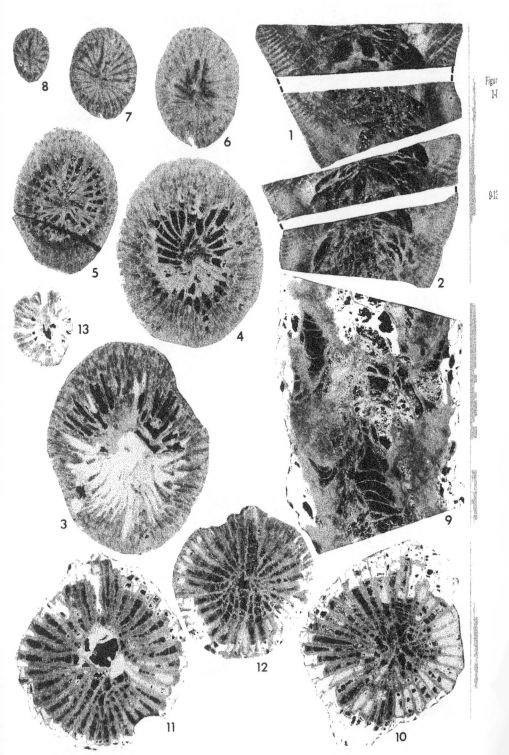

EXPLANATION OF PLATE 26
(All figures ×3.5)

EXPLANATION OF PLATE 27

(All figures ×3.5)

    1, 2. Holotype of *Calamopora favosa* Goldfuss, from the Silurian of Drummond Island, Lake *H*uron. Specimen, apparently lacking catalogue number, is from collection of Geological and Paleontological Institution of Bonn, West Germany, and is the original of Goldfuss' plate 26, fig. 2: 1, transverse section; 2, vertical section. Whole specimen is shown on Pl. 8, fig. 4-6, 9.

    3, 4. UCGM 41957, Specimen from locality 46, the Noland Member, Brassfield Formation near College Hill, Kentucky: 3, vertical section; 4, transverse section. While the small diameter of the corallites may preclude this specimen from Goldfuss' species, the good agreement in all other features suggests that it may be part of an ancestral lineage, or a subspecies of *F. favosus* sensu stricto.

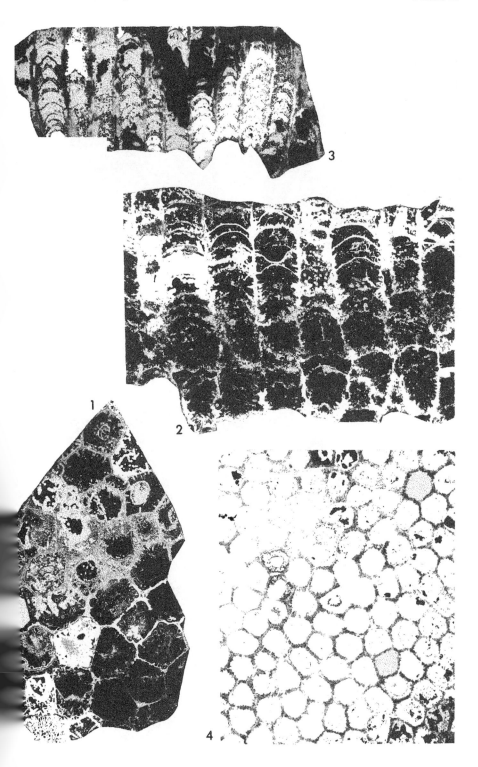

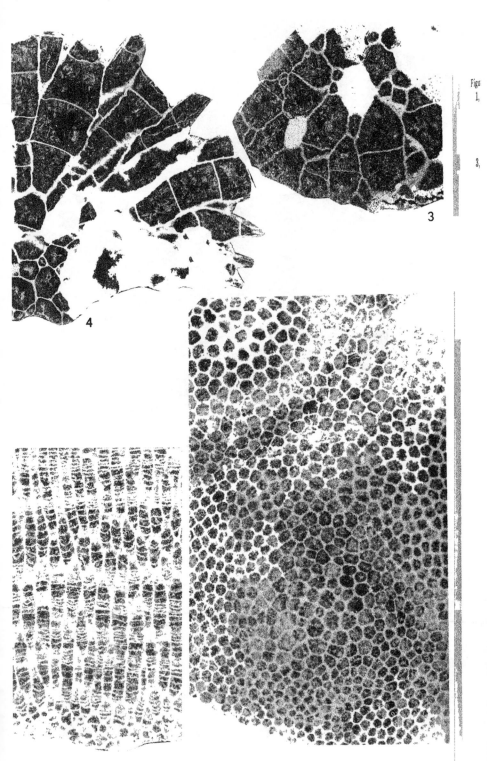

Figu
1,

3,

EXPLANATION OF PLATE 28

(All figures ×3.5)

Explanation of Plate 29
(All figures ×3.5)

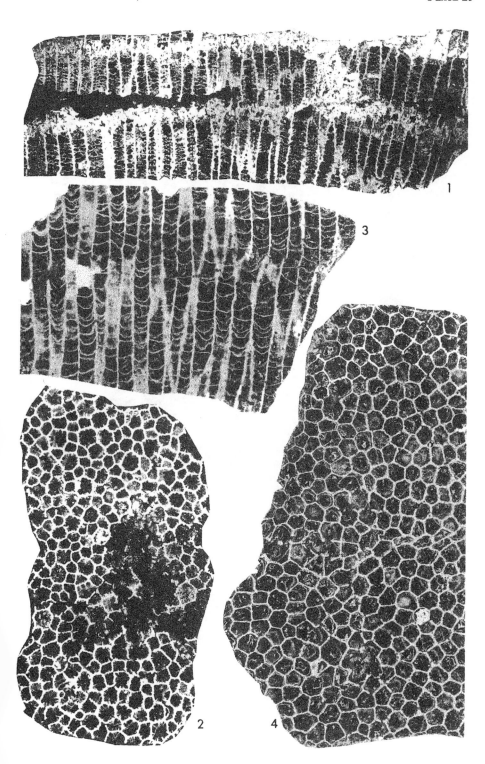

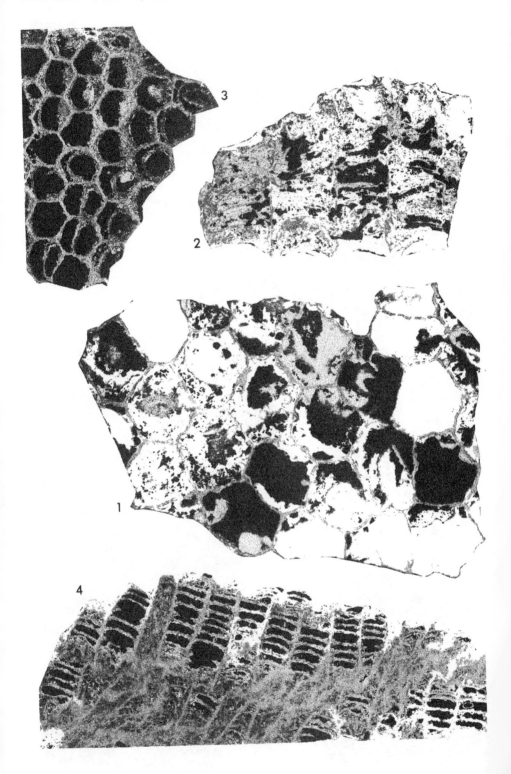

Explanation of Plate 30
(All figures ×3.5)

EXPLANATION OF PLATE 31
(All figures ×3.5)

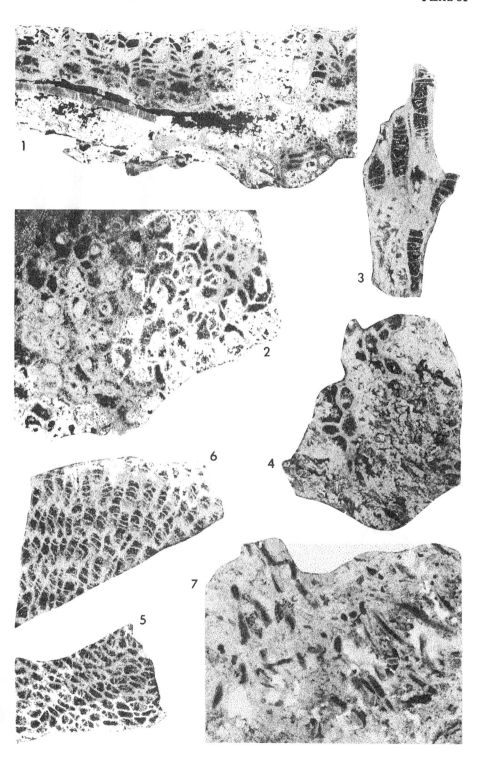

PLATE 32

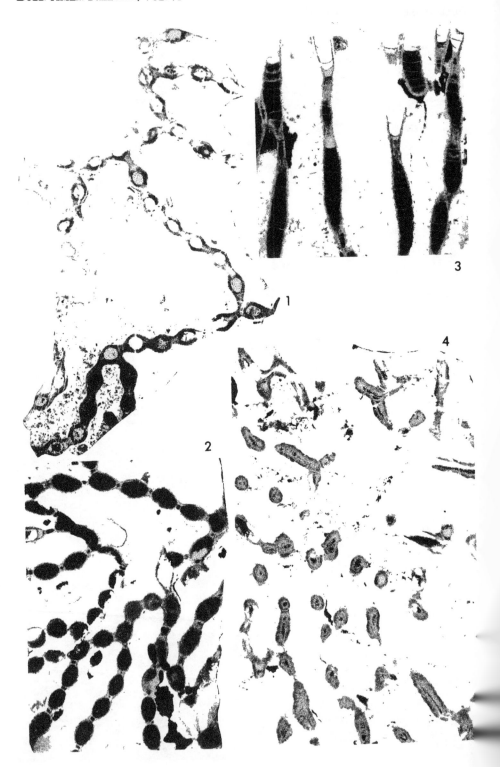

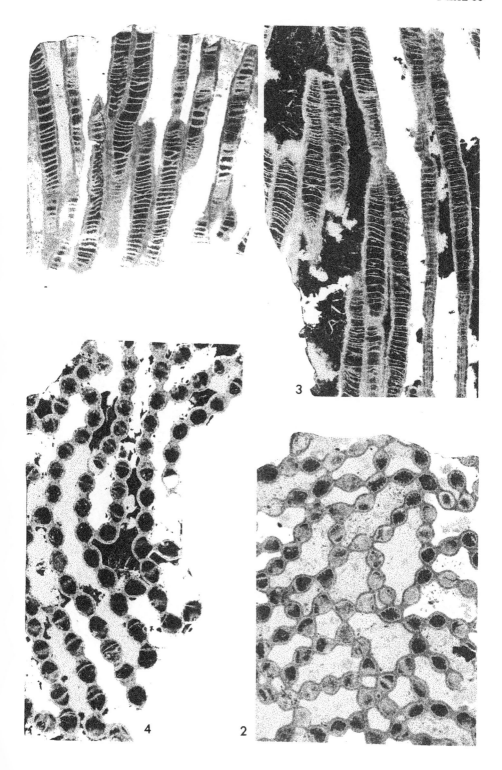

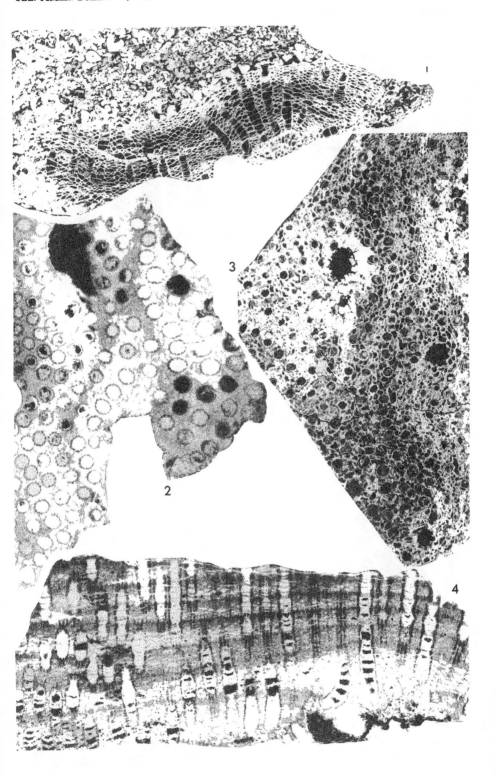

EXPLANATION OF PLATE 35

(All figures ×3.5)

Figure   Page

1, 2. **Palaeoporites** sp. 361
UCGM 41932, Specimen from locality 11, Brassfield Formation
near West Union, Ohio: 1, vertical section; 2, transverse section.

3, 4. **Plasmopora lambii** Schuchert 342
USNM 28140b, Paralectotype, from the Trenton beds of Baffin
Island, Arctic Canada: 3, vertical section; 4, transverse section.
These sections are shown to illustrate the similarities and dif-
ferences between this species and *Propora conferta* (see Pl. 34).
See text discussion of *P. conferta* for details.

5, 6. **Heliolites spongodes** Lindström 353
UCGM 41936, Specimen from locality 1, top of Noland Mem-
ber, Brassfield Formation near Panola, Kentucky: 5, transverse
section (note up-turned spines in corallite axes, appearing as
dots in plane-of-section); 6, vertical section.

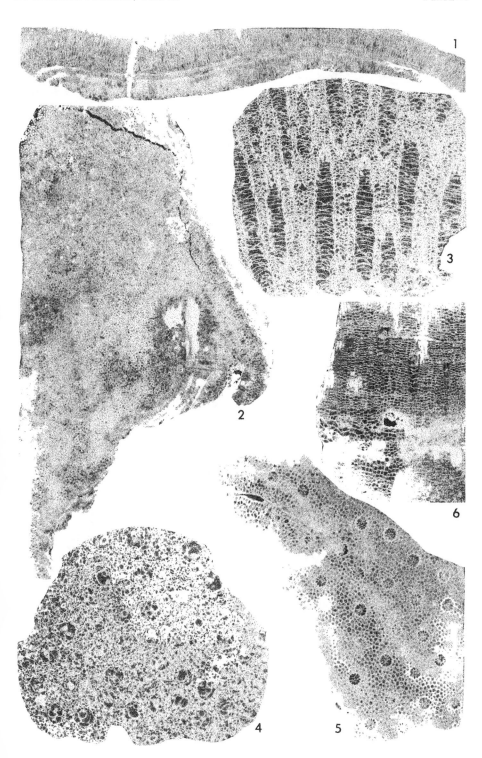

PLATE 36

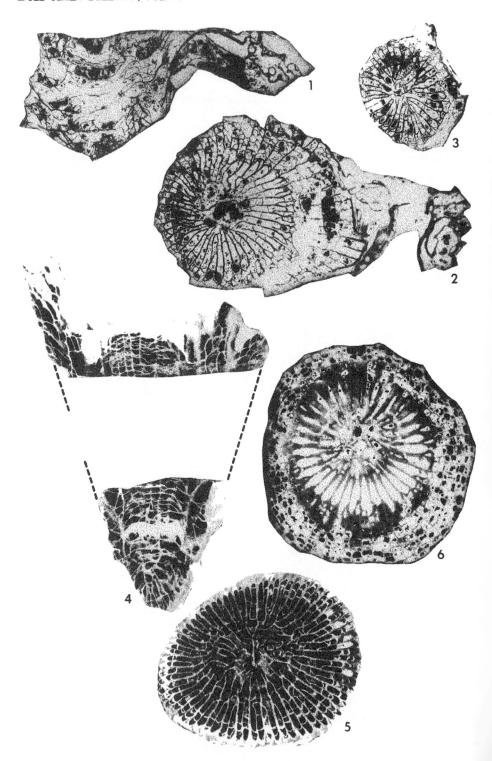

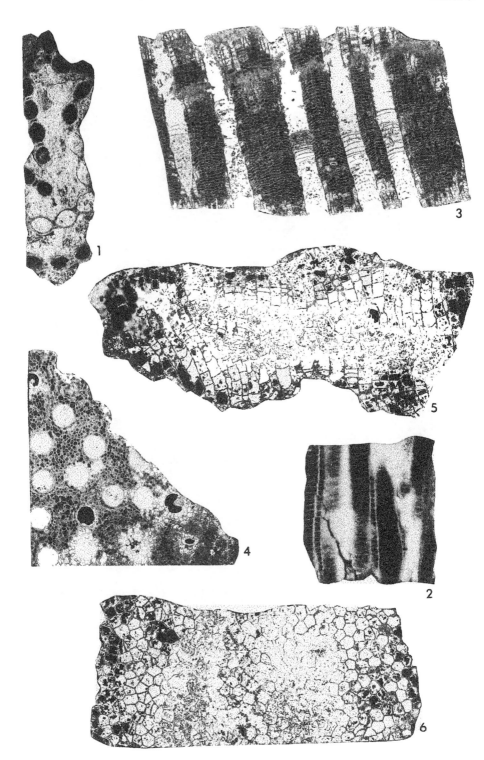

Explanation of Plate 38
(All figures ×3.5)

PLATE 39

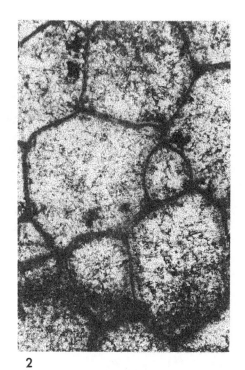

2

4

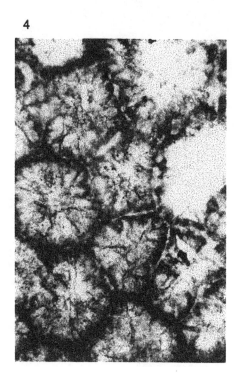

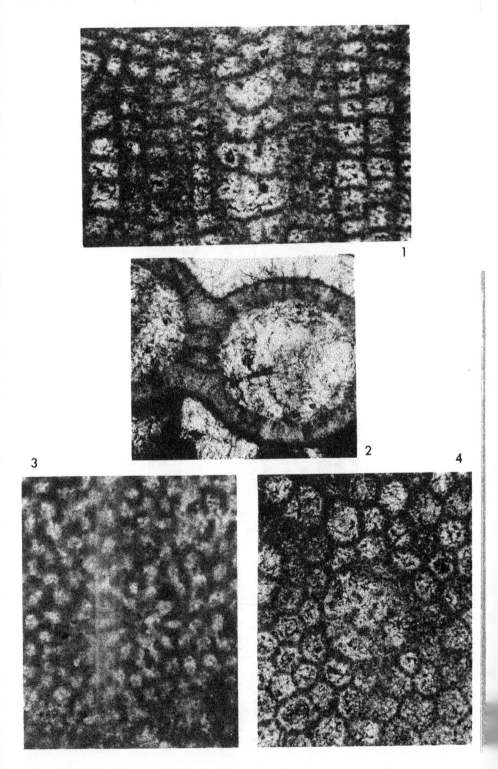

EXPLANATION OF PLATE 41
(All figures approximately ×30)

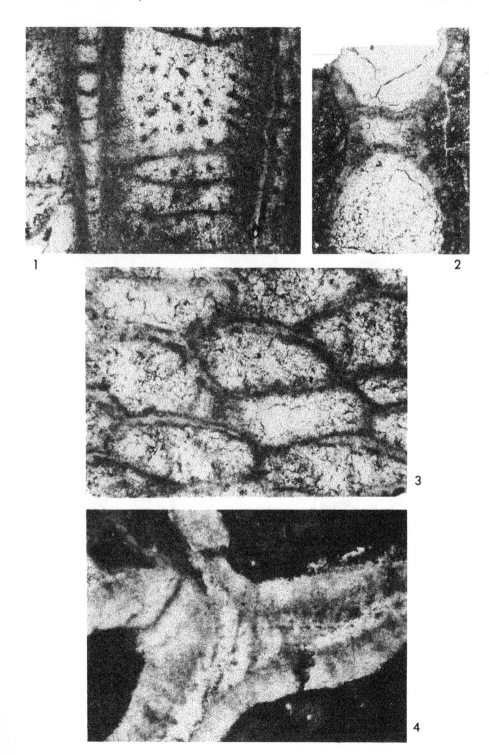

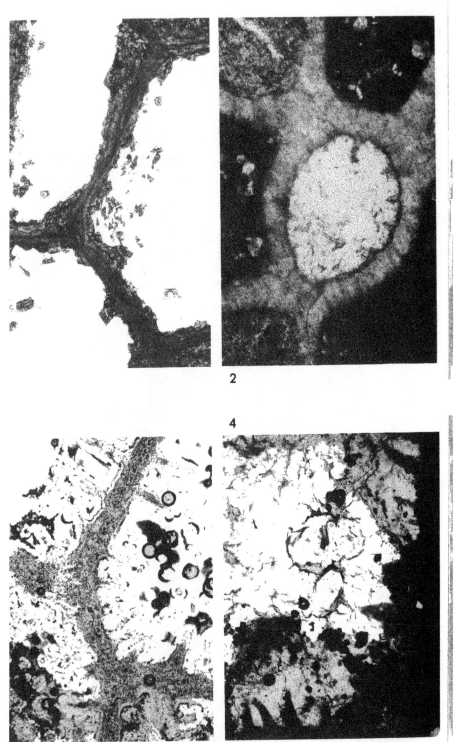

Figu

3,

2

4

EXPLANATION OF PLATE 42

(All figures approximately ×30)

INDEX

NUMBER 305
Note: Page numbers in light face; plate numbers in **bold face** type.

A

INDEX

435

INDEX

436

INDEX

INDEX

INDEX

INDEX

INDEX

INDEX

INDEX

INDEX

INDEX

INDEX

INDEX

447

INDEX

448

INDEX

INDEX

P

INDEX

451

INDEX

INDEX

INDEX

INDEX

INDEX

INDEX

457

PREPARATION OF MANUSCRIPTS

Bulletins of American Paleontology currently comprises from four to six separate monographs in two volumes each year. The *Bulletins* are a publication outlet for significant longer paleontological, paleoecological and biostratigraphic monographs for which high quality photographic illustrations are a requisite.

Manuscripts submitted for publication in *Bulletins of American Paleontology* must be *typewritten*, and double-spaced *throughout* (including direct quotations and references). All manuscripts should contain a table of *contents*, lists of text-figures and/or *tables*, and a short, informative abstract *that* includes names of all new taxa. Format should follow that of recent numbers in the series. All measurements must be stated in the metric *system*, alone or in addition to the English system equivalent. The maximum dimensions for photographic plates are 114 x 178 mm (4½″ x 7″: plate area is outlined on this page). Single-page text-figures are limited to 108 x 178 mm (4¼″ x 7″), but arrangements can be made to publish text-figures that must be larger. Any lettering in illustrations should follow the recommendations of Collinson (1962).

Authors must provide three (3) copies of the text and two (2) copies of accompanying illustrative material. The text and line-drawings may be reproduced xerographically, but glossy prints at publication scale must be supplied for all half-tone illustrations and photographic plates. These prints should be clearly identified on the back.

All dated text-citations must be *referenced*, except those that appear only within long-form synonymies. Additional references may be listed separately if their importance can be demonstrated by a short general *comment*, or individual annotations. Citations of illustrations within the monograph bear initial capitals (e.g., Pla*te*, T*ext*-figure), but citations of illustrations in other articles appear in lower-case letters (e.g., pla*te*, text-figure).

Original plate photomounts should have oversize cardboard backing and strong tracing paper overlays. These photomounts should be retained by the author until the manuscript has formally been accepted for publication. Explanations of text-figures should be interleaved on separate numbered pages within the *text*, and the approximate position of the text-figure in the text should be indicated. Explanations of plates follow the Bibliography.

Authors are requested to enclose $10 with each manuscript submit*ted*, to cover costs of postage during the review process.

Collinson, J.
1962. *Size of lettering for text-figures.* Jour. Paleont., v. 36, p. 1402.